The Individual
Subject and
Scientific Psychology

PERSPECTIVES ON INDIVIDUAL DIFFERENCES

CECIL R. REYNOLDS, *Texas A&M University, College Station*
ROBERT T. BROWN, *University of North Carolina, Wilmington*

DETERMINANTS OF SUBSTANCE ABUSE
Biological, Psychological, and Environmental Factors
Edited by Mark Galizio and Stephen A. Maisto

HISTORICAL FOUNDATIONS OF EDUCATIONAL PSYCHOLOGY
Edited by John A. Glover and Royce R. Ronning

THE INDIVIDUAL SUBJECT AND SCIENTIFIC PSYCHOLOGY
Edited by Jaan Valsiner

THE NEUROPSYCHOLOGY OF INDIVIDUAL DIFFERENCES
A Developmental Perspective
Edited by Lawrence C. Hartlage and Cathy F. Telzrow

PERSONALITY AND INDIVIDUAL DIFFERENCES
A Natural Science Approach
Hans J. Eysenck and Michael W. Eysenck

PERSONALITY DIMENSIONS AND AROUSAL
Edited by Jan Strelau and Hans J. Eysenck

PERSPECTIVES ON BIAS IN MENTAL TESTING
Edited by Cecil R. Reynolds and Robert T. Brown

A Continuation Order Plan is available for this series. A continuation order will bring delivery of each new volume immediately upon publication. Volumes are billed only upon actual shipment. For further information please contact the publisher.

The Individual
Subject and
Scientific Psychology

Edited by
Jaan Valsiner
University of North Carolina
Chapel Hill, North Carolina

PLENUM PRESS • NEW YORK AND LONDON

Library of Congress Cataloging in Publication Data

The Individual subject and scientific psychology.

(Perspectives on individual differences)
Includes bibliographies and index.
1. Single subject research. 2. Psychology—Research. I. Valsiner, Jaan. II. Series.
[DNLM: 1. Individuality. 2. Psychology. BF 697 I39]
BF76.6.S56I53 1986   150'.72   86-18698
ISBN 0-306-42250-6

© 1986 Plenum Press, New York
A Division of Plenum Publishing Corporation
233 Spring Street, New York, N.Y. 10013

Printed in the United States of America

Contributors

ROBERT B. CAIRNS, Developmental Psychology Program, Department of Psychology, University of North Carolina at Chapel Hill, Chapel Hill, North Carolina

BARBARA CARROLL, Department of Psychology, Carleton University, Ottawa, Ontario, Canada

JANE DYWAN, Department of Psychiatry, McMaster University, Hamilton, Ontario, Canada

ISAAC FRANCK, Late of the Kennedy Institute of Ethics, Center for Bioethics, Georgetown University, Washington, DC

MARY GAUVAIN, Department of Human Development, University of Pennsylvania, Philadelphia, Pennsylvania

HERBERT P. GINSBURG, Department of Developmental and Educational Psychology, Teachers College, Columbia University, New York, New York

KLAUS E. GROSSMANN, Institut für Psychologie, Lehrstuhl IV, Universität Regensburg, Regensburg, Federal Republic of Germany

THOMAS R. KRATOCHWILL, Department of Educational Psychology, University of Wisconsin, Madison, Wisconsin

F. CHARLES MACE, School Psychology Program, School of Education, Lehigh University, Bethlehem, Pennsylvania

BARBARA ROGOFF, Department of Psychology, University of Utah, Salt Lake City, Utah

SIDNEY J. SEGALOWITZ, Department of Psychology, Brock University, St. Catharines, Ontario, Canada

EVELYN B. THOMAN, Department of Biobehavioral Sciences, University of Connecticut, Storrs, Connecticut

WARREN THORNGATE, Department of Psychology, Carleton University, Ottawa, Ontario, Canada

JAAN VALSINER, Developmental Psychology Program, Department of Psychology, University of North Carolina at Chapel Hill, Chapel Hill, North Carolina

PETER WALSCHBURGER, Institut für Psychologie, Freie Universität Berlin, Habelschwerdter Allee 45, Berlin, Federal Republic of Germany

Preface

The aim of this book is to bring together scientists from different fields in psychology who are concerned about the gradual disappearance over the past decades of the treatment of individual subjects as viable sources of scientific generalizations in our discipline. Each of the contributors to the present volume brings into it her or his unique (individual!) perspective on the issue of how psychology's inference making based on individual subjects can be constructed. Partially because of the leanings of the present editor, developmental psychologists are particularly well represented in the book. However, this book is oriented toward crossing subdiscipline boundaries in psychology—the contributions coming from researchers in social psychology, behavior analysis, neuropsychology, and psychophysiology illustrate that psychologists working in these different fields have to struggle with the difficulty of thinking about individual subjects in ways that would retain the qualities of the phenomena under study.

Since the beginning of this century, psychologists have tried to capture the individuality of the people they work with in two opposite ways. First, their thinking has often moved toward discarding individuality as "noise" in the process of making generalizations, by averaging it out or replacing it by prototypical individual types of psychological phenomena. Second, differential psychology has turned the vice of individuality into the virtue for interindividual comparative research. In either case, the more relevant question—*How is the psychological functioning of the whole range of different individual subjects organized?*—has largely remained beyond the area of thought in which scientific psychology has been operating. The roots of this difficulty of thinking about individual subjects are, of course, far beyond the boundaries of our discipline—they may be located in the traditions of occidental philosophical thought and in the realm of dominant forms of social life in rapidly industrializing Western societies, in which many of the psychological issues have become "problems" that are given to specially educated experts—psychologists—to solve. Because the education of experts to deal with socially relevant issues is necessarily society-bound, psychologists' thinking about the role of individual subjects in the scientific inference of general psychological principles is guided by certain assumptions that have effectively narrowed down the experts' thinking. The present volume is a collective attempt to deal with some of the basic issues of psychological thought that have complicated psychologists' thinking about individual subjects. The contributors also suggest different solutions to overcome the epistemological hurdles of considering individual subjects in their areas of research. However, the coverage of the issues in the present volume purposefully refrains from direct, cookbook-like directions as to how one *should* study individual subjects in the fields of psychology represented in the book. Instead, the volume is expected to provide intellectual nutrition to those readers who want to know how one could, in principle, construct new methods of analysis of individual-subject phenomena in psychology. Psychology's major problem, in my opinion, has been the trivialization of epistemological problems of research through quick and uncritical acceptance of different formalized systems of data analysis without careful scrutiny of their axiomatic background, and without a thorough evaluation of the fit between these methods with the important aspects of psychological phenomena that psychologists wish to study.

The idea of compiling this book occurred to me a couple of years ago, when I compared the presentation of what was being referred to as "data" in different established child psychology journals of 1930s and 1980s. It became evident from the comparison that, despite cautionary outcries over these decades, the presentation of empirical information as data that preserve the systemic integrity of psychological issues in individual subjects has gradually waned and been replaced by certain abstractions with little connection to their underlying reality. This psychological reality occurs in a variety of *individually structured* forms, whereas the habitual way of the representing these forms as scientific data are often limited to the presentation of group averages, or sometimes even only to statements that a certain "effect" (tested by analyses of variance) appeared to be statistically significant. Empirical psychology, in its striving towards increased accumulation of data, may have actually eliminated the reality that remains behind the data from its consideration. This quite unhappy realization led me to undertake the project that has resulted in the present volume, making both its strengths and weaknesses available publicly to interested readers.

A number of people contributed to the preparation of this volume. First, the enthusiasm of the contributors and their coping with editor-imposed deadlines has made working on the book interesting and not frustrating, and all the contributors deserve the editor's gratitude for their contributions. Second, Robert Brown, the co-editor of the series in which this book appears, was helpful in providing critical feedback to the editor and the contributors. Basil Blackwell Publishers, Ltd., and the editor of the *Journal for the Theory of Social Behaviour*, Paul F. Secord, gave kind permission to reprint the only contribution to the volume (Franck's contribution—Chapter 1) that had been previously published in that journal. The Oxford University Press and the Society for the Advancement of Behavior Analysis are thanked for their permission to reproduce figures used in Chapter 6 (by Mace & Kratochwill). Mrs. Vicki R. Gray's help with typing of different parts of the volume is acknowleged with gratitude.

JAAN VALSINER

Contents

PART II
GROUP VERSUS INDIVIDUAL-BASED INFERENCE IN PSYCHOLOGY:
LOGIC AND PRACTICE

PART III
TOWARD THE STUDY OF INDIVIDUAL SUBJECTS:
CONTRIBUTIONS FROM DIFFERENT FIELDS IN PSYCHOLOGY

Where is the Individual Subject in Scientific Psychology?

JAAN VALSINER

Throughout its history as a science, psychology has been plagued by a double standard in its treatment of the individual subject. In research, psychologists often distrust their observations of an individual subject and strive toward an aggregation of data across many subjects, expecting that idiosyncratic "errors" can be eliminated if a sufficiently large number of subjects is studied. Psychologists justify this expectation by relying on the basic ideas of the statistical world view, which has become the epistemological basis for the activities of most psychologists over the past century.

The research activities of psychologists, however, are aimed at revealing general laws that would adequately explain psychological functioning of individual subjects. Knowledge that results from studies of samples of subjects is expected to be adequate to inform us about individuals—at least in general, abstract terms. This application of knowledge may originate in the specificity of the scope of psychology, which is built on human curiosity about the psychological phenomena of human beings (who, obviously, are individuals). These phenomena exist in the form of particular, concrete cases, and are observed and explained by laypersons as such. However, psychologists' efforts at finding out about principles that govern these phenomena take the form of an aggregation

JAAN VALSINER • Developmental Psychology Program, Department of Psychology, University of North Carolina at Chapel Hill, Chapel Hill, North Carolina 27514.

of information across many individuals within a sample, with subsequent application of the generalized knowledge to populations and to individuals in these populations. Unlike the psychologist-scientist, the layperson-psychologist (and/or clinical psychologist) observes a psychological phenomenon without the luxury of assembling a "representative sample" of a number of the phenomena of the given kind. For the layperson, knowledge taken from scientific psychology is relevant if it is applicable to the particular observed case. Likewise, scientific knowledge in psychology is adequate if it explains not only the majority of the cases of the given kind, but every particular case—even if the latter is unique.

The relevance of generalized knowledge that is applicable to particular individual phenomena is especially important in the applied areas of psychology—where the layperson's and scientist's perspectives cross paths. Successful application of the basic knowledge of psychology in particular concrete situations—be those situations examples of individual or group psychotherapy, of consultation in a business firm, or of dealing with a troubled adolescent—can be consistent only if the basic scientific basis of these applications is adequate to the reality. Certainly it is possible to achieve occasional practical success on the basis of inadequate scientific knowledge—as with the many people who believe in, and try to confirm, predictions made on the basis of horoscopes. Such occasional success, however, would be based on the particular combination of circumstances in the case of a concrete application, and need not follow from any adequate scientific understanding of the phenomenon.

To summarize in psychological discourse (both scientific and applied), the individual subject is constantly given high relevance. In contrast, the individual case is usually forgotten in the practice of psychological research because it is being replaced by samples of subjects that are assumed to represent some general population. Overwhelmingly, psychologists study samples of subjects and often proceed to generalize their findings to the ideal (generic, i.e., average or prototypical) abstract individual. Furthermore, charactistics of that abstracted individual may easily become attributed to particular, concrete individuals with whom psychologists and other laypersons work and interact. The inductive inference from samples of concrete subjects to the abstract individual, and from it (now already deductively) back to the multitude of concrete human beings, is guided by a number of implicit assumptions. The chapters in this volume aim to make some of these assumptions explicit, so that psychologists and laypersons can become aware of the cognitive blinders that obscure insight into the science and hamper its applications.

THE GOALS OF THIS VOLUME

WHAT THIS BOOK IS ABOUT

This volume brings together contributors from different areas of psychology, whose interests have gradually found their way to the reevaluation of the role of the individual subject in psychology in general, and in their own field of research in particular. Each of the participating authors has progressed to the task of understanding the role of the individual subject through different paths, and through the influence of different theoretical backgrounds. The chapters in this volume bear the mark of the authors' personal-scientific life histories in the search for greater adequacy in understanding the individual subject. Because the contributors are all in the prime of their personal scientific development, their contributions to this volume provide a glimpse of where they stand now, and in what directions they might proceed. The contributors differ greatly in their ways of conceptualizing the role of the individual subject in psychology. However, all of them share the determination that the role of the individual subject should be given a more central place in the epistemology of psychology than has been the case thus far if we wish to proceed toward a more adequate state of affairs in our science.

The goal of this volume as a whole is to analyze general epistemological issues that surround the role of the individual subject in scientific psychology. Therefore, the emphasis in the volume—even in those chapters that present empirical examples—is on the general theoretical issues involved in the study of the individual subject in psychology.

WHAT THIS BOOK IS NOT ABOUT

This volume is not a psychological handbook or manual of particular methods that can be applied to the analysis of individual subjects. A reader of this volume who looks for guidance of that kind is looking in the wrong place, and is likely to be disappointed. Quite a few sources that provide concrete advice on methods that can be used with individual subjects have appeared in recent years, and the more immediately pragmatic and application-minded readers are advised to consult those (e.g., Barlow & Hersen, 1984; Hersen & Barlow, 1976; Kazdin, 1980; Kratochwill, 1978; Plooij & Dungen, 1985; Yin, 1984). However, if a reader wishes to become engaged in the intellectual pursuit of theoretical issues concerning the inference from individual subjects to general knowledge in psychology, the content of this volume may provide some interesting leads toward finding novel solutions to our epistemological problems.

The goal of the volume is to analyze the issue of the individual subject
without separating theory from methods and their empirical applications.
This goal obviously leads this volume in a direction that is quite the
opposite of the widespread idea that methods are theory free, and often
independently applicable to diverse phenomena. In contrast to that pos-
itivistic tradition, this volume proceeds to demonstrate that the low priority given to information about the individual subject in psychology follows
from hidden theoretical assumptions in psychologists' thinking about
their research tasks. Therefore, a handbook- or manual-oriented approach to the methods of individual case analyses would leave the origins
of our thinking about individuals concealed in those methods.

WHAT IS THE INDIVIDUAL SUBJECT?

The meaning of the term *individual subject* requires clarification. In
ordinary usage, this term can often refer to an individual *person*. However,
in the present volume the meaning of this term is wider than its commonsense usage implies. The individual subject is a *self-contained whole,
an organism in which psychological phenomena occur, and that functions
as a system.* This definition certainly covers the commonsense usage of
the term—different contributors to the volume discuss how individual
children, neurological patients, or adult subjects in psychophysiological
experiments can be studied as individual persons. However, the term
individual subject as it is used in this volume, also covers systems that
have individual persons as their constituent parts. For example, a social
group, consisting of individual persons who are related to one another,
can be an individual subject. Likewise, a gang of adolescents on a city
street, a group of students attending a certain grade at school, a local
community, a social class (in sociological analysis), a culture (in the thinking of anthropologists), a country, and—ultimately—the whole of man-
kind, can be considered to be the individual subject if our interest in
them emphasizes their functioning as self-contained systems. In the natural sciences, some objects of investigation are necessarily studied as
self-contained systems. For example, an astrophysicist who studies the
planet Jupiter has to work with a "sample" of $N = 1$. The explanation of
how physical and chemical processes on Jupiter are organized necessarily treats those processes as parts of the whole that contribute to the
functioning of that whole. Likewise, a linguist who studies the structure
of a particular language deals with the self-contained system of that
language.

On many occasions, analysis of the functioning of self-contained
systems is based on single examples even if a large number of such
specimens is available. This is the case whenever the structure and func-

tioning of a system, which may be present in many elements (each of which is identical to every other of the kind), is studied. Consider the example of a curious owner of a home computer who tries to understand how a software program works. The particular program can be copied many times—so, in principle, a large sample of specimens of the program is available. However, because it is assumed from the very beginning that every copy of the program works in identical ways, the use of a sample bigger than $N = 1$ is unnecessary. Instead, the computer hobbyist should try to understand the organizational structure of the one copy of the program that he or she possesses, and to understand the systemic functioning of the program. The task—to find out how this program works—is very different from the task of deciding which of the different kinds of software, available on the market, the computer owner would want for his or her particular needs. The former task involves the acquisition of *intrasystemic* knowledge—an understanding of how the particular program (as a self-contained system) works. The latter task is an *intersystemic* comparison in its nature—different available specimens of self-contained systems are compared with one another, on the basis of some criteria, to arrive at the *selection* of one (or more) of those over others. This can be accomplished on the basis of purely formal criteria, and without necessarily understanding anything about the internal organization of the systems compared.

The difference between intrasystemic analysis and intersystemic comparison is a basic one. It largely parallels the distinction between production and consumption. The production task becomes possible when the designer of a new product uses knowledge about the intrasystemic functioning of a system for constructing a new object. Once the new product is ready to enter the market, the characteristics which are similar to, and different from, other available products capable of fulfilling the same function become important. The consumers of the new product will compare the characteristics to other products and make their decisions about obtaining the new one on the basis of that intersystemic comparison. I do not mean that the production side is entirely free of an emphasis on intersystemic comparison, or that the consumers never consider any intrasystemic information. This undoubtedly happens in reality. However, without intrasystemic knowledge, new products cannot be constructed. For example, imagine a self-made engineer who tries to build a TV set using the intersystemic information about the characteristics of different brands of TVs from *Consumer Reports* as his or her sole information source. Without any knowledge about the structure and functions (within the whole of the set) of different parts of the TV, that engineer's construction efforts will end up with a nonfunctional pile of electronic components, each of which may be of superior quality (i.e., as compared to other comparable components used in other TV sets). However, the total system—the planned TV set—will never work, exactly because the "engineer" lacks the intrasystemic knowledge of how to put the parts of superior quality together to make up an adequately functioning whole.

The consumer's perspective is different. He or she does not have to construct the new appliance, but can rely on other people's work. There are a variety of objects that are capable of performing the same function. For example, a buyer in a TV store may compare the characteristics of different TV sets, attempting to buy the best-quality set at the lowest price available. The consumer's task is the opposite to that of the producer, and for that task the intersystemic comparison is eminently suitable.

A simple example of the separation of intersystemic and intrasystemic knowledge in our lives would involve a momentary overview of our own possessions, the many man-made things that we have obtained (TVs, cars, radios, microwave ovens, food processors, computers, stereo systems, etc.). Can we say that we have a clear understanding of how they function? Many of us would not be able to say that we know much about the intrasystemic aspect of these objects. However, whenever we make decisions about obtaining any of these objects, we may become involved in extensive intersystemic comparison of the functional output characteristics, prices, and so forth, of different kinds of objects usable for the same purposes.

Psychology has developed in ways that largely accommodate the needs of consumers rather than to those of the producers. This is not surprising given the dominant role that applied-selection objectives (e.g., application of standardized tests to select the most suitable persons from a pool of applicants) have played in the promotion of psychology's role in Western industrialized societies. As a natural outcome of that emphasis, much of psychology's analytic methodology is intersystemic in its nature. At the same time, many aspects of psychologists' discourse is aimed at getting to understand the intrasystemic nature of psychological phenomena. The opposition between the intersystemic nature of much of psychological research practices and the intrasystemic goals of some of the epistemological efforts in psychology constitutes the background of my effort to put together the present volume.

RELATED ISSUES: AN OVERVIEW

At different times in the history of psychology, a number of issues pertaining to the role of the individual subject in our discipline have been discussed. Quite often, these discussions have related to the sensitive issue of the status of psychology as a science. Because the term *scientific psychology* appears in the title of this volume, and is used throughout the text, it is necessary to make its meaning explicit.

THE MEANING OF SCIENCE IN PSYCHOLOGY

Psychologists have often been engaged in proving to themselves and others that their activities amount to science. Support for such claims is obtained in many different ways. For example, many psychologists consider quantification of psychological phenomena to be a necessary (and—for some psychologists—sufficient) criterion for the scientific status of the discipline. This belief can also be supported by common sense, which may support the notion that anything quantified is scientific. Likewise, the adoption of theoretical models and methods for empirical investigation from some well-established area of science—for example, from physics—is sometimes believed to establish the scientific status of psychologists' endeavors. Finally, the increased computerization of Western industrialized societies, and the dissemination of computer jargon from among computer users to the wider public, may lead to the adoption of "computerese" as a linguistic means to assure the status of psychologists' efforts as science. However, none of these adopted ways of acting, thinking, and speaking—quantification, physicalistic models, and computer jargon—need lead to the actual status of psychology as science.

The Status of Science as a Social Category

Within a society, the term "science" can be defined in many different ways, all of which belong to the system of meanings that is adopted by the society at the given time. In this sense, what is labeled "science" in a society determines what is considered to be science, at least in that society at that time. When one approaches the issue of the definition of science from a sociolinguistic standpoint, it is difficult to define that domain of human activities from any perspective other than one based on connotative meanings of terms in the culture. An example of such efforts to make science by extending the boundaries of meaning of the term *science* is the social movement in the contemporary United States that promotes the inclusion of "creation science" among sciences, on the grounds that this framework of thinking should be given equal representation in a democratically organized society. It is easy to see how decisions made with the help of labeling (and relabeling) about what is, and is not, science are open to influences from different social processes that take place in a society.

The bases for labeling some area of knowledge "science" can vary, and usually concentrate on the object of investigation, or methods, or the nature of generalization—or all three. Thus, somebody may consider a chemist a scientist because the object of investigation for the chemist is objective—it occurs in the form of chemical substances. Or a psychologist may feel proud applying a fancy satistical method to his or her data, because the statistical method itself is labeled as "scientific." It need not matter that the particular investigator does not exactly understand how the method works; as long as it is accepted as scientific by significant others (supervisor, journal editors, colleagues), it establishes the scientific status of the endeavor through social consensus. Finally, a linguist may build a complex formal system of a generative grammar of a language that may look extremely scientific to the uninitiated reader.

None of these three aspects of knowledge—what it is about, the methods by which it is obtained, and its form—are sufficient separately to define science. Instead, science can be defined through looking at the relationships of these aspects. In this volume, science (and scientific psychology) is distinguished from other kinds of knowledge on the basis of *the adequacy of its explanations from the perspective of the reality that is being explained.* The relevance of this criterion of defining science for psychology is discussed by Robert Cairns in Chapter 4. According to this criterion, an investigator's account of a certain psychological phenomenon is scientific if it adequately represents the phenomenon. No single criterion can be set up so that a certain perspective taken by a psychologist can be declared automatically scientific. For example, a researcher's effort to quantify a psychological phenomenon that, in its nature, defies quantification, is not considered scientific if we apply the criterion used in this volume to that case. Likewise, another psychologist's refusal to quantify a psychological phenomenon that is quantitative in its nature, is likewise not a scientific enterprise. Quantification in the case of any psychological phenomenon cannot automatically guarantee that the investigator's efforts are scientific, despite the widespread belief that quantification is science. In a similar vein, the straightforward adoption of theoretical models from physics or of jargon associated with computer science does not necessarily belong to the realm of science as it is defined here, even if these other disciplines have earned the social prestige of being scientific in their treatment of their knowledge domains.

Further Social Differentiation of Science: "Hard" versus "Soft" Science

Within science, a further division is quite often applied—sometimes we may hear psychologists talking about "hard" (implicitly meaning "good") and "soft" (implicitly dubious) approaches in psychology. Not surprisingly, we can often discover that a psychologist's separation of these two subclasses of science proceeds along the lines of the psychologist's personal likes, dislikes, and socialized ideals. Very often, consensus among psychologists on what is "hard" versus "soft" in psychology is used to make one's verdict about other psychologists' work. Social consensus, however, is an epitome of the nominalistic approach to defining the truth through labeling. It is a temporary, albeit widely used, way of constructing solutions to some pressing problems in society or within the human mind. It is not, however, universal for all societies. Consensual decisions may have an important role to play in those societies where what the majority decides is accepted (at least temporarily) by a minority. However, in many societies the role of popular consensus is of low social relevance. In such societies social consensus about what science is, or how "hard" science differs from its "soft" counterpart, is likewise of no importance. Instead, in the scientific establishments in these societies it is the minority of individuals in key power positions who determine what qualifies as a "hard" or "soft" approach in psychology. Neither social consensus, nor the decisions by dominant individuals in the scientific establishment, can determine the "hardness" of one or another approach in psychology.

In the context of this volume, no approach to the study of psychological phenomena is considered "hard" or "soft" from the beginning. Instead, how "hard" an approach to the study of individual subjects in psychology is depends on the adequacy of the approach to the reality, rather than on any value-laden meaning attached to the given approach.

DICHOTOMIES IN THE HISTORY OF PSYCHOLOGY

Psychology is a by-product of the history of the European philosophy of the 19th century. As such, it bears the marks of different conceptual dichotomies that have prevailed in the philosophic categorization of the world and science.

The Erklären–Verstehen Controversy

According to Apel (1982, p. 19), the terminological distinction between *Erklären* and *Verstehen* was introduced by J. G. Droysen in 1858 in his work *Grundriss der Historik*. It was later followed by Wilhelm Dilthey, who in 1883 published his *Einleitung in die Geisteswissenschaften*. The goal in the introduction of this dichotomy was to set historical sciences and arts apart from natural sciences. The former—labeled *Geisteswissenschaften* ("the sciences of the mind")—were considered to be based on the understanding (*Verstehen*) of their subject matter. The latter—*Naturwissenschaften* ("the sciences of the nature")—were supposed to explain (*Erklären*) their phenomena in strict causal terms. The terminological distinction between the two kinds of sciences was introduced in 1851 by French economist A. A. Cournot (1956).

The *Erklären–Verstehen* dichotomy was originally a reaction of thinkers in the mind-oriented domains to prevent the positivisitic philosophies of Auguste Comte and J. S. Mill from infiltrating the domain of learning that attempts to capture the functioning of the mind. In this respect, the introduction of this dichotomy historically served the social function of maintaining a boundary between different sciences. The emergence of psychology coincided in time, and in spirit, with the introduction of the *Verstehen–Erklären* dichotomy in 19th-century German thought. From its beginning, psychology has included the dualism of explanation of psychological functions in general, and the understanding of the psychological life of individual human beings in particular, in its domain of knowledge. From time to time, the explanation–understanding dichotomy has surfaced in the thinking of major theoretical thinkers, who have tried to overcome it. Lev Vygotsky's (1962, 1978) fundamental contribution to psychology began with his analysis of the "crisis" in psychology in mid-1920s, which he directly viewed in the light of the persistence of that dichotomy (Vygotsky, 1982; see also Van IJzendoorn & Van der Veer, 1984). Kurt Lewin's dissatisfaction with the traditions in psychological research that disgarded the study of individual persons as a "lower-status" scientific endeavor was expressed in his criticism of the Aristotelian mindset of his time (Lewin, 1931). On the other hand, efforts by the contemporary humanistic psychologists tend to keep the dichotomy alive as an insurmountable barrier that artificially separates understanding of other human beings from scientific research of their psychological functioning.

The Idiographic-Nomothetic Controversy

The idiographic-nomothetic distinction developed historically from the *Erklären–Verstehen* dichotomy, and has prevailed in psychology ever since Wilhelm Windelband introduced it at the end of the last century. Its history is analyzed in detail in Klaus Grossmann's contribution to this volume (see Chapter 2). This dichotomy continues to be present in psychology, and it occasionally surfaces in the form of specific discussions in different areas of psychology. For example, personality researchers tend to confront this issue at times, the most recent case being evident on the pages of the *Journal of Personality* (Bem, 1983; Kenrick, & Dantchik, 1983; Runyan, 1983; West, 1983). The continuation of this dichotomy has been facilitated in part by controversies in Gordon W. Allport's writings.

Issac Franck's contribution to this volume (Chapter 1) analyzes the conceptual background of Allport's work and its relation to the dichotomy.

Case Study versus Group Study

The third dichotomy in psychology—which stems from the other two, and is relevant in the context of the present volume—is that between studies of single cases and samples of subjects. The study of individual cases has always been the major (alboit often unrecognized) strategy in the advancement of knowledge about other human beings. Medical science has largely depended on the accumulation of intrasystemic knowledge, which in the present context amounts to case studies and their replication by other case studies. In everyday life, all human beings are forced in certain situations to analyze, diagnose, and treat "cases" of different kinds—ranging from physical events (e.g., impending or actual natural disasters) to psychological issues (e.g., how a parent could handle the particular temper tantrum that his or her child is displaying now; or whether it is promising to invest one's money in a certain way at the given time, with all the risks and benefits involved). The handling of such unique problems in everyday life is of utmost importance for the problem-solving individuals, who have to adopt a case-study approach to them. However, case studies in the social sciences have often been labeled "soft" science because there are indeed aspects to them that do not satisfy the requirements of the so-called "hard" sciences: existence of controls, replicability of measurement, separation of independent and dependent variables, and so on. The controversy about the case-study method, and of the idiographic-nomothetic dichotomy, has historically centered around the problem of statistical (actuarial) and clinical prediction in psychology (see Allport, 1940; Beck, 1953; Falk, 1956; Freyd, 1925; Meehl, 1954; Sarbin, 1944; Viteles, 1925). Proponents of the case-study method have their ways for explaining the adequacy of the method for scientific purposes (see Runyan, 1982). However, the established "inferior" status of the case-study method in psychology is slow to disappear. Even the elaboration of statistical methodology for the analysis of case-study data (Davidson, & Costello, 1969; Kazdin, 1980, 1981) is slow in eliminating the prejudicial treatment of the case study as a scientifically viable method.

OVERVIEW OF THE CONTENTS OF THIS VOLUME

The present volume is divided into three parts. In Part I, Isaac Franck (Chapter 1) and Klaus Grossmann (Chapter 2) provide analytic overviews of the history and epistemology of the role of the individual subject in psychology. Grossmann emphasizes the historical connection between Wilhelm Windelband, William Stern, and Gordon W. Allport in the treatment of the issue of the individual subject. Franck analyzes Allport's implicit understanding of the idiographic (morphogenetic) research in psychology. Warren Throngate (Chapter 3) introduces the readers into the realm of inference making from particular individual subjects to an individual subject in general, instead of populations. The basic characteristic of psychological phenomena—divergent amplification—serves as his theoretical basis in that endeavor. Grossmann's chapter includes a number of empirical illustrations of how divergent amplication works in some concrete cases of infants' attachments to parents, and their relationships with siblings.

Part II is introduced by the chapter by Robert Cairns (Chapter 4), which outlines the consequences of research procedures that promote distance between the investigator and the phenomena to be explained. Although the chapter illustrates the problems encountered in developmental research, the issues recur in several areas of psychological study. My chapter on how psychologists and laypersons interpret group-based correlational data (Chapter 5) illustrates how extrascientific cognitive operations are implicitly brought into the scientific inference making. These operations, largely prescribed by our everyday language, facilitate the inference from groups of subjects to individual subjects. Charles Mace and Thomas Kratochwill (Chapter 6) provide an overview of how single-subject research can benefit from the applicaton of adequate statistical methodology to case study data. Evelyn Thoman (Chapter 7) emphasizes the role of the time dimension for the analysis of individual subjects' development, and illustrates that role with empirical examples from her work with newborns. Finally, Warren Thorngate and Barbara Carroll (Chapter 8) describe some novel mathematical techniques that can be used to evaluate the adequacy of different hypotheses that can be tested on the basis of individual subjects' data.

Part III of the volume includes chapters from different areas in psychology, where the authors have tried to develop new approaches for analyses of their particular empirical data in ways that would preserve the integrity of the individual subject data. Developmental, neuropsychological, and psychophysiological approaches to psychological phenomena are represented in this final part of the book. Herbert Ginsburg (Chapter 9) outlines some possibilities for the development of individually centered diagnostic methodology for the study of cognitive development. Barbara Rogoff and Mary Gauvain (Chapter 10) describe their rocky road to the need for finding methods that preserve the systemic nature of individual cases. They provide an overview of their approach to the study of individual mother–infant dyads. Jane Dywan and Sid Segalowitz (Chapter 11) analyze the ways in which inference from single neurological cases to general knowledge can be made in neuropsychology, where quite often single cases constitute the only available basis for generalizations. Peter Walschburger (Chapter 12) outlines the problems of the idiographic-nomothetic controversy in psychophysiological stress research, and describes his empirical methodology, which can integrate the idiographic emphasis of single-subject research with the nomothetic emphasis on generality. Finally, my analysis of the open-systems nature of psychological phenomena (Chapter 13) is illustrated by the method of analysis of recurrent sequences of events in the behavior of individual subjects.

The central issue in this volume—the role of the individual subject in scientific psychology—is a fundamental one in our discipline. It would be overly optimistic to expect that these chapters can solve problems that have been a part of human thinking since at least the time of the Greek philosophers. I hope that this work will provoke new questions and lead to new efforts to solve these problems. It is offered as a "progress report," or, perhaps, a new point of departure in our quest for basic knowledge applicable to the study of the individual subject in psychology.

REFERENCES

Allport, G. W. (1940). The psychologist's frame of reference. *Psychological Bulletin, 37,* 1–28.

Apel, K-O. (1982). The Erklären–Verstehen controversy in the philosophy of the natural and human sciences. In G. Fløistad (Ed.), *Contemporary philosophy: Vol. 2. Philosophy of science* (pp. 19–49). The Hague: Martinus Nijhoff.

Barlow, D. H., & Hersen, M. (1984). *Single case experimental designs.* New York: Pergamon Press.

Beck, S. J. (1953). The science of personality: Nomothetic or idiographic. *Psychological Review, 60(6),* 353–359.

Bem, D. J. (1983). Constructing a theory of the triple typology: Some (second) thoughts on nomethetic and idiographic approaches to personality. *Journal of Personality, 51(3),* 566–577.

Cournot, A. A. (1956). *An essay on the foundations of our knowledge.* New York: Liberal Arts Press.

Davidson, P., & Costello, C. (1969). N = 1: *Experimental studies of single cases.* New York: Van Nostrand Reinhold.

Falk, J. L. (1956). Issues distinguishing idiographic from nomothetic approaches to personality theory. *Psychological Review, 63(1),* 53–62.

Freyd, M. (1925). The statistical viewpoint in vocational selection. *Journal of Applied Psychology, 9,* 349–356.

Hersen, M., & Barlow, D. H. (1976). *Single case experimental designs.* New York: Pergamon Press.

Kazdin, A. (1980). *Research designs in clinical psychology.* New York: Harper & Row.

Kazdin, A. (1981). Drawing valid inferences from case studies. *Journal of Consulting and Clinical Psychology, 49,* 183–192.

Kenrick, D. T., & Dantchik, A. (1983). Interactionism, idiographics, and the social psychological invasion of personality. *Journal of Personality, 51(3),* 286–307.

Kratochwill, T. R. (Ed.). (1978). *Single subject research.* New York: Academic Press.

Lewin, K. (1931). The conflict between Aristotelian and Galileian modes of thought in contemporary psychology. *Journal of General Psychology, 5,* 141–177.

Meehl, P. E. (1954). *Clinical versus statistical prediction: A theoretical analysis and a review of the evidence.* Minneapolis, MN: University of Minnesota Press.

Plooij, F. X., & Dungen, M. van den. (1985). *Hulpverleningspraktijk en dienstverlenend onderzoek: Handelingsplannen en directe observatie van opvoerder-kind interaktie.* Lisse: Swets & Zeitlinger.

Runyan, W. McK. (1982). In defense of the case study method. *American Journal of Orthopsychiatry, 52(3),* 440–446.

Runyan, W. McK. (1983). Idiographic goals and methods in the study of lives. *Journal of Personality, 51(3),* 413–437.

Sarbin, T. R. (1944). The logic of prediction in psychology. *Psychological Review, 51,* 210–228.

Van IJzendoorn, M. H., & Van der Veer, R. (1984). *Main currents of critical psychology: Vygotskij, Holzkamp, Riegel.* New York: Irvington.

Viteles, M. S. (1925). The clinical viewpoint in vocational psychology. *Journal of Applied Psychology, 9,* 131–138.

Vygotsky, L. S. (1962) *Thought and language.* Cambridge, MA: M.I.T. Press.

Vygotsky, L. S. (1978). *Mind in society.* Cambridge, MA: Harvard University Press.

Vygotsky, L. S. (1982). *Sobranie sochineniy: Vol. 1. Voprosy teorii i istorii psichologiy.* Moscow: Pedagogika (in Russian).

West, S. G. (1983). Personality and prediction: An introduction. *Journal of Personality, 51(3),* 275–285.

Yin, R. K. (1984). *Case study research: Design and methods.* Beverly Hills, CA: Sage.

PART I

Individual-Based Inference Methodology: Past, Present, and the Future

CHAPTER 1

Psychology as a Science
Resolving the Idiographic-Nomothetic Controversy

ISAAC FRANCK

I

"What is the subject matter of this science?" is a proper methodological question to ask about any empirical science. The late Gordon W. Allport asked this question about the science of psychology in a special way, and thereby raised the old philosophical problem of whether scientific knowledge is knowledge of particulars or of universals. It was Allport's oft-repeated complaint that psychology has given its attention only to universals, has neglected to study the individual, and has therefore been guilty of a serious failure in the fulfillment of its scientific task. Allport's complaint and prescription are reflected in the following two brief quotations:

> As long as psychology deals only with universals and not with particulars, it won't deal with much—least of all human personality. (Allport, 1960, p. 146)

> Psychology will become more specific, i.e., better able to make predictions, when it has learned to evaluate single trends in all their intrinsic complexity, when it has learned how to tell what will happen to this child's I.Q. if we change his environment in a certain way. (Allport, 1940, p. 17)

Adapted from the *Journal for the Theory of Social Behaviour*, 12, (1), 1982. Copyright 1982 by Basil Blackwell, Ltd. Reprinted by permission.

ISAAC FRANK • Late of the Kennedy Institute of Ethics, Center for Bioethics, Georgetown University, Washington, DC 20057.

Accordingly, Allport distinguished between two methods in the study of man, for which he borrowed a pair of terms from Windelband (Allport, 1942, p. 53; 1961, p. 9; Windelband, 1921). He called the study of broad, general, universal laws, and the methods employed in this kind of study, *nomothetic,* from the Greek word *nomos,* meaning *law,* and the study of individuals and the methods employed by this kind of study, *idiographic,* from the Greek word *idios,* meaning *one's own, private.* Allport saw contemporary psychology caught in the "the dilemma" between "science and uniqueness," and stated the "quandry which confronts us" as follows (Allport, 1961, pp. 8–9):

> The individual, whatever else he may be, is an internally consistent and unique organization of bodily and mental processes. But since he is unique, science finds him an embarrassment. Science, it is said, deals only with broad, preferably universal, laws. Thus science is a *nomothetic* discipline. Individuality cannot be studied by science, but only by history, art, or biography whose methods are not nomothetic (seeking universal laws), but idiographic. Even the medieval scholastics perceived the issue, and declared *scientia non est individuorum.*

In further amplification of his version of the idiographic horn of the dilemma posed by him, Allport quoted from Kluckhohn and Murray (Kluckhohn, Murray, & Schneider, 1959, p. 53) the observation that "Every man is in certain respects a. like all other men, b. like some other men, c. like no other man" (Allport, 1961, p. 13).

It is with the respects in which every man is "like no other man" that this horn of Allport's dilemma is concerned. "Individuality," he insisted, "is a prime characteristic of the human nature" (Allport, 1961, p. 21). Moreover, Allport reminded us:

> The application of knowledge is always to the single case....With all its weaknesses the case study remains the preferred tool of all clinicians, psychiatrists, personnel officers, and consulting psychologists. They did find that the single case cannot be reduced to a colligation of scores. Here, then, we encounter a pragmatic reason why idiographic procedures must be admitted to psychological science: practitioners demand them. (Allport, 1942, p. 58)

Allport thus apears to have been asking for what might be called a science of individuals, of single cases, of unique beings, of insightful understandings of each individual personality. He pointed out that the social caseworker takes an idiographic approach in his work with his client, and that no nomotheist can be helped by the general laws of psychology to know what his wife would like for a gift. "He can make this prediction correctly only by knowing his wife's particular pattern of interests and affection" (Allport, 1942, p. 58).

In propounding this view Allport revived an old controversy, and evoked the expected spectrum of reactions. The humanistic psychologists, the individual psychologists, the existential psychologists, and others like them, have welcomed Allport's call for an idiographic psychology with jubilation and uncritical acclaim. On the other hand, disdain and derision have come from the camps of positivism and those with a physical science orientation. Still others, dismissing both the dogmatic idiographicism typified by Allport, and the doctrinaire nomotheticism of the scientifically minded, have maintained that psychology as a science is and must be *both* idiographic and nomothetic, and that the putative dichotomy between the two methods is spurious.

The latter view is exemplified on the one hand by the noted psychologist, Kurt Lewin, who said:

> Even if all the laws of psychology were known, one could make a prediction about the behavior of a man only if in addition to the laws, the special nature of the particular situation were known....A task of equal importance...involves the task of representing concrete situations in such a way that the actual event can be derived from the laws according to the principles which are given in the general laws. (Lewin, 1936, p. 11)

A more recent psychologist, Merle B. Turner, associating himself with Lewin's view in contrast to Allport's, observes (Turner, 1967, p. 290):

> On closer inspection...it has appeared that the issues between nomothetic and idiographic methods in psychology as propounded by Allport are not at all clearly drawn. Both idiographic and nomothetic methods of description and understanding are essential. One is to supplement the other.

Manifestly, then, clarity about the relationships between idiographic and nomothetic methods in the science of psychology is not available in abundance. In the pages that follow I shall endeavor to contribute to the clarification of the issue by analyzing some of the contending views referred to previously, and by exhibiting the inextricable interdependence between psychological inquiry when it is pursued in a theoretical-conceptual framework, and psychological inquiry when pursued within an operational or clinical matrix by psychologists as practitioners. I hope that additional light will thus be shed on what I may venture to call the idiographic-nomothetic symbiosis in psychology as a science.

II

In exposing his position Allport seems to have gone back, at least so far as knowledge of *man* is concerned, to a prevalent interpretation of Duns Scotus's doctrine of *haecceitas—thisness*. Duns Scotus (1265–1308) argued that, because the human intellect discovers that individual things are distinct, it follows that in each thing there must be some inherent, fundamental characteristic that differentiates it from all others. Individual things are distinguished from each other in *essence* as well as in existence. Thus, according to Richard McKeon on Scotus's ontology and theory of knowledge,

> the principle of individuation is in the form, not merely in the matter; the essence of each individual contains the principle of contradiction and limitation which restrains the universality of the species: the ultimate reality of the thing which is, contracts the specific form. This is the doctrine of *haecceity*, according to which the characteristics of individuation are not to be found in the quantity or in any other attribute of body . . . but in a formal distinction derived from the thing. . . . Duns Scotus . . . considered that which is as particular, and as known in each case by the haecceity peculiar to the particular thing. (McKeon, 1930, pp. 305–306)

What Scotus himself tells us about the nature of *haecceitas*, in one of his texts, is as follows:

> If you ask "what is this individual thing from which difference is taken? Is it not matter, or form, or the composite?" I reply that every quidditive entity, whether partial or total of any kind, is of itself indifferent, as quidditive entity, to this entity and that one, so that the quidditive entity it is naturally prior to this entity as this. . . . And just as the composite insofar as it is a nature does not include the being by which it is "this," so neither does matter insofar as it is nature, nor form. Therefore, this being is not matter, nor form, nor the composite, insofar as any of these is a nature; but it is the ultimate reality of the being which is matter, or which is form, or the composite. . . . (Scotus, 1967, p. 589)

However, another commentator, Anne Fremantle, after quoting from D. J. B. Hawkins's *A Sketch of Medieval Philosophy* appends her own view, which is typical of a widespread misapprehension of Scotus's doctrine of *haecceitas:*

> Hence, beyond all that in reality corresponds to universals . . . [Scotus] claims that things exhibit a principle of individuality, a *thisness*, which is not reducible to any other factor. The singular adds an entity over and above the entity of the universal. . . . We cannot understand anything until we have understood its *thisness*, and the difference between *thisness* and *thatness*. . . . In fact, for Duns Scotus, the individual is the only existing thing, and it is not the being of *being* but the being of the individual which is investigated by philosophy. (Fremantle, 1955, pp. 183–184)

Thus, to Scotus's claim that each individual thing has a *haecceitas* that could never be resolved into a class concept has been superadded a second doctrine and (incorrectly, I shall argue later) ascribed to Scotus by some commentators, Anne Fremantle among them. This doctrine claims that we can have no understanding of anything until "we have understood its *thisness*," and that only the *individual* is to be studied in our pursuit of knowledge.

III

If this latter doctrine is what Allport meant, namely that because every individual has a *thisness* that cannot be resolved into a class concept, the study of man ought to concentrate on the study of these unique and irreducible individualities—and, as I shall try to show, it is not at all clear that this is what he meant—then he would have support from various quarters in current and recent thought. His strongest support may be found in the existentialist school. Allport pointed out that "at bottom the existentialist approach to man is utterly idiographic" (Allport, 1961, p. 557). He sometimes sounded like an existentialist, as when he uttered pronouncements like the following:

> *The outstanding characteristic of man is his individuality.* He is a unique creation of the forces of nature. (Allport, 1961, p. 4)

> Every person deviates in thousands of ways from the hypothetical average man. But his individuality is not the sum of these separate deviations. (Allport, 1961, p. 7)

> Existence ultimately resides nowhere except in the individual's point of view. (Allport, 1961, p. 557)

But some existentialists go much further than Allport. They deny the very possibility of science and of systematic thought in general. They appear to say that the only significant kind of human knowledge is the knower's intuitive, insightful apprehension of the individual thing, or of the individual person in his pure individuality, in his utter uniqueness. All other knowledge claims distort and tend to destroy both the knower and the known. Allport himself pointed out that for the existentialists,

> to force existence into a theoretical system is to destroy it. . . . Another's existence cannot be pinned down or communicated by devices; an extrascientific grasp is the best we can hope for. (Allport, 1961, p. 557, note 6)

He made these comments in a footnote, in connection with his reference to a statement by Soren Kirkegaard, which is quoted by Jean Wahl. Kirkegaard stated (Wahl, 1949, p. 4), "One might say that I am the moment of individuality, but I refuse to be a paragraph in a system."

But Allport's ascription to the existentialists of the view that "an extrascientific grasp is the best we can hope for" is a misleading understatement that, under critical inspection, reveals that there is in fact a considerable gap between Allport's own views and those of the existentialists, in spite of his emphasis on idiographic knowledge and methods. The existentialist would insist that, for authentic knowledge, a direct, intuitive, "extrascientific grasp" of the other person's concrete uniqueness and individuality is the epistemological ideal and goal we should hope for and strive for, and that scientific knowledge of generalizations, abstractions, laws, and so forth, is a barrier to such authentic knowledge. On the other hand, Allport's emphasis on idiographic knowledge did not lead him to any rejection of science. At most it resulted in some confusion on his part about the differences between these two kinds of knowledge and the relationships between them, as will become clear in the ensuing discussion. Allport was not without some awareness of the distance between his and the existentialists' views. He pointed out that the existentialist approach to man "as yet ... offers no special methods for representing the unique structure of persons ... the movement has not yet evolved genuinely novel methods for the representation of individuality" (Allport, 1961, p. 557). However, in a footnote to this passage, Allport acknowledges, with a kind of embarrassed self-consciousness, that the existentialists may find his desire for such "methods and representation" quite uncongenial: "Existentialist writers may dispute this implied criticism. They may say that it is not the purpose of existential analysis to become objective and scientific. To force existence into a system is to destroy it" (Allport, 1961, p. 557, note 6).

For, unlike Allport, even when he urged the pursuit of idiographic knowledge, some existentialists seem to be categorical in their rejection of scientific knowledge. This may be seen unequivocally in the epistemology imbedded in Martin Buber's distinction between the I–Thou and the I–It relationship. Maurice S. Friedman's exposition of Buber's theory of knowledge helps throw into bold relief the existentialist rejection of science, and its claim that the direct, intimate, intuitive I–Thou bond with the concrete particular yields the only real knowledge. In Buber's epistemology, Friedman (1955, pp. 172–173) tells us:

> the philosophical anthropologist ... must discover the essence of man not as a scientific observer ... [because] science investigates man not as a whole but in selective aspects.... Scientific method is man's most highly perfected development of the I–It, or subject–object, way of knowing.... Just for these reasons scientific method is not qualified to discover the essence of man. It can compare men with each other and man with animals.... This scale ... can be of aid ... but not in discovering the uniqueness of man as man.

It is, in fact, only the knowing of the I–Thou relation which makes possible the concept of the wholeness of man.

IV

Allport did not reject the generalized knowledge which comes to us from science and scientific method, though other writers, not only existentialists, who are allied with Allport in the emphasis on idiographic knowledge, are either on the threshold of such rejection, or are deeply enmeshed in it. For example, Alderian adherents of the school of individual psychology seem to be implying such a rejection in descriptions like the following:

> Adler's Individual Psychology would be the idiographic science par excellence. . . . Adler was not satisfied with probabilities; he wanted a psychological theory which would be adequate for each individual case, the exceptions as well as the rule. What he aimed at was a truly idiographic psychology. (Ansbacher, 1959, p. 34, 29)

And A. H. Maslow, in his cavalier, swashbuckling manner, announced that

> American psychologists have listened to Allport's call for an idiographic psychology but have not done much about it. Not even the clinical psychologists have. We now have an added push from the phenomenologists and existentialists in this direction, one that will be very hard to resist, indeed, I think, theoretically impossible to resist. If the study of the uniqueness of the individual does not fit into what we know of science, then so much the worse for the conception of science. It, too, will have to endure re-creation. (Maslow, 1961, pp. 53–54)

A clear warning against this extreme idiographic antiscientism that, paradoxically, some psychologists and other social scientists have wished to insinuate into the social sciences, comes from the pen of Alfred Schutz, a keen student of the methodology of the social sciences (Schutz, 1963, p. 232):

> It has been maintained that the social sciences are idiographic, characterized by individualizing conceptualization and seeking singular assertory propositions. . . . Some proponents of the [above view] . . . were inclined to identify the methodological situation in one particular social science with the method of the social sciences in general. Because history has to deal with unique and non-recurrent events, it was contended that all social sciences are restricted to singular assertory propositions.

Arguing by anology, the idiographic antiscientists contend (and sometimes Allport appears to have contended) that because psychotherapy, social casework, and the like, have to deal with the unique and (in many respects) nonrecurrent individual person, psychology must work toward singular assertory propositions. That the history of the individual case is of crucial importance for the work of the psychotherapist, case workers, or guidance counselor, is of course axiomatic. In psychoanalysis in particular, the unique history of the patient is of greatest importance. This was shown by Hans Meyerhoff in an interesting paper in which he charted the large areas of overlap between the disciplines of psychoanalysis and history, and in which he made the keen observation that "psychoanalysis 'works' only insofar as we reconstruct the history of the individual case" (Meyerhoff, 1962, p. 12).

Thus, for certain purposes, the direct, idiographic knowledge of the unique, nonrecurrent individual person, the person's case history, or the I–Thou insight and *Einfühlung* that take place between close friends, lovers, husband and wife, therapist and patient, are irreplaceable. Moreover, though the existentialists are profoundly wrong in assuming or implying that this kind of idiographic, I–Thou knowledge of individual persons is all the knowledge we need, and that it can somehow constitute a knowledge of man, it also seems clear that the generalized, nomothetic knowledge of man through the science of psychology can never reach or exhaust the rich uniqueness and individuality of the person. The existentialists are profoundly right in insisting that a kind of knowledge other than the generalizations of the science of psychology is needed for the "knowledge from within" of the unique individual person. However, by the same token, one has to respond to Allport that this knowledge of the full uniqueness of the individual person "from within" is by itself uncodifiable into a science of psychology. This has to be said, because Allport appears to have been confused about aspects of the distinction between idiographic and nomothetic knowledge, and therefore, in the final analysis, his point is unclear.

As was noted above, Allport neither denied nor objected to "the proposition that psychology seeks general laws" (Allport, 1961, p. 572). His insistence that psychology must deal with individuals was based, as was pointed out earlier, on two considerations, one substantive, the other methodological. The substantive point is his belief in "the real possibility that no two lives are alike in their motivational processes" (Allport, 1942, p. 57). The methodological point is that even abstract, general, nomothetic science can never escape the individual, because its findings must be applied to the individual object.

> The application of knowledge is always to the single case. We apply the science of engineering only in building particular conduits or bridges. In the human realm we have to particularize our nomothetic knowledge before it is of any value. (Allport, 1942, p. 58)

> No general principles can ever be applied except to concrete and particular objects. The individual case stands at the gateway and terminus of generalized knowledge. (Allport, 1942, p. 150)

So far, nothing in these passages suggests any departure from the assumptions and procedures of the sciences generally, including the physical sciences. However, Allport appears to have believed that he was pointing to psychology's need for a departure from the nomothetic procedures of science, when he directed our attention to the fact that

> general laws of human behaviour known to us are altered and sometimes negated by the idiographic knowledge available to us concerning the personality we are studying. (Allport, 1942, p. 58)

But this is also true of the application of the science of engineering to bridge building. If, upon idiographic examination, a fault is discovered in a beam of structural steel, or some other significant deviation from the required general character of the other beams, the beam is discarded. Or if, as happened a number of years ago, a wind of unusually great force blew down a great suspension bridge, and it was concluded that the bridge's rigidity was responsible for the occurrence, the idiographic knowledge about *that* bridge led to the altering of the general laws of bridge building so that future bridges were to be built with less rigidity and more elasticity. Similarly, in automobiles designed in accordance with nomothetic principles defects show up when individual cars are driven by their buyers, and the car design then has to be changed in conformity with revised nomothetic principles of automotive design. New, nomothetically designed airplanes are produced, a crash of an individual plane occurs, and a revised design is substituted in accordance with revised nomothetic principles of aerodynamics. This, therefore, cannot be the point in Allport's insistence that psychology must depart from the nomothetic model of science and deal idiographically with individuals.

The confusion on this point, so often present in Allport's writing, and carried to a thoughtless extreme by some of the existentialists, Adlerians, and Maslow's humanistic psychologists, is in part a result of the telescoping of two contentions that need to be separated out. First, there is the contention that psychology should be concerned with the individual case, with differences between individuals, and so forth, which is certainly arguable, and which seems to be essentially what Allport was intent on saying. But the second contention is that nothing general can be said about individuals and individual differences, a contention that is not plausible, and that in fact was not what Allport meant, though others (e.g., Maslow) attributed this to him.

A similar correction has to be made in understanding Duns Scotus's doctrine of *haecceitas*. Scotus did indeed insist, for metaphysical reasons, that each individual thing could be completely described only if we could know its haecceity intellectually. However, the need for haecceity was, in Scotist doctrine, "a logical one." An individual's *haecceitas* is known only to God, and "can be known by man only in a future life" (Wolter, 1967, pp. 431–432). Scotus did not believe that we could grasp even conceptually any individual's unique haecceity, or know it by acquaintance intellectually. We only know about it logically, by description, as it were. Nor did Scotus think that we could have a science of the individual according to his haecceity. Our conceptual knowledge is of the universal. We arrive at an intellectual awareness of individuals by a combination of our sense knowledge and intellectual knowledge. So far as the conditions for scientific knowledge are concerned, Scotus stressed repeatedly that "*scientia est universalium*" ("science is of the universal").

And finally, there is an instructive lesson in the science of astronomy that illustrates the intersection in science of concern with individuals and concern with generalizations and classes. Astronomy is a physical science, but it certainly makes claims about particulars, the sun, the moon, that planet, or this star, and propounds "singular assertory propositions" about them. But these singular assertory propositions function within a matrix of generalizations about physical objects, their motion, inertia, acceleration, gravitation, speed of light, radiation, mass, energy, chemical interactions and their consequences.

V

The point that Allport seems to have been getting at is something different from the denial that there can be generalizations about individuals, and this one *is* a point about psychology. Each individual human being is unique. The individual person is "a system of patterned uniqueness" (Allport, 1961, p. 9). But it also a fact, said Allport, that "science likes universals and not particulars" (Allport, 1961, p. 9). This, for Allport, seemed to present a problem. Let us look at the problem as stated in Allport's own words (Allport, 1961, p. 9):

> Personality itself is a universal phenomenon though it is found only in individual forms. Since it is a universal phenomenon science must study it; but it cannot study it correctly unless it looks into the individuality of patterning! Such is the dilemma.

It is the responsibility of psychology, according to Allport, to investigate "how uniqueness comes about," to study "those processes that bring about uniqueness" (Allport, 1961, p. 10). Accordingly, the reason conventional science is baffled is that it cannot see how the internal organization of the particular can fit into its nomothetic search for general laws. (Allport, 1961, p. 10)

But Allport strangely failed to see that the investigation of the processes that bring about uniqueness, the quest for "the laws that tell us how uniqueness occurs," is ultimately a nomothetic enterprise, the successful pursuit of which will give us abstract, general knowledge about how these individual systems of patterned uniqueness come about, rather than knowledge of the idiographic kind (though of course much idiographic investigation has to take place in order to lead to the nomothetic goal of formulating the laws of how uniqueness comes about, but this is true in varying degrees of all empirical investigation). The nomothetic character of this part of Allport's quest was betrayed in his own words, when he said that he had shown "that a general law may be a law that tells us how uniqueness comes about," (Allport, 1937, p. 558), or that he had "drawn special attention to those laws and principles that tell how uniqueness comes about" (Allport, 1961, p. 572).

What remain, therefore, out of Allport's advocacy of the use of idiographic methods in the study of man, are two other contentions: (a) that "the personal patterns of individuality are unique," (Allport, 1961, p. 10) that is, that each individual person is unique; and (b) that "the behavior of every individual is lawful in its own right," (Allport, 1961, p. 10) that

each personality is a law unto itself . . . The course of each life is a lawful event, even though it is unlike all others of its class. Lawfulness does not depend upon frequency nor upon uniformity, but upon necessity. There is a necessary patterning in each life. (Allport, 1937, p. 558)

VI

Now, before getting on with the main argument of this section, one feels constrained to digress, and inquire what Allport meant by the claim that the behavior of every individual is "lawful" in its own right; that each personality is a "law unto itself"; that the course of each life is "a lawful event"; that this lawfulness depends "upon necessity"; and that there is a necessary patterning in each life. There is no trouble with the obvious meanings of this claim, which Allport makes clear, namely, (a) that there is a consistency, a predictability, even a necessity in the behavior pattern of each individual; and (b) that the better you know an individual person, that is, the more you have studied or experienced the person, the more accurately you will be able to predict that person's behavior.

However, the less obvious and the more knotty problem, namely, what are the causal explanations of this consistency and lawfulness, Allport did not probe. He disposed of this problem by brief, almost casual, and dogmatic references to some concepts out of the writings of Aristotle, St. Thomas Aquinas, Spinoza, and Leibniz.

When a body of phenomena is said to display lawful behavior, it is natural to seek explanations for this, to ask what the causes are for this lawful behavior. As to what the causes are for the lawfulness of the behavior of each person, Allport would of course have rejected the behaviorist attempt to give explanations exclusively in terms of physiological causes, such as drives and tensions, and of the "law of effect" that refers to reward-and-punishment conditioning. He would also have rejected the Freudian answer that would explain the lawfulness of each person's behavior in terms of such psychological causation as early childhood experiences, the oedipal phase and toilet training.

What Allport seems to have suggested is the rather vague theory that there is a kind of *immanent* purposive force within each person that explains the person's lawful, consistent behavior. He invoked Aristotle's doctrines of *orexis* and *entelechy*, Aquinas' doctrine of *intention*, Spinoza's concept of *conatus*, and Leibniz's notion of the intellect as the source of its own ideas; and his own notion of the person as the source of his own acts. This seems to be a doctrine of immanence with respect to the behavior of the discrete individual which would make each person a *causa sui*. Such a doctrine is at least dubious, and would require considerable critical scrutiny.

VII

To return now to the main argument, it is difficult to understand why the two contentions mentioned earlier, namely, that each person is unique, and that his behavior is lawful, require additional idiographic knowledge of individuals. These contentions are themselves generalizations, general laws, and therefore by definition nomothetic in character. Either these two generalizations have been reasonably well established on the basis of empirical evidence, in which case no additional idiographic evidence is needed for their support (except in the sense that the evidence for empirical generalizations of this type is never complete, and the generalization or law or theory always retains the character of a "conjecture," cf. Popper, 1962, pp. 33–66) or, if the generalizations have not been established well enough, then additional idiographic evidence for their support has to be sought; but the objective of such a quest would be a nomothetic one, that is, the establishment of a general law, and would not differ essentially from the procedures of other scientific investigations. On the other hand, if the therapist, or social caseworker, or personnel director needs for his special purposes genuinely idiographic knowledge *about this or that person,* the acquisition of this knowledge would not require extensive idiographic studies of other individuals. Allport himself understood this, as we can see in his (Allport, 1961, p. 10) assertion that

> we do not need to understand every life in order to discover the lawful regularities in one life. If you have an intimate friend, you may know very well why he behaves as he does . . . because you know the lawful regularities in his life.

Let us, however, further examine Allport's emphasis, and the emphasis of others, on the need for "an idiographic psychology" by making the extreme assumption that, pursuant to this goal, every human being on earth were to be subjected to the suggested idiographic analysis; what kind of knowledge would this hypothetical undertaking yield, and to what uses could it be put? What, as a result of this undertaking, would the science of psychology get that it now does not have? What would the results of this undertaking contribute toward the three contentions stressed by Allport, namely, that the personal matters of individuality are unique; that the behavior of every individual is lawful in its own right; and that psychology must discover the laws of how uniqueness comes about?

First we must note that because, *ex hypothesi,* every human being on earth would be studied, we would have an instance of what the logic textbooks call "perfect induction," or "complete induction." It would then seem that, if the idiographic studies gave proper attention to properly selected variables, the consolidated results should give us complete and certain knowledge of the laws of how uniqueness comes about. However, as was pointed out above, this would be essentially nomothetic knowledge, knowledge of general, abstract laws of certain aspects of human nature, abstracted from the totality of all of the individual human beings. Moreover, there is no reason to suppose that, for the discovery of the laws of how uniqueness comes about, a study of all members of the class of human beings is needed, and that reasonably well-established laws could not be arrived at on the basis of studies of a sample of the class. Indeed, if only perfect induction could give us these laws, then all future generations of human beings would have to continue being subjected to the idiographic studies advocated by Allport. Thus psychology would finally achieve a full knowledge of these laws of how uniqueness comes about only in the ultimate moment of collective self-contemplation immediately before the cataclysm that destroys the entire human race. This seems to me not too unfair a *reductio ad absurdum* of the call for an idiographic psychology.

As to Allport's contentions that each individual is in some respects unique and that each person is a lawful system, either these have already been established as sound empirical generalizations, or as in the case of other scientific disciplines, psychology must continue its investigations by studying individual human beings (idiographically, if you will) to see whether the two nomothetic contentions under discussion are supported by the evidence. Here again, it would hardly require idiographic studies of all human beings in order to establish satisfactorily these two generalizations. At most, what it would require is more careful or thoroughgoing study of individuals in order to have a more solid foundation of evidence for the nomothetic generalizations. This does not add up to an idiographic psychology.

It would seem, however, that the knowledge obtained from such idiographic studies of all human beings could have two kinds of uses. First, it would be useful to therapists, social case workers, guidance counselors, or personnel directors in helping them to a more thorough understanding, and to more accurate predictions of behavior, of those in the total population who are their patients or clients. In other words, it would be useful to have this idiographic knowledge about those individuals about whom it is important to have individual knowledge for such specific purposes. But this seems hardly more than a tautology, and one is therefore tempted to ask what need there is for universal idiographic studies of individuals other than the patients and clients concerned, and, by the same token, to ask what need there is for an idiographic psychology. What *is* needed is careful idiographic study of each individual who, for one reason or another, is in need of the attention of the type of practitioner mentioned above, so that the practitioner would know as much as possible about the laws of his patient's or client's inner personality system. If Allport meant that there ought to be more of this done, and that it ought to be done better than it is being done now, then his admonition is probably justified, but, from the point of view of the systematic psychological study of human nature, relatively trivial.

The other use to which idiographic studies of all human beings could be put is the prediction of the behavior of human beings generally. We can predict the behavior of an intimate friend more accurately than can be done by another person who does not have our idiographic knowledge of our friend; and it is so with each person of whom we possess idiographic knowledge. Accordingly, where the prediction of individual persons' behavior is important, idiographic knowledge of these persons is important. Allport complained that science is greatly disadvantaged in the prediction of the behavior of individuals, "because the best source of that prediction is the past behavior of the individual" (Allport, 1942, p. 155). His complaint about science continues this theme:

> Where this [scientific] reasoning seriously trips is in the prediction applied to the single case, instead of to a population of cases. A fatal non-sequitur occurs in the reasoning that if 80% of the delinquents from broken homes are recidivists, then *this* delinquent from a broken home has an 80% chance of becoming a recidivist. The truth of the matter seems to be that *this* delinquent has either 100% certainty of becoming a repeater or 100% certainty of going straight. If all the causes in his case were known, we could predict for him perfectly (barring environmental accident). His chances are determined by the pattern of his life and not by the frequencies found in a population at large. (Allport, 1942, pp. 156–157)

Here Allport was once again saying that if you wish to know about the probable behavior of an individual, you will be better off if you have as much knowledge as possible about this individual's personality system, than if you have only statistical generalizations about a class of persons to which he belongs. This is probably right, and it means that if you wish to make predictions about recidivism in the case of 1,000 specific delinquents, your predictions will be better if you have idiographic knowledge of each one of them. But you will then in fact be making 1,000 disparate predictions, and idiographic studies of other individual human beings, whether all of them or only a sample of them are likely to be irrelevant to these 1,000 predictions. Again we end up with the trivial conclusion that it is useful to have idiographic knowledge about those individuals about whom it is important to have individual knowledge for specific reasons or purposes. And again, this does not add up to an idiographic psychology.

VIII

To sum up then, in the study of the nature of man we may acquire knowledge of individuals in their unique and irreducible individuality, and we may acquire general knowledge of aspects of human beings, for example, traits, or propensities, which are abstracted from the concrete individuals. This is true in all other empirical sciences. However, in the case of human beings, the knowledge of the individual is of special importance, because the behavior, the future, the fate or destiny of fellow humans in general and of certain individual humans in particular is of importance to us. Idiographic knowledge of the individual person is of crucial importance in the work of therapists, caseworkers, guidance counselors, and the like, because it is with the individual that these psychologists and practitioners deal. Knowledge of the uniqueness of the unique individual in the sense of *Einfühlung*, or the I–Thou relationship between persons, is of transcendent importance in interpersonal relations, but this is of course the kind of knowledge that one can only *experience* directly and ineffably; and it is hardly possible to talk about it. At best one can try to describe it elliptically or metaphorically. In itself, it is not systematically codifiable.

Manifestly, the study of the nature of man also requires the acquisition of general, abstract knowledge of human behavior; general laws about the psychological constitution of human beings, and about probable human actions under certain conditions. This is the kind of knowledge Allport called nomothetic. Without this kind of knowledge, a science of psychology, and more generally, a science of man, would not be possible. Allport was no doubt right when he said that *"psychological causation is always personal and never actuarial"* (Allport, 1942, p. 157), but it is through the study of many instances and many kinds of personal causation, and through codification of them into nomothetic generalizations, that our systematic knowledge of man will increase and deepen. Indeed, just as new idiographic knowledge of individuals may serve as a point of departure for extending and enriching our knowledge of laws of human nature, so may nomothetic knowledge of general laws of human behavior make possible deeper and more trustworthy idiographic insights into individuals by facilitating our understanding of the respects in which the individual under study is "like (some or all) other people," and the respects in which he is "like no other men."

Historical knowledge, and the methodology of historical inquiry, have often been said to be strictly idiographic. However, Ernest Nagel rejects such rigid exclusivity. Approaching this issue in connection with his examination of problems on the logic of historical inquiry, Nagel (1961, p. 548) concludes that (natural) science is not purely nomothetic, and that history is not purely idiographic discipline. "It would be a gross error," Nagel tells us, "to conclude that singular statements play no role in the theoretical sciences or that historical inquiry makes no use of universal ones." As to the natural sciences, Nagel reminds us that even they can assert their general statements as empirically warranted only on the basis of concrete factual evidence, and therefore only by making use of singular statements. Moreoever, many . . . laws of "pure" science have a generality that is at least geographically restricted . . . Furthermore, some branches of natural science . . . are primarily concerned with spatiotemporal distributions and the development of specific individual systems, and are therefore engaged in establishing statements singular in form. (Nagel, 1961, pp. 548–549, emphasis added)

Returning to psychology now, the interplay of singular statements and universal statements in the systematic discourse about human nature and conduct is in fact quite familiar. But what may be worth recalling here, in connection with Allport's call for an idiographic psychology, is the extent to which even therapists depend on nomothetic generalizations to help them better to understand idiographically their patients or clients. Anthony Kenny's observation is helpful here, when he says that "Being told that a man acted out of vanity helps us to understand his action . . . because we say to ourselves: 'Yes, of course, men *often* act like that' " (Kenny, 1963, p. 95). Moreoever, the therapist is helped in making predictions about individuals, that is, idiographic predictions, by the nomothetic knowledge he has of regularities and of generalizations concerning the class of similar cases. Jacob A. Arlow's clinical illustration is instructive in this connection, and it is worth quoting from it at length (Arlow, 1960, pp. 206–207):

During an initial interview I asked a patient how long he had been married. He answered, "Sixteen months, three weeks." The overly exact quality of this response aroused in me the suspicion that I was dealing with a person whose character structure was colored by obsessional thinking and compulsive traits. To confirm my suspicion I asked further, "How long did you know your wife before you married her?" He answered, "Two years, three months." At this point, inwardly, I make a further set of predictions concerning this individual's mental traits. I guessed that he would be especially concerned with money. . . . A further set of predictions concerned his relationship to cleanliness. I could guess that he would be excessively neat regarding his person and his clothes, tidy in his surroundings, orderly in his manner, and rigorously punctual regarding appointments and the fulfillment of financial obligations. Questioning confirmed each of these predictions in minute detail. But even further predictions can be made on the basis of the minimum hints given by this patient. Such predictions in psychoanalysis . . . have been validated regularly, hundreds of times in psychoanalytic investigations.

The various illustrations in the preceding pages exhibit the paradigmatic interplay and reciprocal illumination between the psychologist's idiographic knowledge of the individual person and his nomothetic knowledge of generalizations about all human beings or about classes of human beings. The illustrations also must lead to the conclusion that the idiographic-nomothetic controversy has been misconceived by the doctrinaire and dogmatic partisans on both sides, as well as by those— most often nomotheticists—who cavalierly and impatiently dismiss the whole matter as a "pseudoissue" (cf. Marx, 1964, p. 313). It is not a pseudoissue. The fact that, as was argued previously, there is no mutually exclusive dichotomy between these two methods and two kinds of knowledge in psychology, and that they are in fact inextricably intertwined, does not mean that they may not be distinguished from each other in discourse, or that there is no need to understand the nature of the relationships between them. This has been exhibited previously, and two additional illustrations should help make the relationships clear.

The nomothetic generalization that all human acts, no matter how trivial, capricious, or inexplicable they may appear to be, acts such as slips of the tongue, memory lapses, or dreams, are in fact explicable in terms of unconscious processes in which they originate, does not by itself tell us either that this patient's lapsus linguae is of any special significance, nor does it by itself explain what its significance may be in this patient's symptomatology. It is only in the context of and in relation to much other idiographic knowledge about this patient that his slip of the tongue may be seen to have a cause or a reason that is explanatory of this and other elements of his behavior. When, in the light of the additional idiographic knowledge of the patient, the slip of the tongue is indeed discovered to be of significance in this particular case, this in turn may be seen as another confirmatory instance of the nomothetic generalization, and thus the nomothetic knowledge is enriched.

On the other hand, the intensive idiographic study of each of the 500 juvenile offenders in a particular institution would no doubt help the parole authorities to hazard a prediction of future behavior in each case when an offender is up for parole. But such idiographic knowledge and succession of individual decisions will not advance our knowledge of delinquency, confinement, parole, and recidivism, unless it is supplemented by the nomothetic search for trends, invariances, or correlations that may help discover possible relations between certain fators and recidivism. The resulting nomothetic generalization, though only of a certain degree of probability, may lead to the application of this nomothetic knowledge idiographically in the form of revised treatment of this or that offender who has displayed the given characteristic. In short, if psychology as a science pursued nomothetic generalizations exclusively, without interest in the individual case, or if it pursued only the idiographic knowledge of individuals with no interest in nomothetic generalizations, the hoped for achievement of explanation and understanding of human behavior would remain a will-o'-the-wisp.

ACKNOWLEDGMENTS

A substantially abbreviated verison of this paper was presented at the Sixth Annual Meeting of the Society for Philosophy and Psychology on March 15, 1980, at the University of Michigan, Ann Arbor, Michigan. The author is grateful to Susan B. Haack (University of Warwick) and Allan B. Wolter (Catholic University of America) for their critical reading of this paper, and for their helpful suggestions, a number of which have been incorporated in this revised version; to Kent Bach (San Francisco State) who served as Commentator on this paper at the Ann Arbor meeting; and to Laurence B. McCullough (Kennedy Institute of Ethics) for making available to me his dissertation, "The Early Philosophy of Leibniz on Individuation" (unpublished), which was helpful in the clarification of some ideas. The author is especially grateful to Paul F. Secord for his criticisms of the original manuscript and his editorial suggestions.

REFERENCES

Allport, G. W. (1937). *Personality: A psychological interpretation.* New York: Holt.
Allport, G. W. (1940). The psychologist's frame of reference. *Psychological Bulletin, 37, 1–28.*
Allport, G. W. (1942). *The use of personal documents in psychological science.* New York: Social Science Research Council.
Allport, G. W. (1960). *Personality and social encounter.* Boston, MA: Beacon Press.
Allport, G. W. (1961). *Pattern and growth of personality.* New York: Holt, Rinehardt & Winston.
Ansbacher, H. L. (1959). Causality and indeterminism according to Alfred Adler, and some current American personality theories. In K. A. Adler & D. Deutsch (Eds.), *Essays in individual psychology* (pp. 27–40). New York: Grove Press.
Arlow, J. A. (1960). Psychoanalysis as scientific method. In S. Hook (Ed.), *Psychoanalysis, scientific method, and philosophy: A symposium* (pp. 201–211). New York: Grove Press.
Fremantle, A. (1955). *The age of belief: The medieval philosophers.* New York: Mentor Books.
Friedman, M. S. (1955). *Martin Buber: The life of dialogue.* Chicago, IL: University of Chicago Press.
Kenny, A. (1963). *Action, emotion, and will.* London: Routledge & Kegan Paul.
Kluckhohn, C., Murray, H. A., & Schneider, D. M. (Eds.). (1959). *Personality in nature, society, and culture.* New York: Knopf.
Lewin, K. (1936). *Principles of topological psychology.* New York: McGraw-Hill.
Marx, M. H. (1964). Confusion in attitudes toward clinical theory. In M. H. Marx (Ed.), *Theories in contemporary psychology* (pp. 311–323). New York: Macmillan.
Maslow, A. H. (1961). Existential psychology: What's in it for us. In R. May (Ed.), *Existential psychology* (pp. 49–57). New York: Random House.
McKeon, R. (Ed.). (1930). *Selections from Medieval philosophers* (Vol. 2). New York: Scribners.
Meyerhoff, H. (1962). On psychoanalysis as history. *Psychoanalysis and the Psychoanalytic Review, 49 (2),* 3–20.
Nagel, E. (1961). *The structure of science: Problems in the logic of scientific explanation.* New York: Harcourt, Brace & World.
Popper, K. R. (1962). *Conjectures and refutations.* New York: Basic Books.

Schutz, A. (1963). Concept and theory formation in the social sciences. In M. Natanson (Ed.), *Philosophy and the social sciences: A reader* (pp. 230–249). New York: Random House,

Scotus, Johannes Duns (1967). *The Oxford Commentary on the Four Books of the Sentences* (Book II, Dist. III) In A. Hyman & J. J. Walsh (Eds.), *Philosophy in the Middle Ages* (pp. 560–604). New York: Harper & Row.

Turner, M. B. (1967). *Philosophy and the science of behavior.* New York: Appleton-Century-Crofts.

Wahl, J. (1949). *A short history of existentialism.* New York: Philosophical Library.

Windelband, W. (1921). *An introduction to philosophy.* London: Unwin.

Wolter, Allan B. (1967). Article on John Duns Scotus. In P. Edwards (Ed. in Chief), *Encyclopedia of Philosophy* (pp. 427–436). New York: Macmillan.

CHAPTER 2

From Idiographic Approaches to Nomothetic Hypotheses

Stern, Allport, and the Biology of Knowledge, Exemplified by an Exploration of Sibling Relationships

KLAUS E. GROSSMANN

INTRODUCTION

The apparent dichotomy between attempts to establish psychology on the basis of universal laws on the one hand, and psychology as an area that tries to concentrate on individual life courses on the other hand, has had a long tradition in philosophy. The dichotomy has been called spurious by Franck (Chapter 1, this volume), because one side must supplement the other for the only possible "idiographic-nomothetic symbiosis in psychology as a science" (Franck, p. 19).

The idiographic-nomothetic controversy is not a pseudoissue (Franck, p. 34) because of the different traditions in scientific thinking. In my own idiographic history I have participated in all three positions. I was very

A portion of this chapter was delivered in symposium at the Biennial Meeting of the Society for Research in Child Development, Boston, 1981. The German quotations have been translated by the author.

KLAUS E. GROSSMANN • Institut für Psychologie, Lehrstuhl IV, Universität Regensburg, D-8400 Regensburg, Federal Republic of Germany.

much in favor of the personalistic system represented by William Stern in Hamburg. At the time I was a student of Kurt Bondy's, who had reopened the Institute of Psychology originally established by William Stern when the University was founded in 1919 (Stern, 1927). Later I was very much in favor of a purely nomothetic psychology when I eagerly absorbed *Psychological Research* (Underwood, 1957) as a Fulbright Student in the United States. And, later still, I was very much in favor of the position that "there is no mutually exclusive dichotomy between these two methods and two kinds of knowledge in psychology and that they are in fact inextricably intertwined" (Franck, p. 34) when I tried to compensate for the grave errors of behaviorism as a student of ethology back in Germany.

In retrospect, the bias toward one end or the other of a proposed continuum between the individual and his integrative organizational powers (as exemplified by the ability to plan for the future), and those universal laws that seem to govern most of our lives, appears to be a matter of preference. Allport's truism, borrowed from Kluckhohn and Murray, was that every man is in certain respects (a) like all other men, (b) like some other men, (c) like no other man. That distinction was known much earlier and it was made the very basis of a "differential psychology" by William Stern as early as 1911 (Stern, 1921). Allport seems to have imported the issue from Hamburg into the American literature with some noteworthy distortions by emphasizing a spurious contradiction at the expense of dialectic unity.

I shall first present a brief account of Stern's personalistic psychology as expressed essentially by Stern himself (Stern, 1927). Second, I shall discuss some of the issues that Allport's writings raised for psychological objectivists (behaviorists) before the rise of humanistic psychology. Third, I shall try to view the issue from a comparative-ethological perspective, where observation, categorization, and interpretation are the main tools of (heuristic) knowledge production. The resulting insights, however, are merely hypotheses and expectancies, open for the methodological, theoretical, and probabilistic challenges of modern psychology. Fourth, and finally, I shall present an example in the making: the abductive categorization of different types of sibling relationships on the basis of interviews, narrative reports, and audiotapes, and its consequences for meaningful psychological conceptualizations (see Eco & Sebeok, 1983).

IDIOGRAPHY AND STERN'S PERSONALISTIC SYSTEM

Stern was born in 1871, founded the famous Hamburg Psychological Institute in 1919, escaped from the Nazis to Holland and later to Duke

University, where he died in 1938. Stern was among the first psychologists who were interested in the single, unique personality, and who, for that reason, considered themselves as "idiographically" oriented. They contrasted themselves to those who, because they were interested in groups of individuals, considered themselves as nomothetic. Stern used these terms in his *Differentielle Psychologie*, first published 1911 (Stern, 1921). Stern's main concern was to clarify the confusion that existed in the trait concept, and he emphasized an idiographic *and* a nomothetic dimension in differential psychology.

Differential Psychology

Stern presented a diagram to explain his ideas (see Figure 1). Located along the vertical axis (a ... z) are traits, whereas individuals are located along the horizontal axis. Four different kinds of research can be conducted, two of the nomothetic variety, and two of the idiographic.

A. Nomothetic
   1. One horizontal level means that one trait is investigated across many individuals for its variability.
   2. Two or more horizontal levels mean that two or more traits are investigated for their correlation across many individuals.
B. Idiographic
   1. One vertical section means that one individual is described on several traits and their interrelationship (*Zusammenhang*) with regard to its uniqueness.
   2. Two or more vertical sections mean that two or more individuals are compared with each other on the basis of many traits.

Stern argued in favour of an idiographic approach, but he was far from constructing any unwarranted dichotomy between the two concepts. Because traits were always phenomenological, they had to be described idiographically. The dangers involved, however, were that such description might yield a sample card of innumerable traits without a center point ("*Gefahr der mittelpunktlosen Musterkarte zahlloser Merkmale*"). This problem, according to Stern, must be solved by the traits themselves, because they are "relatively constant forms of behavior of an individual." The danger, as we will see in the following, has not been totally avoided.

Method B.1 seems to be idiographic, but what about B.2.? Is there any standard of a meaningful comparison of traits and trait interactions in single individuals? In order to understand uniqueness, a nomothetic basis is needed, as Windelband said in a famous speech (Windelband,

1919). In order to apply precise and general trait concepts to an idiographic representation of a unique individual, we need the correlations of characteristics defined "horizontally" as traits. Stern used the idiographic method with the intention of finding the traits that actually represent "dispositions for specific behaviors." This is important because it leads directly to his later *Konvergenz-Theorie* (Stern, 1911/1921), an attempt to unify the nativists' and the empiricists' prejudices concerning the individual.

The important point is that once traits are nomothetically defined by methods A.1 and A.2 (Figure 1), they provide the basis for an understanding of the individual. From here on, Stern believed, the idiographic method provides the best understanding on the basis of the unique interaction of nomothetically defined traits. We will see later that this does not necessarily have to be a mysterious type of understanding. In fact, in order to become scientifically testable, it has to be handled with the tools of nomothetic approaches.

Because William Stern is probably the most outstanding figure in a long history of attempts to create a psychology centered around the individual, a brief account of Stern's attempt, based on Stern's own autobiographical sketch (Stern, 1927b), will be presented.

Unitas Multiplex—Convergence and Dialectics

Stern's book on differential psychology superseded his previous ideas on the nature of individual differences and the methodological founda-

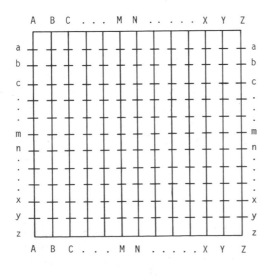

Figure 1. Diagram depicting the idiographic and the nomothetic approach in differential psychology. From *Die Differentielle Psychologie in ihren Methodischen Grundlagen* (Differential Psychology in Its Methodological Foundations) by W. Stern, 1921, p. 17. Copyright 1921 by Barth. Reprinted by permission.

tions of their study. It owes its special character to the peculiar halfway position that differential psychology occupies between classical psychology on the one hand, and individual diagnosis and description on the other. Whenever differential psychology seeks lawfulness in psychic variations, establishes relationships between types, or measures variability or correlational ties—it is *generalizing*. Clearly this generalizing function of the discipline is performed in ways very distinct from the traditions of classical psychology. However, whenever a differential psychologist tries to grasp the psychical make-up of an individual, his character, the degree of his intelligence, his total personality pattern, the psychologist is *individualizing* and is thereby coming close to the historical sciences (Stern, 1930b, p. 360). Stern admits that "real individuality" (i.e., "the understanding of which I had set my goal"—Stern, 1927, p. 14) cannot be approached through differential psychology because that discipline dissects the "unity of spiritual life" and generalizes—just like general psychology does, only to a lesser degree (Stern, 1930b, p. 347).

Stern views the individual as a *unitas multiplex*.

> The wealth of phenomena concomitantly or successively observable arrayed themselves in a unified life-line of the developing individual, and received their significance directly from this.

Stern discovered at this point the "fundamental form of personal causality": the "*convergence* of the stirring character traits in the developing child, with the totality of environmental influences" (Stern, 1911/1921).

Convergence as the basic principle of adaptation to life was a view of psychology most prominent in the days when Allport visited Hamburg. It is a most important perspective in modern behavioral ontogeny (Plomin, 1981, 1983; Scarr & Kidd, 1983). It is based on an evolutionary perspective of genetically determined differences that during the life course lead to an active organizational structure. For any given individual, this structure is ultimately quite unique, although the general laws of interaction of genetic propensities, developing preferences, and the interests toward particular aspects of one's life space are responsible for producing the individual in the first place. Sandra Scarr's model of ontogeny is a modern example of Stern's old convergence concept (Scarr & MacCartney, 1983). Scarr, on the basis of two adoption studies carried out from early infancy and across races well into adolescence, has good data to assert that after an early phase of passive influence, the adolescent's main task is to find the best fit between those aspects of the environment that suit their motives and interest (and values?) best. The same basic idea, essentially, is expressed in modern cognitive psychology. Highly gifted mathematics students, for example, see essential abstract features of complex patterns faster and more reliably than highly gifted psychology students. The psychology students can be trained, but they do not transfer their skills into new situations, whereas with the mathematics students the gift seems to be present in all situations that are suited to this particular ability for abstract thinking (Klix, 1984).

In contrast to Allport (see the following) Stern felt that he had to find a dialectic solution for the law in science versus the private or particular phenomenon. Stern

> found it as impossible to maintain a one-sided point of view, which should simply ignore the other, as to asquiesce to an eternal dualism such as had been repeatedly attempted from Kant to Münsterberg.

The dialectic solution Stern was striving for was not intended to "be a compromise, but a genuine *radical synthesis* of teleology and mechanism" (1930b, p. 353). Stern thought it possible "to render a picture of the personality of an individual psychographically . . . through observation, experiment, etc. . . . A synthetic process was to yield the result." Later, however, he realized that "a genuinely personalistic psychology must transcend not only analytic but even synthetic method" (1930b, p. 361). This expectancy was not fulfilled in Stern's time nor, perhaps, has it been today. Stern's ideas on a dialectic psychology have been most recently pushed considerably by the late Klaus Riegel during his influential career as a psychologist in the United States (Riegel, 1979). Riegel, in the late fifties, was exposed to Kurt Bondy's presentation of William Stern's psychology.

Karl Bühler struggled with a similar problem in a book published the same year that Stern's autobiographical essay appeared. Bühler tried to unite three different perspectives: the behavioral, the functional, and the understanding psychologies ("*Verstehende Psychologie*"; Bühler, 1927, Grossmann, 1983). There is no reason to assume that either Stern or Bühler have failed. On the contrary, they have dealt with issues that were and still are pertinent when a number of individual people are the basis for the abductive construction of hypotheses to explain their past, present, and future behavior (Pierce, 1955).

Intelligence, Adaptation, and Values

A few more things should be mentioned about Stern's personalistic psychology. He continued working on problems in intelligence and special abilities. Ten years of research led to the following concept: intelligence was conceived "as a purposively oriented, personal disposition," that of "general intellectual adaptability to new tasks and conditions of life." However, it was not known "through what psychical processes the solutions were attained, and in what way the special contributive factors: attention, memory, thought, etc. were involved." And in the course of such specialized treatment, the factor of intelligence certainly was presented "too much in isolation, as an apparently independent disposition, without due regard to its original membership in the personality as a whole." Therefore, Stern argued, "we must pay more attention to its personalistic anchorage" and therewith to its fundamental relatedness to the impulsive, volitional, and practical life (Stern, 1930b, p. 365), or "*Interessenleben*," the development and pursuit of one's interests (Stern, 1927b, p. 31). This issue is virtually paraphrased by Sandra Scarr in her conclusions on individual differences in intelligence:

> The child's current intellectual level is a function of a motivationally determined history of learning in which motivation has played as much or more of a part than cognition in helping or hindering the child's intellectual development." (Scarr, 1981, p. 1160)

> Intervention that address the motivational and adjustment aspects of learning may well be more effective with these children than those that primarily address the cognitive lags. (Scarr, 1981, p. 1165)

Stern developed his personalistic ideas during the years when he compiled questionnaire data on taste and distaste for curricular activities, trends of interest, and motivation behind likes and dislikes. In addition, together with his wife Clara, he compiled diaries about the earlier development of "three essentially very different children from birth well up into the school years." Of the originally intended six monographs three were published, one about children's language (1907/1928) and one about memory, statement, and lying (1909/1931). A summary of the psychology of early childhood until the age of six was finally published in 1914 (Stern, 1927a). Stern recalls: "but for my own real development [in] these studies of my children... I observed *concrete* spiritual (*seelisches*) life and was thereby safeguarded against those false schematizations and abstractions which we meet all too often under the name of psychology. Here I became aware of the fundamental personalistic factors of *unitas multiplex*." Not only practitioners demand idiographic procedures, but also any research-oriented psychologist who wants to generalize from a descriptive base of individuals in general cannot accomplish much without them.

Adaptation, then in Stern's mind, is accomplished by the living individual that keeps its own tensions in proper equilibrium with those of the environment. Intelligence, therefore, is a personal disposition, that of "general intellectual adaptability to new tasks and conditions of life" (Stern, 1927a, p. 30). Stern believed that values were the core matter of individuality. Stern's "Wertphilosophie" (Stern, 1924) exceeded a philosophy of value, it was, more profoundly, a theory of meaning. Charlotte Bühler, Karl Bühler's wife, came toward the end of her life as one of the world's few ingenious developmental psychologists to very similar conclusions (C. Bühler, 1976).

TOWARD A PERSONALISTIC EPISTEMOLOGY

To summarize, the individual as a person conceptualizes his or her world view, establishes values, interests, and is motivated to search for these aspects of the world. The understanding of this personalistic concept is a mental construction of the psychologist. This is the problem that actually underlies the many disputes centering around Allport's claims. It is an epistemological problem, which was anticipated by Stern (1927b, p. 34).

> The oft-repeated demand to let critical thought take the place of creative thought in philosophy is distinctly repudiated. Never can criticism precede creativity, let alone forbid it (i.e., deny us the right of metaphysical attempts altogether); but it must follow in the footsteps of invention, must control, justify, or rectify it. (Stern, 1930b, p. 369)

Stern, as I will argue, was entirely correct.

Stern's notion is that the person develops fundamental types of introception, that is, that one creates values that guide one's future life. Stern examines loving and understanding, aesthetics, sensitivity and practical activity, creativity, and finally also worshipping. At this point his philosophy, admittedly, became detached from known empirical bases. Methodological foundations had not been worked out, and, as in the case of the Bühlers in Vienna, a most original way of conducting psychological research within a constant tension between (metaphysical) meaning and empirical rigor was suffocated by the murderous political events that eventually forced Stern out of Germany in the early thirties (Grossmann, 1983). A new understanding for Stern's ethical imperative may perhaps still develop. Some of the flair of Stern's outlook is captured in the following citation:

> The ethical imperative cannot be individualistic: live out your life, for this ipsification would ignore the objective values. It cannot be universalistic: subject yourself to the general ethical law, for that would be de-ipsification, denial of the uniqueness and therewith of the unique task of every personality. It can only be: "introcept!" or: "mould your ego microscopically into a personality, in that you raise all service to the non-ego-values to essential traits of the

individual intrinsic value. . . . The content-giving ethical imperative is: "live up to your vocation." (Stern, 1927, 53, 54)

THE IDIOGRAPHIC-NOMOTHETIC CONTROVERSY IN (PAST?) EXPERIMENTAL PERSPECTIVE

THE MISCONCEIVED DICHOTOMY: FROM WINDELBAND TO ALLPORT

Traditional experimental psychology was—and still may be—mainly concerned with the discovery and variation of universal laws. Underwood (1957), in agreement with many other researchers, pointed out that his approach "deals almost exclusively with the classical nomothetic approach in which groups of organisms are used." Underwood contrasted, as Allport did, his approach with an "idiographic psychology," which is supposed to deal with individual organisms only. We now know, however, that universal laws can be studied very well with single organisms. Single-subject designs, repeatedly applied, have amply demonstrated that this is not the true contrast between the idiographic and nomothetic orientation in psychology. Underwood pointed out that

> those who champion idiographic analysis are trying to say that there is no communality of laws from one organism to the next, in which case we will have as many sets of laws as we have people. (1957, p. 88)

As pointed out in the previous section, this is a semantic problem, as well as a matter of the level of analysis.

The distinction between idiographic and nomothetic orientation has often been linked to the German distinction between *Naturwissenschaften* (sciences) and *Geisteswissenschaften* (humanities). These terms had been used by Wilhelm Windelband in 1894 (Windelband, 1919). Windelband's main argument was that an empirical discipline of such important subject matter as psychology has no place between *Naturwissenschaft* and *Geisteswissenschaft*. In order to appreciate Windelband's idea it may be essential to understand his distinction. *Wissenschaft* or *scientia* have been defined as "objectively, the systematic unity of knowledge that belongs together in principle and which forms an area of its own" (Eisler, 1930, p. 617). There exist two concepts of science—*Naturwissenschaften* are the sciences that are concerned with objects that belong to nature, whereas *Geisteswissenschaften* constitute those disciplines that deal with "spiritual concepts" ("*Geistige Gebilde*"), functions, and processes. Fechner believed that *Naturwissenschaft* abstracts from all qualitative aspects of objects and makes objective only those aspects of our perception that can be perceived in a quantitative fashion. Münsterberg thought that *Naturwissenschaft* deals only with things *made* objective, and not with the immediately real or true things. These, "immediately true things," are of course accessible through a phenomenological approach in which idiographic methods are often used. Wundt considered the separation between *Natur-* and *Geisteswissenschaft* to be in the fact that all experience contains an objectively given object of perception (*"einen objektiv gegebenen Erfahrungsinhalt"*) as well as experiencing subject (Eisler, 1929, p. 226).

Eisler's analysis of the issues seems to be fairly representative. All the thinkers whose views were mentioned seemed to be in favor of a science of the phenomenology of the "truly given" reality. The mental structures (*Geistige Gebilde*) are embedded within a normative-teleological reference system of values. To some extent they belong to the *Kulturwissenschaften* of the systematic and historical kind. Wundt argued that

> the problems of *Geisteswissenschaft* begin where man as a thinking being with his own will is an essential factor. All *Geisteswissenschaft* deals with the immediate experience as determined by interaction of objects with recognizing and acting subjects. (Wundt, 1880, p. 1ff)

He actually collected data and analyzed the *Kulturwissenschaft* side of human psyche in his *Völkerpsychologie* (published in ten volumes between 1910 and 1920) in ways that were meant to parallel his experimental approach in psychology. Dilthey, who is considered to be the founder of "understanding psychology" (*Verstehende Psychologie*) and who, through his disciple Eduard Spranger, has had a considerable influence on Gordon Allport's thinking, held the notion that "we explain nature but we understand mental life" (*"die Natur erklären wir, das Seelenleben verstehen wir,"* Dilthey, 1924, p. 144). Stern acknowledged these ideas but insisted on the correctness of his own conceptualization of the issues (Stern, 1927, p. 19).

The elaboration of these two aspects of thinking within the German tradition reached a dichotomy that for Windelband actually were unacceptable. From the beginning, Windelband had in fact substituted the factual dichotomy between *Natur-* and *Geisteswissenschaften* by a methodological dichotomy. He asserted that psychology, as far as its subject matter is concerned, was obviously *Geisteswissenschaft*, whereas the entire procedure of psychology, its methodological basis, was from beginning to end that of the *Naturwissenschaften*. Psychology collects facts, and uses them to understand the general laws that regulate these facts; knowledge in form of a natural law is the goal of the nomothetic *Ges-*

etzeswissenschaften (lawful sciences). The idiographic *Ereigniswissenschaften* (event sciences) strive for the unique in the historically determined form (*"nach dem Einzelnen in der geschichtlich bestimmten Gestalt"*).

The implications of Windelband's new distinctions were that behavior outside of "uniqueness in its historically determined form" is to be submitted to a nomothetic approach. The differentiation has already been conceived of as a relative differentiation between general and special cases. Whether the one or the other approach becomes dominant is entirely an aspect of the research perspective. In fact, Windelband himself pointed out that "the idiographic approaches need the nomothetic. They need the general laws which they can borrow only from the nomothetic disciplines" (Graumann, 1960, p. 104).

On the other hand, there exists no common source for these two aspects of knowledge. The unique cannot be derived from the general, nor the general from the unique. "The totality of givens in time seems to exist in underivable independence parallel to the general lawfulness according to which it actually happens" (Graumann, 1960, p. 104).

The meaning of the two words idiographic and nomothetic, then, appears to be somewhat as follows: Investigation leading to laws, with control and prediction, is and has to be nomothetic. Idiographic uniqueness does exist and, further, cannot be inferred from nomothetic relations. It cannot be understood, however, without such relations. Any focussing on the individual with the intention to find laws in his behavior no longer confines itself to "uniqueness in the historically determined form." It must eventually leave the frame of event science and extend to the science of lawful relations.

ALLPORT'S NONDIALECTIC STAND

The influence of Windelband's distinction on the psychology of personality is not easy to judge because Stern did not adhere to it. Personality research and personality theory were the areas in which Windelband's distinction, at least in retrospect, seem to have exerted some influence. Gordon W. Allport is mainly responsible for having introduced Windelband's distinction to American psychology. He mentioned it in a short paragraph but asserted at the same time that this distinction was too sharp, because it required a psychology divided against itself. He regarded the two methods as overlapping and as contributing to one another, as, in fact, Windelband had already done. Psychology, Allport declared, can treat both types of subjects. In refering to the term *Scientia*, which prescribes no method but which signifies simply knowledge, Allport asserted:

> If there are psychologists who in the face of this growing movement still
> declare that the study of the individual is not and never can be part of science,
> they must now be left alone with their views. (Allport, 1937, p. 22)

Allport confronted what he held to be the entirely idiographic the subject
matter of literature with the type of psychology that exhibits a persistent
interest in the discovery of laws of human behavior (1937, p. 61). Allport's
main idea, as was Stern's and Windelband's, was to integrate the two
approaches for a true understanding of the individual.

Allport, a B. A. from Harvard (1919) in economics and philosophy,
was a teacher of sociology and English in Istanbul in 1920. He received
his Ph.D. in psychology from Harvard in 1922 and returned for another
2 years to Europe. During that time Allport spent some time in Berlin
(with Max Wertheimer and Kurt Lewin) and in Hamburg, where William
Stern was professor of psychology. Allport's orientation was guided by a
strong interest in sociological and ethical problems; he synthesized the
empiricist tradition of Locke, the quantitative methods of psychology,
and Leibniz's ideas of the unique spiritual activity with a qualitative
psychology (Heiss, 1936).

Allport rejected Stern's (Stern, 1911/1921) attempts to create an "idiographic differential psychology" because, to him, it was not idiographic.
Stern's proposal (see Figure 1) was too limited to him because "we learn
nothing about Sam that has not previously been chosen by the investigator for testing," and: "the profiles tell us nothing about the organization
of the qualities in question" (Allport, 1961, p. 15). To Allport, personality
was not a problem for science or for art exclusively, but for both together.
To him, each approach had its merits, but both were needed for even
an approximately complete study of the infinite richness of personality
(Allport, 1950). The psychologist has to learn from literature "something
about the nature of the substantial and enduring dispositions of which
personality is composed (Allport, 1950, p. 203) and "it has been handled
more successfully through the assumptions of literature than through
the assumptions of psychology." Furthermore, the psychologist has to
learn from literature the "self consistency of its products"; and elsewhere:

> No one ever asked their authors to prove that the characters of Hamlet, Don
> Quixote, Anna Karenina, Hedda Gabler, or Babbitt were true and authentic.
> Great characterizations by virtue of their greatness prove themselves. (Allport,
> 1950, p. 203)

Skaggs attacked Allport's "bold stand for the broadening of the concept of science," which enabled him to include literature into scientific
psychology as "idiographic methods" (Skaggs, 1945, p. 234). Skaggs asked
what actually was different and unique in individuals. The learned traits

were different from one individual to another, whereas the unlearned primary processes were alike. Stern had known better already! Skaggs's idea was that the investigation of secondary differences of one individual as the characteristic of uniqueness was not scientific, but that the investigation of the acquisition of secondary characteristics was. Skaggs thought that Allport's laws concerning personality were as nomothetic as any of the older principles of Titchener and Wundt, and he concluded:

> We believe that the philosopher Windelband arrived at the proper conclusion, namely, that the knowledge of psychology must be classified into two groups, nomothetic and idiographic. The nomothetic class of knowledge is scientific.

Here is the old untenable dichotomy again! When these discussions are seen in light of diaries like those of Tiedemann, Pestalozzi, and Preyer, or in light of modern video analyses (Grossmann, Grossmann, & Schwan, 1986) or on the basis of narrative reports as used most successfully by Ainsworth and others (Ainsworth, Blehar, Waters, & Wall, 1978; Grossmann & Grossmann, 1984; Grossmann, Grossmann, Spangler, Suess, & Unzner, 1985), one starts wondering about the old dispute. Perhaps it all boils down to the gift of being able to represent keen observations by— among other things—highly differentiated prose at some stage of one's observational research on similarities and differences between people. As stated before, this process remains on the side of heuristics, discovery, and knowledge production. The true difference seems to rest with the question whether this is the end ("understanding") or the beginning ("best available hypothesis") of a process of substantiating the groupings on the basis of perceived similarities and of finding the reasons for them, up to the level of hypothesis testing and proof. For this reason Stern's epistemological notion has been chosen as a guide line.

WHAT IS THE SOLUTION OF THE OLD PROBLEM?

Stern's first ideas concerning differential psychology (1911/1921) have been taken up by personality theorists many years later. Guilford (1959, p. 23) stated that "in the typical study using the idiographic approach . . . the person provides his own reference point, which is his own average, if a reference point is wanted." Guilford suggested a resolution. The "technologist" (i.e., the clinician working with an individual) needed more information of the type he already had to make his diagnoses. This scientific, nomothetic information went beyond the single case; it had to be derived from many individual studies.

Cattell (1944) provided some further insights from the psychometric point of view on normative, ipsative, and interactive measurement. Although he argued mainly in favor of interactive measurement ("it recognizes the oneness of the organism-environment"), Cattell presented ideas for measurement using the individual as a standard. These "ipsative" (*ipse* = he, himself) measurements can be expressed:

1. In standard scores either from a population of measurements
2. As ratios of a single measurement (e.g., Vincent curves)
3. As "fractional ipsative measures," in which the units are fractions of some total possessed by the individual
4. As normative-ipsative units, in which interactive measurements, which have first been made normative, are themselves used as a basis for ipsative recalculation

We have encountered the problem and its rejection by Allport already in connection with Stern's method as shown in Figure 1.

There is, indeed, some semantic confusion when Windelband's terms are used. Many arguments have greatly diverged from the original issues, which in any type of observational and abductive research are always with us. Windelband's distinction must not be misunderstood as if *idiographic* meant those events that are inaccessible for psychology, a domain that simply provides pleasure for the mind interested in the arts, in philosophy, or in esthetics. The problem, as we have already seen, is still with us: the mind's integrative power occasionally provides the very concepts that may become an important issue of further nomothetic research. These may be creations of writers and poets. The same creations may be achieved by scientists as a first, but most important step. The difference is that the scientist ultimately has to find the criteria that make up the constructs of similarities and differences, the way they belong together, and the procedures or methods that may render his invention a more or less useful one.

Allport's "true and authentic characters" are true and authentic indeed as creations of their poets. This does not mean that they are not convincing, but it certainly means that creation of a character is not the same as idiographic description of a character. That leaves us with the same epistemological problem already inherited from William Stern. It is time finally to confront it from a position that owes much to the psychology of Karl Bühler and to the ethology of Konrad Lorenz.

THE ETHOLOGICAL APPROACH IN PSYCHOLOGICAL RESEARCH

OSCILLATION BETWEEN THE UNIVERSAL AND THE INDIVIDUAL

There have been a number of collections on infant and child observation in as early as the 18th century and even before. Pestalozzi as well as Jean Paul wrote such diaries about individual infants and children, and the idea of their ultimate benefit was widely accepted in those days. Tiedemann, in 1787, published a child's diary, but around 1800 the diary movement was dead. According to Ch. Bühler and H. Hetzer (1929) the most likely reason was that one did not know how to use the observational information properly. The art of inducing order in a multitude of facts and arriving at an immediate interpretation did and does not seem to come easy, nor does it seem to be very helpful if it is attempted without some integrating organizational structure.

The German philosopher F. A. Carus characterized naive observation as early as 1808 as follows: "One always oscillates from the essential to the accidental, from the universal to the individual" (Bühler & Hetzer, 1929; Carus, 1808). This insight is known to anyone having slaved through video tapes. The process of discovery of the scientist working on the basis of repeated viewing of childrens' behavior in unmanipulated situations can easily seem like a rather fruitless enterprise. On the other hand, it is a necessary way for creating the most insightful characterization necessary for later critical analyses. Actually, Karl Bühler, in 1925, thought that film and audio recordings, luckily, had not been available in Preyer's and Tiedemann's days, because knowledge from observation depends on both "on the context of the psychological consideration and on the exactness of the observations" (K. Bühler, 1925, p. 33).

There is often a cumbersome interplay between misleading and productive speculation in the researcher's mind about the observed facts. An easy solution has often been looked for. The easiest way out is to concentrate on prematurely defined variables before one has attempted to account for the observed events in an exhaustive and integrative (understanding, insightful) manner—such an account is a necessary but insufficient step toward understanding complex interrelated events. The Cattell–Guilford approach was presented as an example for an attempt to solve the problem on methodological grounds but that, concurrently, was conceptually inadequate for a solution of the old idiographic-nomothetic controversy. The process of creating (understanding) insights sometimes is called speculation and is meant to mean *unscientific*. However, as Tinbergen & Tinbergen have pointed out, progress in science depends as much on creative observations and intuitive interpretation as on verification (1972, p. 50).

Oscillations between observations of many single events and their integration on the basis of some similarities may produce many incompatible conceptions among different observers. This may just indicate competition for the most appropriate hypothesis, or a still pending breakthrough. It is a normal state of affairs in natural science, and it may cause less damage than an unjustified application of single variable experimental results with no apparent basis for rational transfer to real life events.

AN EXAMPLE OF PREMATURE DEFINITIONS

It would be no surprise if some readers feel tempted to abandon the views presented here as old-fashioned, prescientific European ways of thinking, which was luckily overcome by the experimental paradigm, by operational definitions, and by rational methods of proof. Such was the tenor of the discussions around Allport's notions. Although there were not as yet really many experiments in developmental psychology that may fit to Wundt's criteria established about 100 years ago, experimental thinking is nonetheless quite dominant. This implies the "necessary" clear definition of the concepts and variables used. But this, as the following example will show, may come much too early, before even a rudimentary understanding is available.

The example is from a piece of observational work of a research group that had conducted hundreds of hours of observations on children in a kindergarten. The researchers were trying to identify rank order. In their minds, rank order was defined by a concept of attention structure (Chance, 1967): That child who was looked at by at least three other children most often was ranked the highest, the second most often looked at child was second in rank order, and the children looked at only rarely or not at all by three other children at the same time of observation were considered lowest in rank order. The rank positions for some children changed considerably over a year's time, but for others the rank positions remained stable. Rank order correlated with popularity in the United States (Vaughn & Waters, 1980) but not in Germany. Aggressiveness was only moderately correlated with rank order as determined by attention structure.

The researchers in my example thought it to be important to determine the correlations between rank order, friendship, and prosocial behavior. This resulted in the following premature definitions. Friendship (or playmate preference) was operationalized by the number of nonantagonistic interactions between any two children. Prosocial behavior was conceptually confined to, first, helping: for example, assisting a given child to complete a given intended act when hindered in doing so by some obstacle or barrier; and, second, supporting: for example, joining in when another child was involved in an aggressive act towards a third child, with the hope of helping the supported child to win. Group statistics—correlation coefficients and significance testing—provided some meager and uninteresting results. When, however, one looked at the individual children who stood in high esteem, that is, who held high ranks, a fascinating story revealed itself. It went far beyond the operationally defined variables and is perhaps somewhat closer to what Allport and Stern had in mind: Some children were really aggressive, noisy, obnoxious bullies and they were looked at by the other children with a kind of monitoring and vigilant self-guarding attitude. Others were quiet and creative in their activities, and they were sometimes admired and imitated. Others had access to valued resources, for example, they dominated the private dollhouse or the construction corner, and they were looked at longingly. Still others were just cute and charming and they were smiled at more often. Still others were rather unpredictable, and these, sometimes, seemed to be objects of a watchful concern to some of the others. These differences had not even entered the process of statistical analyses. A proper ethological approach in the tradition of Tinbergen, Lorenz, and others would have given the process of discovery a much longer time to develop. An exhaustive description would perhaps have prevented any artificial concepts and premature definition of rank order and other related variables. It also would have been better suited to account for the intricacies of human interactions in the special kindergarten setting. Luckily the situation has been remedied in the meantime.

OSCILLATION BETWEEN IDIOGRAPHIC DESCRIPTIONS AND CREATION OF HYPOTHESES—THE EXAMPLE OF AINSWORTH'S STRANGE SITUATION

During our longitudinal research in developmental psychology we observed 49 German children at 12 months of age with their mothers in Ainsworth's Strange Situation. At 18 months we saw 46 of the children in the same situation with their fathers. During the first year we observed typical home interactions at 2, 6, and 10 months of age. At 2, 3, and 6 years we saw various forms of interplay with the mothers, fathers, and a highly cooperative participant observer. For the analysis of the childrens' behavior in the Strange Situation we used Ainsworth's criteria. These had been developed on the basis of an interesting epistemological interplay between induction and theory. Close inspection of our children's behavior, particularly of a few individual children, revealed that many of them failed to express relief, enjoyment, and happiness upon the parents' return after brief separation. Attachment theory holds that close or tender bodily proximity and contact eases separation-induced insecurity most efficiently. On the basis of the classification criteria developed in Baltimore, which worked well also with other American samples, more than two thirds of our children were classified as insecurely attached to their mothers and, independently of that classification, also to their fathers (Grossmann, Grossmann, Huber, & Wartner, 1981). But were they really? Some of the children definitely avoided their parents. But most of them—most of the distinct and clear cut avoiders (A_1 in the Ainsworth's terminology)—displayed rather well-conducted self-control. Their facial expressions were calm, sober, well-behaved, and so were their movements. Their voices sometimes let their feelings leak through, but rarely their faces—they were not dead pan or poker faces, but rather serious, sober ones.

What was the matter with our many children whose behavior toward their parents would have to be labeled insecure-avoidant on the basis of Ainsworth's criteria? No operational definition and no premature conceptualization would have helped us to make psychologically meaningful statements about *our* children's quality of relationships to their parents. Some of the children may not expect anything special from their parents, and some may not need to. Still others may already know that it is inappropriate to show affection out of the proper context (Grossmann, Grossmann, Spangler, Suess, & Unzner, 1985).

Can anything, for solving the problem, be learned from an ethological approach? What can be gained from oscillation between idiographic understanding and nomothetic data treatment? This is not at all an argument in favor of another artificial dichotomy between the method of induction versus deduction. Although even Darwin wrote in his autobiography that he "worked on true Baconian principles, and without any theory collected facts on a wholesale scale," Stephen Jay Gould, in an analysis of Darwin's notebooks, clearly showed that Darwin was never just an inductivist, nor was he just an eurekaist (Archimedes shouted "Eureka," running naked through the streets of Athens after having discovered that water displaced by his body suggested a method for measuring volumes like that of the king's crown.) Gould concluded, after carefully documenting Darwin's struggle for understanding the origin of species, that

> inductivism is inadequate, that creativity demands breadth, and that analogy
> is a profound source of insight—and also that great thinkers cannot be di-
> vorced from their social background. (Gould, 1979, p. 31)

Of course, the source of an idea is one thing; its truth or even fruitfulness is another. It has become customary to refer to changes in scientific orientation as changes in the paradigm which are supposed to underlie scientific revolution (Kuhn, 1962). Ainsworth suggested that the different paradigmatic orientations may indeed be operating in the most experimentally-oriented learning theorists as compared to most observation-oriented ethnologists. The difference in thinking and conceptualization is profound, as exemplified by the dispute between Ainsworth and Gewirtz on the issue of whether infants, during their first year of life, learn to cry less or more when paid attention to by significant others. The learning paradigm postulates a direct reinforcing effect of the cry response by the responding adult (Gewirtz & Boyd, 1977a, b). The ethological orientation postulates a match between an inner organizational structure called attachment, and its increased state of arousal as a function of any attachment figure's unresponsiveness (Ainsworth & Bell, 1977; Sroufe & Waters, 1977; Waters, 1981). Learning theorists would count frequencies of cries as a function of contingent reinforcements by responding to the cries, and would venture to show that by not responding the cries would eventually be "extinguished." Ethologically oriented observers would attempt to understand the role of crying within the system of developing emotional communication by describing the function within an evolved infant–mother system as some kind of gestalt or qualitative pattern, as done by Ainsworth *et al.* (1978).

Empirically, it has been shown beyond any reasonable doubt that for an understanding of the functions of infants' cries during the first year the conditioning by reinforcement orientation is wrong, and it precludes any ultimate understanding (Tinbergen, 1963), whereas the ethological gestalt orientation has developed a solid empirical basis for its qualitative conceptualizations. This is, however, not to say that infants who have been deprived of an opportunity to learn the subtle ways of prespeech communication based on emotional expressions of the voice, of the face, and of the whole body, would not eventually be reinforced for their demanding cries, which then would be the result of a dramatically reduced process due to social incompetence, an entirely different notion than to say crying has been reinforced by social conditioning. How, then, can the recognition of such qualitative behavioral patterns on a systematic level be achieved?

The context in Ainsworth's argument was the fulfillment of the basic biological function of an infant's cry: to bring another person close enough for further interaction in terms of need fulfillment. Crying, if appropriately and promptly responded to, is subsequently replaced by more differentiated and efficient communication. Gewirtz, on the other hand, was convinced that adult responding reinforced crying. Therefore children paid attention to when crying would learn to cry more. Both contingencies do in fact occur, at different age levels and in different situations, and in different ecological contexts. Gewirtz did indeed go about demonstrating that he was right, but the argument was similar to that of Watson, who—after having conditioned Albert to be afraid of a rat—

claimed that by producing fear in this manner he had in fact proven that this was how fear was acquired by children in general, which of course he had not done at all. Infants whose cries remain unanswered show despair and finally exhaustion. Older children may indeed use obnoxious cries for getting what they want if they have not acquired better means of social competence.

RE-COGNITION IN THE CHILD AND IN THE SCIENTIST

At this point ethological thinking in developmental psychology touches very closely on major issues of epistemology on two levels: on the level of the child and on the level of the recognizing scientist. A child develops to adapt to caretakers first and to the wider world eventually. Infants build up expectancies that are by no means identical with conscious planning strategies. The avoidant behavior of some infants in Ainsworth's Strange Situation and children in Main's work (Main, Kaplan, & Cassidy, 1985) are easily recognized as learned expectancies toward the specific parent who returns to the child after brief separation.

The re-cognizing scientist also builds up expectancies that are by no means identical with planning strategies. When he calls these expectancies facts, he may be as mistaken as the developing child, because they may not account for the whole context, but only for a selected fraction of it, and they may be wrong too. In this sense, too early a concentration on attention structure as an operationalization of rank order, as outlined previously, may prevent us from discovering the actual events, particularly from the child's own individual perspective. Furthermore, a premature transfer of mother-avoidant behavior beyond the Strange Situation, and even into other industrial cultures, may prevent the discovery of different paths to either eudemonic happiness or to an externally enforced dutiful life, in agreement with strong self- or ego-control and, perhaps lower ego-resiliency (Block & Block, 1980).

Stern's hope for a "radical synthesis of teleology and mechanism," the dialectic solution he was striving for has been very close to most recent developments in the biology of knowledge (Lorenz, 1973; Riedl, 1980; Vollmer, 1975). The dialectic unity, implicit in Stern's personalistic epistemology was designed to overcome an "eternal dualism"; it actually consists of an unending process of discovery and proof on the scientific level. On the phylogenetic level it consists of selecting out those individuals of a given species who do not adapt to a given environment because they have not inherited the necessary information to adjust their behavior in such a manner as to allow for survival and procreation. Discovery, on the scientific level again, very often uses idiographic strategies, and thereby

provides the necessary basis for actuarial proof strategies as will be outlined in the following. This is true for very specific observations, as in
the case of Ainsworth's Strange Situation (Grossmann & Grossmann, 1984).
It is also true for the general case that we otherwise may learn nothing
"that has not previously been chosen by the investigator for testing" (Allport, 1961, p. 15).

CHILD ETHOLOGY AS A COGNITIVE STATE OF MIND

Hypothetical realism, a term coined by Donald Campbell (1966), refers
to a philosophy of science that is able to account for the processes of
understanding through guided observation and of successful (and unsuccessful) approximation to given complex facts. Perhaps at the lowest
level this is all there is behind observation (Lorenz, 1973; Riedl, 1980;
Vollmer, 1975). Each individual has to be prepared to judge the relevance
of his or her ideas. Or, as one might say with Pasteur, "fortune favours
the prepared mind" (quoted by Gould, 1979, p. 31). The child and the
scientist both may severely distort what they see on behalf of ideologies
that may take the place of reality. But reality is, from all we know, not
constructible *ad libitum* (despite Berger & Luckmann's, 1966, assertions).

An exclusively experimental orientation, concentrating on isolated
variables while ignoring the context at any given moment of observation
and data analysis (deplored by Bühler & Hetzer as early as 1929) ignores
the mind's integrative powers for qualitative pattern recognition so vitally
important for the behavioral sciences. It would only be acceptable if the
theories from which the variables were derived were comparable to the
mathematical theories of mechanical physics from where the model was
originally copied. But because the analogy does not hold for the status
of theories in the behavioral sciences, with perhaps a few exceptions in
psychophysics, the resulting loss for discovery and ultimate (Tinbergen,
1963) understanding is to be greatly deplored. The apparent poverty of
some behavioral research caused by the obvious prejudices in calling the
mechanistic model "scientific," and the pattern recognition model "subjective" has been successfully challenged by ethological thinking. There
is a fair chance that thereby the obviously false dichotomy between ethology versus experimentation (see the following) may indeed be overcome.

Lorenz's epistemological prologomena provided a hypothetical realism founded on biology (Lorenz, 1973), an orientation lacking in Tiedemann's days 200 years ago. The notion that perceiving and conceiving
have evolved to match reality casts a sense of interpretative unity on the
minds of the ethologically-oriented child-watchers: the object of knowl-

edge and the instrument of knowledge cannot legitimately be separated, but must be taken together as a whole (Bridgeman, quoted by Lorenz, 1973, p. 12). In addition, as Karl Popper noted: "The thing in itself is unknowable: we can only know its appearances which are to be understood (as pointed out by Kant) as resulting from the thing in itself, and from our own perceiving apparatus" (quoted by Lorenz, 1973). Thus the apparatus results from a kind of interaction between the things in themselves and ourselves. Again, its truth or fruitfulness is another matter, because intellectual insights can be utterly wrong. But discovery is not favored by avoiding possible errors. The child and the ethological child-watcher create knowledge only through "the relationship between all of our innate forms of potential experience and the fact of objective reality which these forms of experience make possible for us to experience" (Lorenz, 1973, p. 21).

ETHOLOGY VERSUS EXPERIMENTATION: ANOTHER FALSE DICHOTOMY

Modern ethology's insights into the evolution of hypothetical realism stands in some contrast to the seemingly rational process of variable testing in traditional experimental research. Furthermore, it appears to be a way of making the process of hypothesis generation equally reputable as, and actually more important than, the process of hypotheses testing. This is perhaps the dialectical process of understanding as intended by Stern. Not much emphasis, however, has been devoted to the process of adaptation towards life goals, which is the utmost task of an individual, and a major issue of developmental psychology. In ethology the false opposites were well integrated by Tinbergen more than 20 years ago (Tinbergen, 1963). The problem of meaning and the problem of generalizability in psychological research remain unsolved when the rational proof strategy is overemphasized at the expense of the documenting and interpretative understanding of differential, sometimes idiographic strategies. On the contrary, the breadth that creativity demands is the very context into which detailed variables must be placed.

Description in some cases may even be sufficient for correctly identifying what is going on, provided one has learned to develop those perceptual functions that convey to us the "experience of qualities constantly inherent in certain things in our environment" (Lorenz, 1973, p. 21). It means, of course, learning to see and to create interpretations. Provided that these interpretations can be meaningfully identified by others who have achieved the same breadth of understanding, they may eventually lead to common understanding (or, for that matter, to common misunderstanding).

Perhaps ethology's main contribution is to teach in a new way how to learn to see, and to have provided the foundations for more confident uses of our potential powers to perceive in a rather integrated manner. A good example is Tinbergen's analysis of infantile autism. Although it may not exhaustively explain the clinical state of affairs associated with that anomaly, it does, nonetheless, provide an insight into a helpful kind of interaction analysis.

The consequence would be, of course, a heavy emphasis on the representativity of the observational data for a given ecology and on the appropriateness of the interpretative frame. The paradigm change, sometimes mistaken for a method, sometimes referred to as just those aspects of behavior that can be explained on phylogenetic grounds, proved actually to be a change in the care with which lawful order can be discovered at a highly integrated level by the observer. Here again Ainsworth's Strange Situation may serve as an example. Frequency counts of perfectly well-defined movement patterns, such as smiles, looks, and movements of the infants upon the return of the attachment figures meant nothing, because most infants were flexible enough to use different strategies to reach the same goal. The appropriate means of recognition for the researcher, then, is not frequency counts, but qualitative description of all possible occurrences during the process of mutual signaling and responding within a given context for individual cases or dyads, such as "sinking in," cuddling, appropriate changes in emotions, and so on. It is, in E. v. Holst's words, a matter of "niveauadäquate Terminologie" (terminology appropriately used for the relevant level of description and analysis), which, of course, requires all the insights outlined above before the research process may ever become fruitful (v. Holst & St. Paul, 1960, p. 409ff).

The change in paradigm, then, is actually a change in state of mind. That may create tensions not only in those researchers who have to come up with insights, but also in those who not only mistrust their own cognitions—which is scientifically sound—but who would not let these insights come between them and what functions without them (objectively) in terms of manipulated functions between variables. This, as it turns out, is critical for (nomothetic) science but unpractical without (idiographic) understanding insights. The application of these dialectic principles is somewhat difficult in longitudinal developmental research. In many discussions with fellow developmentalists it is often stated that there is actually no discrepancy at all, no difference between an ethological approach as presented here and the traditional psychological one. Perhaps this will come to be true in the near future. Judging from the present literature, however, there are not yet many descriptive, contextual, and well-documented observations with careful formulations of hypotheses about the development of complex questions of life, and their influence of conceptualization, goals, strategies, and means of adaptation in an ontogenetic and phylogenetic perspective on an individual level. Definition, systems, and methods come second. They trail after the insights of long and tedious observations. Ethology, then, to repeat, demands a state of mind that oscillates indefatigably between what one sees and what one makes out of it. What one makes out of it always settles with the best interpretation available for things seen at any given time. This principle of mutual elucidation (Lorenz, 1973) is capable of increasing the probability of a given hypothesis or of its ultimate falsification. It is an interplay between the perceiving subject and the perceived object, a struggle for understanding that is evident to anyone who has ever tried to account for what children do in direct observation on videotape, in narrative reports, or even in interviews. For this kind of insight, analogy is a profound source of knowledge (Lorenz, 1974). It is the ethologist's best way to discover similarities. For analogies he sharpens his eyes and his wit by looking at man and animal wherever possible. And sometimes analogies help discover true similarities. Lorenz is correct in stating that exhaustive observation and description, which are necessary for understanding behavior development, are still held in low esteem when compared to operational definitions and experimental methods. This may well have an ultimately devastating and demoralizing effect on some child ethologists.

In summary, the ethologically oriented child observer develops an understanding thinking pattern that he constantly sharpens during the process of interpretation of individuals' behavior. A close connection between the historical-philosophical concern with nomothetic and idiographic methods has been demonstrated. Although they are "inextricably intertwined," (Franck, this volume) their flexible use has added a new dimension to researchers' exploration in the field of developmental psychology. The researcher is concerned with qualitative descriptions of acquisition of life strategies that are flexible, that is, goal corrected. With such an orientation toward understanding the developmental researcher attempts to document widely and tries to give exhaustive interpretations in the light of what he has learned to see. The interpretations are treated as the best available approximation to reality for anyone at any given moment. They are tested usually on the basis of systems analysis (e.g., Bischof, 1975), or on the basis of active search for other bits of knowledge. What is to be avoided by all means is succumbing to the "causal-mechanistic and to the accidentally single variable" (Bühler & Hetzer, 1929, p. 223).

The argument, in line with Stern's and perhaps even with Allport's intentions, but based on the methods of modern ethology may be stated as follows: The particular individuality of a given child, in comparison to a number of other children, will emerge only as a creative act of a knowledgeable person. The way to prepare one's mind is to watch constantly and describe and to try to account exhaustively for the things one learns to see. It is a way of access to employing the prepared mind's integrative powers as a means of discovering laws, similarities, causes, and purposes (Riedl, 1980). It means venturing speculation within the naturalistic context of ethological thinking. Proof, the dialectical counterpart of discovery, is then no longer a matter of just testing null hypotheses, but testing knowledge obtained on the basis of the principle of mutual elucidation.

In simpler terms it can be said that knowing many individual children in many situations, to the child ethologist, is perhaps more important for future understanding than keeping up with the incoherent accumulation of a multitude of unrelated statements on the basis of precisely defined variables.

And in terms of the original issues of the individual subject it may mean that by knowing more about the individual child the observer can develop insightful understanding of each individual personality. By shifting his emphasis between similarities and differences of many infants and parents, he may discover the principles which, after critical scrutiny, may acquire the status of psychological laws within a differential perspective.

EXPLORATION OF SIBLING RELATIONSHIPS: HEURISTICS AND EXAMPLES

About 2000 publications have appeared between 1946 and 1980 on birth order and later social behavior (Ernst & Angst, 1983). Research has concentrated on self-esteem, sex, age differences, family size, and other topics. Some authors have related behavioral differences between siblings to differences in parental behavior. Dunn (1983) has added the interaction patterns of family members. Dunn & Kendrick (1979) have given examples of elder siblings providing security for the younger one and of the influence of the behavior of siblings on mothers. There has not been, however, any systematic research on the quality of sibling relationships *per se.* Welz (1984) in close cooperation with Karin Grossmann, used an approach that is based on a variety of qualitative materials. The purpose was to find a base for describing individual patterns of sibling relationships, and, as a second step, to look for different types or groups of sibling relationships. A third step, typical for an idiographic approach, was to get some insights about the nature of mothers' influences on the development and regulation of the sibling relationships.

DATA BASE AND METHOD

The material consisted of narrative reports containing very careful and minute observations of 49 babies at 2, 6, and 10 months of age and their interaction with family members present (mostly mothers, of course). These were the same infants later seen in Ainsworth's Stress Situation as previously discussed. In addition, lengthy and exhaustive interviews had been conducted with the mothers, so that some detailed knowledge existed about the families' situations (Wachinger, 1981) and about the mothers' views of their infants (Ziemer, 1982). Twenty-three of the 49 infants had elder siblings; 17 had 1, 3 had 2, 1 had 3, and 2 had 4. Four elder siblings were hardly ever present during the observations, therefore only 19 families could be analyzed. Of these, 13 families had two children, that is, there was one elder sibling. Only these will be considered here.

For any actuarial research this would have been a rather small basis indeed. But for gaining some possible understanding of sibling relationships the idiographic-descriptive-abductive approach proved to be invaluable. The most detailed data base consisted of the narrative reports of the younger siblings' ongoing interactions. They were prose that had been produced by well-trained observers. Mothers were mostly, but not always, the main interacters; sibling interactions were always recorded whenever they occurred during the observational periods. Interviews were used to find additional cues for the best possible descriptions and interpretations. After all sibling encounters and mothers' descriptions of the siblings' behavior had been compared four salient issues emerged: (a) animosity and jealousy, (b) defense of property, (c) quality of interaction of the elder toward the younger sibling, and (d) quality of interaction of the younger toward the older sibling.

There were, as a final result, 13 individual descriptions of our target infants' encounters with their older siblings during their first year. Some were unique, some were paralleled by others. The procedure of marking similarities and differences in qualitative terms is, of course, open to criticism, but first it has to be achieved.

FOUR TYPES OF SIBLING RELATIONSHIPS

Welz came up with four types of sibling relationship that were almost exhaustive for the material on hand: (a) a hostile and rival sibling relationship ($n = 2$); (b) an ambivalent sibling relationship ($n = 3$); (c) a co-operative relationship ($n = 4$); and (d) an emotionally supportive sibling relationship ($n = 4$) (total $= 13$). The elder siblings of a *hostile* sibling relationship were characterized by a very selfish attitude, by jealousy, protest, by seemingly friendly, but covertly antagonistic and even aggressive behavior, by a lack of any cooperative advances, and by limiting the baby's range, ignoring (purposively) his or her bids for attention. The younger siblings withdrew, the mothers were unreliable in regulating the siblings' behaviors, they did not cooperate with both at the same time (lack of integrative ability), and they often ignored the elder child.

Elder siblings of an *ambivalent* sibling relationship often began or ended play or interaction sequences rather abruptly and without motivation. They too showed jealousy and protest, but without bodily aggression; they did not share their own play things. The younger siblings were very interested in and growingly assertive toward the elders and during periods without rejection some tendencies toward a cooperative relationship were discovered. The mothers treated the elder siblings as interfering with the ongoing observations and often sent them away; they regulated both children by supporting the younger, while inappropriately admonishing, even threatening the elder and ordering them around. They did not establish any triadic interactions and hardly accepted, respected, or even payed attention to the elder siblings' bids and needs.

Elder siblings of a *cooperative* sibling relationship accepted the younger ones as playmates. They rarely showed jealousy, shared their playthings, were invited to participate in the younger ones' games, and allowed the younger ones to play with the toys alone. The interactions were often joyful, they showed understanding and empathy and tender bodily contacts. They were hardly ever jealous or antagonistic and showed no protest, but consoled the younger ones instead. The mothers appeared to be the center of the sibling relationship. In most instances the mothers were present and initiating. In all cases there was at least some evidence for a harmonious triadic interaction pattern. The elder siblings were always included in the ongoing events. The mothers were not dominating, but rather negotiated between the siblings by modeling and by explaining; they were proud of both children, they invited the elder siblings to participate responsibly in the younger one's care, they enjoyed both and talked positively about them, they were interested in both, included both in joint games, and they encouraged exchanges between the two children.

Elder siblings of *emotionally supportive* sibling relationships actually spent more time with their younger brothers or sisters than their mothers. They showed motherlike behaviors. They consoled, cared for responsibly, were proud of and interested in them and enjoyed them. There were long empathetic interactions, sharing, and joint games; they were quite sensitive to the younger siblings' signals, and tender bodily contacts were frequent. The younger siblings stayed close to the elders; they protested and became dissatisfied when they left; they greeted them happily upon their return. In general, they showed more interest in them than in their mothers. There was no jealousy, no protest, no hostility and no possessiveness. The tendency of the elder sibling toward an emotionally supportive relationship was already recognized in the narrative reports of the observations performed when the younger sibling was only 2 months old. They were already very interested, participating, empathetic, and responsible. The mothers of siblings who had an emotionally supportive relationship interacted infrequently with the younger siblings, nor did they noticeably integrate games between them, and they were relieved when the two played together alone. The mothers were not part of their togetherness. They used the elders as a mediator between themselves and the youngest infants.

What Can Be Done with the Descriptions?

Each of the brief characteristics above is superficially plausible. This is a danger for drawing inferences. Each one is also amply founded by an exhaustive documentation of all examples available from the narrative reports, from the interviews, and from the audiotapes. That is a strength for drawing inferences. The greatest danger is to use the plausibility as if it were an explanation. It just suggests certain hypotheses, for which, however, some support is found in the literature. One attempt, for example, was to apply David and Apple's (1969) perspectives on the mother–child relationship to the interaction patterns between siblings. Although helpful it was just an attempt at sharpening possible hypotheses.

The greatest gain, in light of the problem of understanding the single case on the basis of general psychological laws, may result from broadening the inductive base. Perhaps more types will be discovered, perhaps some exceptional or individual cases, perhaps more similarities and more differences. Perhaps another typology or classification eventually may become more suitable. And, of course, as we proceed up the steps of Allport's truism, any typology ends up with more or less unrelated individual cases. There is no escape from the need to search the appropriate level for one's descriptive and categorial terminology, as pointed out by von Holst and St. Paul (1960). In any case, more and different kinds of information may demand a change in kind of idiographic understanding. Eventually, however, some kind of understanding will turn out to be more useful, more integrative, and more exhaustive than others, and they will persist until we know better.

The four types of sibling relationships outlined here are not distinctly separate, defined classes; there are in fact various overlaps. The descriptive prose used here—intended to create some insightful understanding in the reader—is far from the exactness required for research aimed at falsification in the sense discussed by Popper. But it is a step on a way to that ultimate goal. Dunn has started in that direction:

> In families with a particular intense and playful relationship between mother and firstborn girls, before and immediately after the birth of a sibling, not only was the behavior of the elder sibling relatively hostile to the younger over the next year but by 14 months the younger sibling was also particularly negative to the older; ... [and]

> in families where the mother engaged in particularly warm and playful exchanges with her 8-month-old second-born, both children were particularly hostile to one another six months later. (Dunn, 1983, p. 804)

But, whereas Dunn reported that a poor sibling relationship will result from a warm and close mother–child relationship, this, we know already, is not true for some of our cases, in which the mother integrated the elder siblings, rather than isolating them, into a warm and emotionally supportive relationship.

CONCLUDING REMARKS

The case of sibling relationships was not intended primarily as a research report but rather as an example of the necessity, as a first step, of an idiographic, somewhat personalistic understanding of infants as individuals in a particular life context. It was introduced as a piece of empirical reality after an excursion into history and the evolutionary theory of knowledge. The idiographic-nomothetic controversy, it seems, has been overcome by an open-minded, unprejudiced use of the full powers of one's (empirical) mind for discovery and for critical analysis. Perhaps time was not ripe for William Stern to push his personalistic psychology to that point of dialectic unity.

It is a relief to see the narrowness and constraints imposed by the old Allport controversy vanish in history. It is, for an observing mind, invigorating to perceive, at the same time, the role of discovering intricate behavior patterns and the role of inventing integrating concepts as utterly vital for any kind of deeper psychological understanding. It is a fundamental pleasure (and one supported, I believe, by the historical development outlined earlier) to enjoy the wider range of curiosity and freedom gained by modern ethology in its attempt to understand more of the integrated aspects of the lives of individuals from their own perspective—even if they are infants not even one year old. The next step is always, as Stern anticipated, the nomothetic, critical support of such personal insights, their change, correction, and possibly their improvement.

REFERENCES

Ainsworth, M. D. S., & Bell, S. M. (1977). Infant crying and maternal responsiveness: A rejoinder to Gewirtz and Boyd. *Child Development, 48(4)*, 1208–1216.

Ainsworth, M. D. S., Blehar, M. C., Waters, E., & Wall, S. (1978). *Patterns of attachment: A psychological study of the Strange Situation.* Hillsdale, NJ: Erlbaum.

Allport, G. W. (1937). *Personality: A psychological interpretation.* New York: H. Holt.

Allport, G. W. (1950). Personality: A problem for science or a problem for art? In G. W. Allport, *The nature of personality* (Selected papers). Cambridge, MA: Addison-Wesley.

Allport, G. W. (1961). *Patterns and growth in personality* (2nd ed.). New York: Holt, Rinehart & Winston.

Berger, P. L., & Luckman, T. (1966). *The social construction of reality.* New York: Doubleday.

Bischof, N. (1975). A systems' approach toward functional connections of attachment and fear. *Child Development, 46*, 801–817.

Block, J. H., & Block, J. (1980). The role of ego-control and ego-resiliency in the organization of behavior. In W. A. Collins (Ed.), *The Minnesota symposia on child development: Vol. 13. Development of cognition, affect, and social relations* (pp. 39–101). Hillsdale, NJ: Erlbaum.

Bühler, C. (1976). *Die Rolle der Werte in der Entwicklung der Persönlichkeit und in der Psychotherapie* (The role of values for development of personality and in psychotherapy). Stuttgart: Klett-Cotta.

Bühler, C., & Hetzer, H. (1929): Zur Geschichte der Kinderpsychologie (On the history of child psychology). In *Beiträge zur Problemgeschichte der Psychologie. Festschrift zu Karl Bühlers 50.Geburtstag* (pp. 204–224). Jena: Verlag von Gustav Fischer.

Bühler, K. (1925). *Abriß der geistigen Entwicklung des Kindes* (Summary of the mental development of the child, 2nd ed.). Leipzig: Quelle & Meyer.

Bühler, K. (1965). *Die Krise der Psychologie* (The crisis of psychology, 3rd ed.). Stuttgart: Fischer. (Original work published 1927)

Campbell, D. T. (1974). Evolutionary epistemology. In P. A. Schilpp (Ed.), *The philosophy of Karl R. Popper* (pp. 413–463). La Salle, IL: Open Court. (Originally prepared 1966)

Carus, F. A. (1808). *Psychologie* (Band 2). Leipzig: Johann Ambrosius Barth & Paul Gotthelf Kummer.

Cattell, R. B. (1944). Psychological measurement: Ipsative, normative, and interactive. *Psychological Review, 51*, 292–303.

Chance, M. R. A. (1967). Attention-structure as the basis of primate rank orders. *Man, 2*, 503–518.

David, M., & Apple, G. (1969). Mother–child relationship. In J. G. Howells (Ed.), *Modern perspectives in international child psychiatry* (pp. 98–124). Edinburgh: Oliver & Boyd.

Dilthey, W. (1924). Ideen über eine beschreibende und zergliedernde Psychologie (Ideas on a descriptive and analytic psychology). In W. Dilthey, *Gesammelte Werke* (Band 5, pp. 139–240). Leipzig: Teubner. (Original work published 1894)

Dunn, J. (1983) Sibling relationships in early childhood. *Child Development, 4*, 787–811.

Dunn, J., & Kendrick, C. (1979). Interactions between young siblings in the context of family relationships. In M. Lewis & L. A. Rosenblum (Eds.), *The child and its family* (pp. 143–168). New York: Plenum Press.

Eco, U., & Sebeok, T. A. (Eds.). (1983). *The sign of three: Dupin, Holmes, Peirce.* Bloomington, IN: Indiana University Press.

Eisler, R. (1927–1930). *Wörterbuch der philosophischen Begriffe* (Vols. 1–4). Berlin: E. S. Mittler und Sohn.

Ernst, C., & Angst, J. (1983). *Birth order, its influence on personality.* Berlin: Springer Verlag.

Franck, I. (1982). Psychology as a science: Resolving the idiographic-nomothetic controversy. *Journal of Theory of Social Behavior, 12*(1), 1–20.

Gewirtz, J. L., & Boyd, E. F. (1977a). Does maternal responding imply reduced infant crying? A critique of the 1972 Bell and Ainsworth Report. *Child Development, 48*(4), 1200–1207.

Gewirtz, J. L., & Boyd, E. F. (1977b). In reply to the rejoinder to our critique of the 1972 Bell and Ainsworth Report. *Child Development, 48*(4), 1217–1218.

Gould, S. J. (1979, December). Darwin's middle road. *Natural History*, pp. 27–31.

Graumann, C. F. (1960). Eigenschaften als Problem der Persönlichkeitsforschung (Traits as a problem of research in personality). In P. Lersch & H. Thomae (Eds.), *Handbuch der Psychologie: Band 4. Persönlichkeitsforschung* (pp. 87–154). Göttingen: Verlag für Psychologie-Hogrefe.

Grossmann, K. E. (1983, April). *Historical contributions of German developmental psychology.* Paper presented at the Symposium on International Perspectives on Research in Child Development, Society for Research in Child Development, Detroit, MI.

Grossmann, K. E., & Grossmann, K. (1984). Discovery and proof in attachment research. *The Behavioral and Brain Sciences, 7*, 154–155.

Grossmann, K. E., Grossmann, K., Huber, F., & Wartner, U. (1981). German children's behavior toward their mothers at 12 months and their fathers at 18 months in Ainsworth's Strange Situation. *International Journal of Behavioral Development, 4*, 157–181.

Grossmann, K., Grossmann, K. E., Spangler, G., Suess, G., & Unzner, L. (1985). Maternal sensitivity and newborns' orientation responses as related to quality of attachment in Northern Germany. In I. Bretherton & E. Waters (Eds.), Growing points in attachment theory and research. *Monographs of the Society for Research in Child Development* (Nos. 1–2, Serial No. 209), pp. 233–256.

Grossmann, K. E., Grossmann, K., & Schwan, A. (1986). Capturing the wider view of attachment: A re-analysis of Ainsworth's Strange Situation. In C. Izard & P. Read (Eds.), *Measuring emotions in infants and children* (Vol. 2, pp. 124–171). Cambridge: Cambridge University Press.

Guilford, J. P. (1959). *Peronality.* New York: McGraw-Hill.

Heiss, R. (1936). *Die Lehre vom Charakter* (About character). Berlin: DeGruyter & Co.

Holst, E. von, and St. Paul, U. von. (1960). Vom Wirkungsgefüge der Triebe (On the effect network of instincts). *Naturwissenschaften, 47*, 409–422.

Klix, F. (1983). *Erwachendes Denken* (The dawn of thinking, 2nd ed.). Berlin: VEB Verlag der Wissenschaften.

Klix, F. (1984, September). *Parameters of cognitive efficiency: A new approach in measuring human intelligence.* Paper presented at the 23rd International Congress of Psychology, Acapulco, Mexico.

Kuhn, T. S. (1962). *The structure of scientific revolutions.* Chicago, IL: University of Chicago Press.

Lorenz, K. (1973). *Die Rückseite des Speigels. Versuch einer Naturgeschichte menschlichen Erkennens* (Behind the mirror. A search for a natural history to human knowledge). München: Piper.

Lorenz, K. (1974). Analogy as a source of knowledge. *Science, 185,* 229–234.

Main, M., Kaplan, N., & Cassidy, J. (1985). Security in infancy, childhood and adulthood: A move to the level of representation. In I. Bretherton & E. Waters (Eds), *Growing points in attachment theory and research. Monographs of the Society for Research in Child Development, 50*(1–2, Serial No. 209), 66–103.

Pierce, C. S. (1955) Abduction and induction. In J. Buchler (Ed.), *Philosophical Writings of Pierce.* New York: Dover.

Plomin, R. (1981). Ethological behavioral genetics and development. In K. Immelmann, G. W. Barlow, L. Petrinovich, & M. Main (Eds.), *Behavioral development: The Bielefeld Interdisciplinary Project* (pp. 252–276). Cambridge: Cambridge University Press.

Plomin, R. (1983). Developmental behavioral genetics. *Child Development, 54,* 253–259.

Riedl, R. (1980). *Biologie der Erkenntnis: Die stammesgeschichtlichen Grundlagen der Vernunft* (Biology of knowledge: The phylogenetic roots of reason). Berlin: Paul Parey.

Riegel, K. (1979). *Foundations of dialectical psychology.* New York: Academic Press.

Scarr, S. (1981). Testing for children: Assessment and the many determinants of intellectual competence. *American Psychologist, 36,* 1159–1166.

Scarr, S., & Kidd, K. K. (1983). Developmental behavioral genetics. In M. M. Haith & J. J. Campos (Eds.), *Handbook of child psychobiology: Vol. 2. Infancy and developmental psychobiology* (pp. 345–433), New York: Wiley.

Scarr, S., & MacCartney, K. (1983). How people make their own environments: A theory of genotype environment effects. *Child Development, 54,* 424–435.

Skaggs, E. B. (1945). Personalistic psychology as a science. *Psychological Review, 52,* 234–238.

Sroufe, L. A., & Waters, E. (1977). Attachment as an organizatorial construct. *Child Development, 48*(4), 1184–1199.

Stern, C., & Stern, W. (1928/1931). *Monographien über die seelische Entwicklung des Kindes.* I. Die Kindersprache, eine psychologischen und sprachtheoretische Untersuchung (3rd ed., original work published 1907). II. Erinnerung, Aussage und Lüge in der ersten Kindheit (3rd ed., original work published 1909). Leipzig: Barth.

Stern, W. (1921). *Die differentielle Psychologie in ihren methodischen Grundlagen* (Differential psychology in its methodological foundations, 3rd ed.). Leipzig: Barth. (Original work published 1911)

Stern, W. (1924). *Person und Sache: System des kritischen Personalismus: Band 3. Wertphilosophie* (Person and matter: The system of critical personalism, Vol. 3. Philosophy of value). Leipzig: Barth.

Stern, W. (1927a). *Psychologie der frühen Kindheit bis zum sechsten Lebensjahr* (mit Benutzung ungedruckter Tagebücher von Clara Stern) (Psychology of early childhood up to the sixth year of age, using unpublished diaries of Clara Stern, 4th ed.). Leipzig: Quelle & Meyer. (Original work published 1914)

Stern, W. (1927b). William Stern. In R. Schmidt (Ed.), *Die Philosophie der Gegenwart in Selbstdarstellungen* (pp. 1–56). Leipzig: Verlag von Felix Meiner.

Stern, W. (1930a). *Psychology of early childhood up to the sixth year of age, supplemented by extracts from the unpublished diaries of Clara Stern* (2nd ed., Anna Barwell, Trans.). New York: H. Holt. (Original work published 1914).

Stern, W. (1930b). William Stern. In K. Murchison (Ed.), *A history of psychology in autobiography* (Vol. 1, pp. 335–388). Worcester, MA: Clark University Press.

Tinbergen, N. (1963). On aims and methods of ethology. *Zeitschrift für Tierpsychologie, 20,* 410–433.

Tinbergen, E. A. & Tinbergen, N. (1972). *Early childhood autism: An ethological approach.* Fortschritte der Verhaltensforschung, Beiheft 10 zur Zeitschrift für Tierpsychologie. Berlin: Paul Parey.

Underwood, B. J. (1957). *Psychological Research.* New York: Appleton-Century-Crofts.

Vaughn, B. E., & Waters, E. (1980). Social organization among preschool peers: Dominance, attention, and sociometric correlates. In R. Omark, F. F. Strayer, & D. G. Freedman (Eds.), *Dominance relations: An othological view of human conflict and social interaction* (pp. 359–381). New York: Garland Press.

Vollmer, G. (1975). *Evolutionäre Erkenntnistheorie* (Phylogenetic theory of knowledge). Stuttgart: Hirzel.

Wachinger, A. M. (1981). *Untersuchung zur Lage junger Familien mit Säulingen und Kleinkindern anhand von Interviews* (Explorations on the situation of young families with infants and small children on the basis of interviews). Diplomthesis, Universität Regensburg.

Waters, E. (1981). Traits, behavioral systems, and relationships: Three models of infant–adult attachment. In K. Immelmann, G. W. Barlow, L. Petrinovich & M. Main (Eds.), *Behaviorial development: The Bielefeld interdisciplinary project* (pp. 621–650), Cambridge: Cambridge University Press.

Welz, S. (1984). *Analyse der Geschwisterbeziehungen in Familien der Bielefelder Längsschnittuntersuchung* (Analysis of sibling relationships in families of the Bielefeld longitudinal study). Diplomthesis, Universität Regensburg.

Windelband, W. (1919). *Präludien II. Aufsätze und Reden zur Philosophie und ihrer Geschichte* (Essays and speeches about philosophy and its history, 6th ed.). Tübingen: Verlag von J.C.B. Mohr.

Wundt, W. (1880–1883). *Logik: Eine Untersuchung der Prinzipien der Erkenntnis und der Methoden wissenschaftlicher Forschung* (Vols. 1–2). (Logic: An investigation of principles of knowledge and methods of scientific research). Stuttgart: F. Enke.

Ziemer, G. (1982). *Das Zusammenleben mit 10 Monnate alten Kindern aus der Sicht ihrer Mütter, sowie deren Einstellungen zu ihren Kindern* (Mothers' reports on their lives with 10-year-old infants and their attitudes toward them). Diplomthesis, Universität Regensburg.

CHAPTER 3

The Production, Detection, and Explanation of Behavioral Patterns

WARREN THORNGATE

INTRODUCTION

The essence of science is the detection and explanation of patterns. Physical scientists devote themselves to the detection and explanation of patterns of matter and energy. Biological scientists are concerned with patterns of life. And behavioral scientists are concerned with patterns of human behavior and experience. Despite the differences in subject matter, all attempt to discern regularities in their domains and to analyze why these regularities occur. Few, if any, attempts are made to study irregular or patternless phenomena. Accidents and other unique events are sometimes investigated by scientific means if deemed sufficiently important (e.g., determining a cause of death, a disputed authorship, or the trajectory of an epidemic), and if they can be viewed as products of regularities or patterns. But phenomena that exhibit no discernible regularity or pattern are relegated to the status of anecdote, transience, chaos, or noise. They are acknowledged in statistics by a concept known as error. And they are assumed, by tradition if not definition, to lie outside the domain of scientific enquiry.

WARREN THORNGATE • Department of Psychology, Carleton University, Ottawa, Ontario, Canada K1S 5B6.

Though patterns are the stuff of science, relatively few of them are readily discernible, and even fewer can be easily explained. Often the detection of patterns must await the invention of new instruments of observation and analysis; the telescope, microscope, polygraph, and inferential statistics are notable examples. Often the patterns that are detected resist explanation, at least until new concepts are developed to account for them, and new methodologies are developed to test the validity of the accounts. There is some merit in speculating that scientists have now found and explained most, if not all, of the simple and easily discovered patterns, and that only the complex and elusive ones remain. Physics, for example, has long ceased to be a "naked eye" discipline, and biology has come to concentrate on the elusive complexities of genes. Surely most of us who call ourselves research psychologists can testify that the patterns we seek are usually complex and elusive, and that explaining the ones we do find is usually far more difficult than falling off a log. Anyone who has observed human behavior is led to the inescapable conclusion that humans are never entirely neat, tidy, or regular. For every rule we invent that accounts for patterns found, two or more exceptions are soon discovered. Yet, if we have been frustrated in our attempts to find simple patterns in our subject matter, and discouraged by our attempts to find simple explanations of the patterns we do find, we have not yet abandoned our faith that patterns do exist and can sooner or later be detected and explained.

Our failures to discover simple patterns and simple explanations of behavior have encouraged a strong belief in complexity. It is now almost taken for granted that patterns of human behavior are quite complex, subject to influence by a host of variables, and prone to haphazard fluctuations. In order to make some sense of the beast that would produce such complexity, we have come to rely almost exclusively on concepts and research strategies borrowed from another discipline: statistics. They are by and large the concepts defining the general linear model, and though they remain controversial in their mother discipline, we embrace them with almost religious conviction. We have assumed that patterns of human behavior are infused with random variation that in the long run will cancel itself out to reveal patterns of averages or proportions that will, in their statistically significant differences, mimic the behavioral patterns themselves. We have assumed that the patterns so revealed will be generated by some explicable composite of independent, correlational, or predictor variables and indicated by some composite of dependent or criterion variables of our choosing. We now place so much faith in these assumptions that most of our graduate schools require a course in statistics of all students, and leave other courses, such as those in the history of psychology, methods of theory construction, or philosophy of science, as mere options. When our statistics do not lead us to the patterns we seek, we add more variables to our research designs, deploy fancier statistics in hopes of detecting otherwise hidden patterns, or look for new and simpler topics of enquiry (Newell, 1973; Thorngate, 1976).

I think that our faith in the statistical assumptions just mentioned is misplaced. At the risk of heresy, I should like to argue that the statistical assumptions on which our research practice is so firmly founded are often demonstrably false, and consequently that they misguide much of our efforts to find patterns of behavior, particularly of the behavior of individuals. Furthermore, I should like to argue that it is often reasonable to expect behavior to be quite patternless, or to exhibit patterns that can be explained by nothing other than accident (see Bandura, 1982), history (see Langmuir, 1940; London, 1946), or chance. I hope the arguments will give some indication of why our conceptions of an idiographic science of behavior must depart from those we have previously, and often erroneously, justified as appropriate for a nomothetic science. And I hope to indicate just how these conceptions should change.

PATTERNED FALSEHOODS

ARTIFACTS OF AGGREGATION

It has now been over 30 years since psychologists were shown the dangers of seeking patterns in averages and other aggregates. At least since Robinson (1950) publicized the "ecological fallacy," and Sidman (1952) demonstrated the possibility of misperceiving patterns of learning by averaging learning data across subjects (see also Bakan, 1954), we have had at our disposal clear indications of the possibility that patterns of aggregates have no necessary connection with aggregates of patterns. Alas, it appears that these indications have been largely forgotten or ignored. Our most popular statistics and methodology texts make no reference to the problem. At best, they acknowledge certain difficulties with within-subject designs—those resulting in variance, covariance of other distributional differences that foul up our ability to employ popular statistical procedures properly. These differences are usually treated as nuisances that—if fortune smiles—can be attenuated or alleviated by appropriate data transformation or conservative tests. In this way we pay homage to Procrustes.

Yet stretching or cutting our data to suit our assumptional beds will not in general solve our artifactual problems. Aggregates of transformed

74   WARREN THORNGATE

data are still aggregates, and conservative tests remain tests of aggregations. The possibility of artifacts thus remains. To illustrate this possibility,
consider one of the most common artifacts of aggregation with an example from the cognitive literature. Assume that we are interested in
determining something of the nature of judgment processes employed
by members of a graduate committee faced with the task of allocating
assistantship funds to new students. We employ a popular research design. Each of the three committee members is given a set of descriptions
of hypothetical student records containing information about students'
sex, grade point average, and number of dependents varied in a factorial
design.

    Assume further that each committee member had very fixed opinions
about who should be funded. Member A believed that only males should
be funded; Member B believed only females should be funded; Member
C believed only students with high grades should be funded. Their beliefs
were manifested in their judgments. These are shown in Table 1.

    Table 1 reveals that, on average, the committee members gave $3,000
more to students with high grades than to students without them. Thus,
we may be tempted to conclude that grades were the only consistently
important cues in the committee's judgment processes. Of course, this
is false. Sex was the only important cue for the majority of members.
Grades were important for the third member, and far more important
than the averages reveal. Had a fourth member been added to the committee with judgments opposite to those of member C, the resulting
averages would all have been equal, indicating no consistent judgment
policy. This, too, would be false, as each of four members had a highly
consistent judgment policy. Covariance differences would indicate a vi-
olation of the assumptions of a within-subject design analysis of the

Table 1. Monies Allocated to Students (in Thousands of Dollars)

Sex	Male				Female			
Grades	High		Not High		High		Not High	
Dependents	Yes	No	Yes	No	Yes	No	Yes	No
Member								
A	$9	9	9	9	$0	0	0	0
B	$0	0	0	0	$9	9	9	9
C	$9	9	0	0	$9	9	0	0
Average	$6	6	3	3	$6	6	3	3
SD	5.2	5.2	5.2	5.2	5.2	5.2	5.2	5.2

variance model. But this problem could be circumvented by a Geiser-Greenhouse or similar test (e.g., see Kirk, 1969, pp. 142–143). It would not matter. If four members generated eight averages in this experiment that were all equal, no test would show significant differences in these averages.

The obvious moral of this little exercise is that what occurs on average is not necessarily what occurs in general. But there are less obvious, and equally important lessons to be learned here. The traditional statistical techniques we may employ here are designed to analyze experiments; we use the techniques to analyze responses, but we are really interested in cognitive processes. We have thus confused, and confounded, three different units of analysis. To analyze processes by averaging responses, it is necessary to assume that the processes are equivalent across people and time. In the case of cognitive processes, such an assumption is almost always incorrect (see Einhorn & Hogarth, 1981; Thorngate, 1975). Our first task, therefore, is to determine why processes vary, and to do so we must discover what processes are possible, then find discriminating indicants of each. The values of these indicants generated by each subject can then be analyzed to estimate his or her most likely processes (e.g., see Payne, 1976; Thorngate & Maki, 1976). Only when this is done can we make statements appropriate to the study of process variance—statements about process popularity and variety, not about process averages. In a phrase, psychologists are far better off to analyze before aggregating than to aggregate before analyzing.

These observations have some interesting implications for discussions of nomothetic and idiographic approaches to research. It is tempting to equate the nomothetic approach with the analysis of averages. To do so is to equate statistical models of experiments with models of people (Wilson, 1973). If the equation is unjustified—and it usually is—then what we learn about averages will not be an accurate reflection of what occurs in general. Yet nomothetic laws are supposed to address what people do in general. So, in general, averages will not reflect nomothetic laws. To find out what people do in general, we must first discover what each person does in particular, then determine what, if anything, these particulars have in common. This implies that we pay more attention to case histories, find or develop models sufficient to account for each, then examine the models for common themes or elements. Thus, for example, if models X, Y, and Z each provide a sufficient explanation for the behavior of person A, models X and Y each provide a sufficient explanation for the behavior of person B, and models X and Y and Z do the same for person C, then we should look to X for hints about a nomothetic law. A statistical method for doing so is presented in another chapter of this volume

(Thorngate & Carroll). It is based on a simple premise: Nomothetic laws lie at the intersection of idiographic laws; the former can be discovered only after we find the latter.

PATTERN AND PURPOSE

The notion of purpose, and the allied notions of goals and motivation, seem central to our understanding of human behavior. Alas, these notions have had a checkered past in psychology, and at present are largely ignored while we pursue a strange fascination with cognition as a topic of investigation and as an explanatory concept. There can be little doubt that such concepts as purpose, motivation, and goals are often troublesome and difficult to examine in detail by scientific methods. But the concept of cognition is equally troublesome, and is certainly no more encompassing in explanatory power than many of those we have dropped along the way.

The current dominance of cognition as a topic of investigation and explanatory concept seems to be the result of causes that have less to do with scientific merit than with intellectual history, the sociology of science, and well-developed habits of designing and conducting research. Most of us have been well trained to explore and explain covariations between stimuli and responses. Our statistical methods are designed to detect such covariation. Research results that do not reveal "significant" covariations are usually assumed to be of less importance than those that do. So there is good reason to look for responses that vary with our stimuli and to avoid responses that do not. First, of course, we must get the responses, and to do so we must give our subjects some reason for giving them to us. The easiest way to accomplish this is to provide in our research instructions and tasks some goal for subjects to attain, some purpose for responding. When we hold this purpose constant we cannot use the concept of purpose to explain any stimulus–response covariation we may subsequently obtain. So we look to leftover concepts, particularly those that concern what goes on in the heads of subjects between our stimuli and their responses. Most of these concepts—save perhaps skill, emotion, and the reflex arc—are cognitive. It is thus no accident that cognition is a popular phenomenon and explanatory construct.

In principle, it should be possible to hold cognition constant and explore how changes in purpose affect stimulus–response regulations. In fact, this is virtually impossible to accomplish. It seems almost absurd to tell our subjects, "This is the way I want you to think . . ." and then vary the reasons why they should. At best, we can try to vary stimuli and purpose together, then try to figure out their independent and mutual effects. In doing so, however, we must reconsider the importance of covariation in our research. As it happens, we cannot study purpose properly by always looking for covariation. For, paradoxically, purpose is regularly revealed by the *absence* of stimulus–response relationships, rather than by the presence of them. How can this be? Consider the demonstration of Powers (1978).

Powers created a simple computer program that slowly moved two points of light (A and B) across a CRT screen. As they went from left to right, they also moved up and down in regular (e.g., sinusoidal) and parallel trajectories—A undulated across the top of the screen, B undulated across the bottom. However, the routine that generated the movement of A also regularly sampled the setting of a joystick controlled by a subject watching the screen. Each momentary numerical value of the joystick setting was subtracted from the momentary numerical value of A, and the result was projected on the screen as A's adjusted value. No such substitution was done for point B. In effect, the subject's joystick movement could control the trajectory of A; it could not control B's trajectory. (For a demonstration computer program, see Appendix A.)

In a typical experiment, Powers asked a subject to move the joystick so as to keep point A horizontal and flat as it moved. With a few minute's practice, this was easily accomplished. As a result, A would proceed across the screen in a straight line, B would continue to undulate (as A would have in the absence of joystick movement), and the joystick would wiggle in a mirror image of B's movement.

Consider, now, a psychologist who had no knowledge of the program, and who began to observe the situation in an attempt to determine why the subject moved the joystick. One relationship would be readily apparent: every time the point of light B moved up, the subject would pull back the joystick; every time B moved down, the subject would move the joystick forward. Indeed, the correlation between variance in stimulus B and variance the joystick response would approach -1.00. In contrast, point A would show no variance, and the correlation between the constant vertical value of A and the joystick movement would be zero. It would thus be difficult for almost any psychologist to resist the temptation to conclude that stimulus B caused the subject's joystick movement.

And how wrong the conclusion would be. The relation between B's movement and the joystick movement is entirely epiphenomenal. B is in no way a stimulus, and the joystick movement is in no way a response to it. But is A therefore the stimulus? Paradoxically, from the viewpoint of the psychologist, it is not. Our current research mentality requires that causal connections between stimulus and response exhibit covariation.

Although correlation does not necessarily imply causation, causation does imply correlation—A and the joystick movement correlate no more than the joystick movement and the height of the table on which the CRT screen rests, so neither can be claimed as causal.

What, then, is the stimulus? What causes the joystick to move? There is no cause for the movement (save perhaps the muscles in the subject's hand). Instead, there is a *reason* for the movement: Powers' instruction to the subject, "keep point A flat." This reason is the stimulus. The trajectory of A is called feedback. Let us rephrase the point. The goal, motive, or purpose of keeping A flat stimulated the joystick movement. But this purpose cannot be detected by traditional research designs and statistical analyses.

A necessary, if not sufficient, condition for detecting the stimulus requires that the psychologist observe the movement of A and B in the absence of the subject, then compare them to the movements in the presence of the subject. More generally, we must examine how the person affects the situation rather than how the situation affects the person. To do so is to reverse our standard research procedures, to turn 100 years of research tradition upside down. It will also be necessary to know something of the relations between elements of the situation—by analogy, something of the program—and something of a person's knowledge of these relations, in order to determine what the person is trying to do. Such estimates are tedious, if not difficult, to make. We can, of course, circumvent the tedium by merely asking the person "Why?" This does not challenge our ingenuity as researchers. But it may speed the recovery of purpose as a useful concept for the explanation of behavioral patterns.

METHODOLOGICAL IMPLICATIONS

The artifacts of aggregation, and the paradox of purpose, should give us cause to consider that by employing traditional approaches to statistics and research design we may be seeking patterns in the wrong places. What we may find are patterns that have no necessary connection to the rules, processes, or reasons that in fact govern human behavior. In the words of Mitroff and Featheringham (1974), we may commit Type III errors: solving the wrong problems when we should have solved the right ones.

The lesson we can learn from the preceding discussion is easy to summarize: Patterns appropriate to the domain of psychology are most likely to be found by examining, one at a time, combinations of individuals and the situations in which they behave. Each combination may generate several patterns of interest, and each pattern may be amenable to several sufficient explanations. But as we sample increasing numbers of combinations, certain sufficient explanations may arise far more frequently than others. We can then adjust our degrees of belief in the explanations accordingly (see Levins, 1966).

Such a research strategy has, of course, been part of science for a long time. In medicine it is known as the case history approach. In biology it is the one preferred by naturalists. Psychologists have generally avoided it. Case histories and naturalistic observations are often collected by clinicians, ethologists, and a few followers of Roger Barker. But they remain on the fringe of the discipline, and have not yet been exploited for their full scientific potential. There are, to be sure, several difficulties with case history and observational methods—difficulties ranging from bias in the selection of cases observed to a dearth of standards for recording patterns. But time spent overcoming these difficulties is likely to advance the discipline far more than, say, time spent on developing or deploying yet another omnivariate statistical procedure. Charles Darwin managed to advance biology without the aid of the analysis of variance, canonical correlation, or causal path modeling. It would probably do us little harm to follow his lead. We may at least reduce our Type III errors.

Yet before we rush to collect case histories and observe beyond the confines of our variables, I think it would be judicious to consider again just what we are looking for. Presumably, we are looking for patterns of behavior—patterns generated by individuals, in given situations, over time. But is is possible for behavior to show no pattern, or more often to deviate from usual patterns in unpredictable ways. Our understanding of patterned behavior might well be enhanced if we first considered why individuals would not exhibit pattern. Let us now do so.

PATTERNLESS TRUTHS

In Praise of Inconsistency

People are often inconsistent. They often do not produce the same response to equivalent stimuli. And they often fail to exhibit any logical relation between what they believe, say, and do. These inconsistencies have presented psychology with a major dilemma: What should be done with them? Science is supposed to concern itself with the discovery and explanation of regular patterns in nature and society. Inconsistencies usually generate irregular patterns. Should we therefore assume that the inconsistencies are irreducible perturbations of regularities and ignore them? Or should we assume that they are imaginary, a reflection of our ignorance, something that will be eliminated as we increase our abilities to discover regular but complex patterns, which now appear irregular?

Psychologists have danced around this dilemma for decades. Once we adopted the general linear model as our major data analytic device, we came to believe that we should concentrate on whatever significant differences we could find and ignore what was left. Main effects and interactions constituted our regularities, the stuff worthy of science. What remained was "error"; we needed it for our F-ratios, but we never considered it worthy of serious enquiry. Determinists could say that it would ultimately disappear when we measured enough important variables at once. Indeterminists could say it would always exist. We really did not care. As long as inconsistencies were normally distributed and relatively small, they became the least of our worries. Between-subject inconsistencies could be dismissed as random effects of nature, nurture, and subject selection. Within-subject inconsistencies could be dismissed as products of wandering attention, fatigue, mood, and such. In the end they would all cancel out.

How different has been the biologists' view of inconsistency. Instead of dismissing inconsistency as mere error or noise, they have made it a central concept of evolutionary theory. Instead of attempting to eliminate it through experimental controls, they have sought to explain its pervasiveness. Perhaps we should do the same.

In order to stimulate your mimetic passion I ask you to ponder the following: Biologists consider genetic variability (inconsistency) to be essential for the survival of a species in changing environments (see Maynard-Smith, 1958); does behavioral, cognitive, or motivational inconsistency have a similar functional significance for individuals in similar environments? A psychologist's first impulse is to answer "No." In most cultures, inconsistency is considered a weakness. People who display it in public are often judged as unthinking, addleheaded, perfidious, or hypocritical. Arguments are usually fought by attempting to find inconsistencies in an opponent's statements or behaviors. Most legal systems place consistency of case and statute law above all else; inconsistency is an anathema in the courtroom. And several theories of attitude change assume a natural aversion to attitudinal or behavioral inconsistency within individuals (e.g., see Feldman, 1966).

But our first impulse may well be wrong, or at least shortsighted. There are several occasions when inconsistency may have great advantages in one's life (see Hogarth, 1982). In competitive games, for example, behavioral inconsistency has a definite advantage; more goals are scored by doing the unexpected than by doing otherwise. Inconsistency often helps save face or esteem. When a male, for example, wishes to become more intimate with a female, he may resort to making a serious request in a lighthearted manner. If the woman accepts the request, his wish is granted; if she does not, he can exclaim that he was "only joking" and avoid undue embarrassment. Certainly inconsistency is advantageous to politicians. Telling different things to different people allows a politician to pander to divergent interests, and increase the chances of holding power. Most inconsistencies will never be discovered or understood by members of the electorate. If they are, they will usually be quickly forgotten. And if not forgotten, they can usually be smothered in rhetoric.

Individuals tend to be very forgiving of inconsistency in themselves and others, especially when it leads to greater rewards or fewer punishments than its complement. Men have long profited from the "double standard" of sexual conduct and often continue to promote it. The rich and powerful are usually quite willing to overlook the logical inconsistencies that lead others to give them preferential treatment. Our schools reward children who are able to learn the inconsistent relations between English pronunciation and spelling; those who are consistent are often regarded as retarded. Inconsistency can even be its own reward. It is an excellent antidote for boredom.

There is yet another function of inconsistency that I find of special theoretical interest. Inconsistency is a prime indicant of weak preferences (Fischhoff, Slovic, & Lichtenstein, 1980; Luce, 1959). When we have very strong preferences for outcomes or courses of action, we tend to pursue them consistently; otherwise we do not. If the nature and frequency of outcomes in social interactions are uncertain, and if we have far less than complete power to control them, then it makes good adaptive sense to weaken our outcome preferences (or as Lewin would say, lower our level of aspiration). For if they are not weakened, the strongly preferred outcomes may rarely, if ever, occur and we may starve for the rewards they bring. As one weakens one's preferences, the result will be a decrease in response consistency, and a commensurate increase in the diversity of alternative outcomes or actions that will be sampled over time. The more one samples, the greater the likelihood that one will discover new response–outcome relationships. Some of these may produce the original, strongly preferred outcomes with much greater frequency than before, or they may regularly produce previously unsampled outcomes that are strongly preferred to the originals. If so, an individual will have not only survived, but will have flourished.

Let me simplify my thesis: Inconsistency is an effective strategy for insuring that not all of one's eggs are placed in one basket. It insures that an individual's behavior is neither inexorably linked to the past, nor totally dictated by the present. Social environments are often unpredictable, and one's success in them may depend largely on the maintenance of loose and protean relationships between stimuli and response, or response and response. Haphazard fluctuations in "normal" behavior allow an individual to test the resiliency of his or her environment, and to develop or exercise strategies of coping with future or unforeseen changes in that environment. Inconsistency insures variety, and in unpredictable environments variety promotes survival (Dobzhansky, Ayala, Stebbins, & Valentine, 1977).

If inconsistency does have functional-adaptive significance in environmental and social interactions, then several important research questions can be raised. For example, it is sometimes assumed that cognitive and perceptual deficits produce inconsistencies (see Tversky & Kahneman, 1974). Is the reverse also true? Are the deficits actually adaptions to inconsistency that insure that the advantages of inconsistency will not be spoiled by constant worry about ordering one's beliefs and actions? Can response inconsistency be predicted from stimulus inconsistency, and vice versa? To what extent, and under what conditions, might response inconsistency result in the equivalent of "genetic drift," or adaptively neutral shifts in response norms (see Nagao & Davis, 1980)? Does such drift account for the infusion of American spelling into Canadian English, or the diversity of schools of art? What are the conditions and strategies for tolerating inconsistency in others? Are some strategies more adaptive? If so, can they be incorporated in organizational rules of conduct (see Katz, 1974)?

Inconsistency appears to be worthy of serious enquiry. To conduct such enquiry, it is first necessary to devise means for measuring inconsistency. Once such means are developed we can treat inconsistency as the subject of analysis and examine the extent to which changes in inconsistency may or may not exhibit patterns. Considerable work has already been done on analogous measures in ecology (e.g., see Pielou, 1969, Chapter 18). These measures focus on the dispersion (vs. concentration) or variance of research observations rather than their location or central tendency. Psychologists normally ignore variance differences, or treat them as an anathema, in hot pursuit of significant differences in means. By doing so we may well be overlooking our richest source of patterned phenomena (see also Bem & Allen, 1974).

THE NECESSITY OF CHAOS

I have tried to argue in the discussion above that inconsistency may have adaptive merit, and thus given an individual reason to exhibit patternless behavior. Recent developments in theoretical biology and physics strongly suggest there may also be causes for such behavior—biological and physical conditions that will, by necessity rather than choice, induce an individual to behave in patternless ways. These conditions may be described by very simple equations. Indeed, the equations are so simple it is almost mysterious why they can behave in the complex ways they do. Let us first examine the behavior of these equations, then consider their implications for the study of pattern in psychology.

Assume for a moment that we wish to examine the trajectory of some behavior, X, by measuring the behavior at some fixed interval of time, say once each second, for some relatively long period, say one hour. Assume also that the behavior is generated according to a simple and deterministic rule. The rule states that the behavior at any time, $t + 1$, will be entirely a function of the behavior during the last observation period, t. The function, however, is nonlinear. Several nonlinear functions will serve for illustration, but let us use one of the simplest:

$$X_{t+1} = aX_t(1 - X_t) \qquad (1)$$

where X_t can range from 0 to 1, and a can range from 0 to 4. (See Appendix B for relevant computer program.)

We begin our observations at $t = 1$, setting $a = 0$ and trying assorted starting values of X. At $t = 2$, X_2 becomes $0[= 0 \times .5 \times (1 - .5)]$ for all X_1, and remains there throughout the observation period—a very predictable, and boring, pattern. We now try again setting $a = 1$. This time all starting values of X become predictably smaller on each occasion, approaching 0 by the end of our observations (see Figure 1).

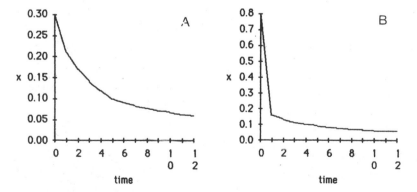

Figure 1. Trajectories of Equation 1 for $a = 1.0$. (A) $X_0 = 0.30$; (B) $X_0 = 0.80$.

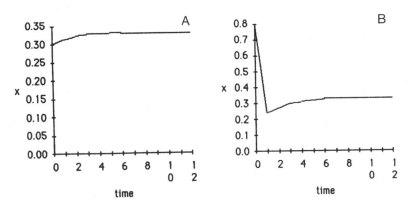

Figure 2.   Trajectories of Equation 1 for $a = 1.5$. (A) $X_0 = 0.30$; (B) $X_0 = 0.80$.

Next we try various settings of a between 1.0 and 2.0. These settings do not allow X to extinguish with time, but instead force X to approach some constant value. The value depends on the setting of a (e.g., for $a = 1.5$, X will approach 0.333; for $a = 2.0$, X will approach 0.500) but all starting values of X will approach it in time, either always from above or from below, according to a regular pattern (see Figure 2). The "homing" tendency is called *convergent amplification* (Langmuir, 1943; London & Thorngate, 1981).

We now try settings of a between 2.0 and 3.0. For each of these settings, we again try different starting values of X. Here too, X will come to approach some asymptotic value—to show convergent amplification. But rather than always approaching the value from above or from below, as occurs when a is between 1 and 2, X will instead approach it as a regular dampened oscillation. We may declare that such a pattern is a bit more complex than the others, but it is still quite regular or predictable (see Figure 3).

At last we try settings of a between 3.0 and 4.0. Suddenly, X begins to behave in increasingly strange ways. At each setting of a between 3.0 and about 3.45, all starting values of X quickly settle on an oscillation pattern that never dampens. For example, with $a = 3.2$, all starting values of X soon settle on the following oscillation: 0.51, 0.80, 0.51, 0.80, 0.51, . . . When $a = 3.4$, all values of X soon settle on 0.45, 0.84, 0.45, 0.84, 0.45, . . . At about $a = 3.45$, the oscillation begins to show a "harmonic": 0.43, 0.84, 0.45, 0.85, 0.43, 0.84, . . . Harmonics are quickly added as a approaches 3.57; when $a = 3.57$, all harmonics that are powers of 2 are present in the trajectory (see Figure 4).

As we increase a from 3.57 to about 3.828, new harmonics appear.

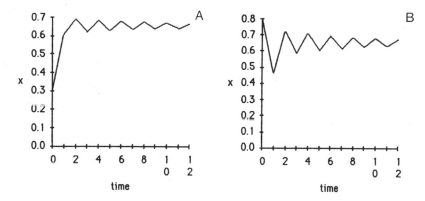

Figure 3.   Trajectories of Equation 1 for $a = 2.9$. (A) $X_0 = 0.30$; (B) $X_0 = 0.80$.

At $a = 3.6786$ harmonics with odd numbered periods appear. And at 3.8284 every harmonic with an integer period emerges (see Figure 5).

Between $a = 3.8284$ and $a = 4.0$ the pattern of X takes on a very special character: it is totally chaotic. To illustrate this, here is the trajectory of X for $t = 1$ to 12, $a = 3.9$ and $X1 = 0.50$:

0.50	0.962
0.975	0.141
0.095	0.475
0.335	0.972
0.869	0.104
0.443	0.363

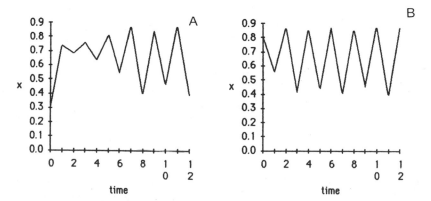

Figure 4.   Trajectories of Equation 1 for $a = 3.5$. (A) $X_0 = 0.30$; (B) $X_0 = 0.80$.

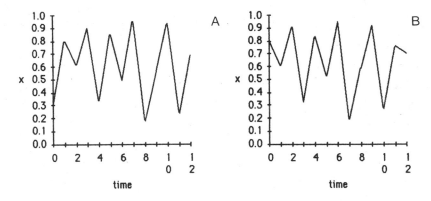

Figure 5. Trajectories of Equation 1 for $a = 3.8$. (A) $X_0 = 0.30$; (B) $X_0 = 0.80$.

In addition, different starting values of X now produce unique chaotic trajectories. Compare the preceding, for example, with the values of X when $a = 3.9$ but $X_1 = 0.52$:

0.52	0.916
0.973	0.298
0.101	0.816
0.354	0.584
0.891	0.947
0.378	0.194

Notice how the two trajectories begin with very similar values, and how the two diverge in their trajectories as they proceed. This is known as *divergent amplification* (Langmuir, 1943; London & Thorngate, 1981). A short computer program that will demonstrate the characteristics of these trajectories is given in Appendix B.

Again, Equation 1 is only one of many simple nonlinear rules that can produce exceedingly complex—even chaotic—behaviors. By way of example, here is another:

$$X_{t+1} = aX_t \quad \text{if} \quad X_t < 0.5 \tag{2a}$$

$$X_{t+1} = a(1 - X_t) \quad \text{if} \quad X_t > 0.5 \tag{2b}$$

The equations and their behaviors have become topics of great interest to theoretical physicists and biologists in recent years because they seem to mimic such phenomena as the transition from laminar flow to turbulence, the often chaotic fluctuations of species populations, and the seemingly spontaneous transitions of chemical compounds. Accounts of their work can be found in Gleick (1984), May (1976), and Prigogine (1980).

What are the implications of the equations and their behaviors for psychology? The answer, of course, depends on the extent to which equations such as these can capture essential features of psychological processes. Good exemplars have yet to be discovered or explored in the discipline; at the moment one can only guess that they will be found at the level of the neuron, and in situations where feedback is delayed (thus necessarily generating oscillation) or where behavioral recovery takes longer than behavioral emission. Anyone who has lived in a poorly constructed apartment complex has probably experienced one analogue in the morning shower. As each dweller enters the shower and turns on the hot water, the hot water available to other showering dwellers diminishes. In response, each of these dwellers will either turn up his or her hot water tap or turn down the cold, or both. These adjustments and their effects are delayed in time by the distance between dwellers. Each causes a new round of tap turning. The oscillations never settle, and the resulting tap dance can proceed for the entire morning in a seemingly chaotic trajectory.

History may show that the equations have no good analogue outside the bathroom. If so, the lessons following will probably be irrelevant to psychology. But consider for at least a moment the paradoxes we will face if several psychological phenomena are governed by the little nonlinear rules.

Perhaps the first lesson to be learned is that many simple, deterministic processes can, under certain conditions, behave chaotically. Indeed, in many cases it is impossible to distinguish the chaos from truly random behavior. (It may be of passing interest to note that the pseudo-random-number generators in most computer languages are variants of Equations 1 and 2.) One implication of this is sobering: If we observe the behavior of such a process in its chaotic state, we will likely conclude (a) there is no pattern to explain, (b) there is a pattern but it is so complex that it must be generated by hundreds of interacting variables, or (c) that whatever pattern exists is entirely stochastic. Each one of these conclusions would be false.

Even if we were clever enough to suspect that the generating function of a chaotic pattern was as simple as that of Equations 1 or 2, we would be unable to reproduce it. There is no known way of "recapturing" the function—as we might by using time series or spectral analyses of linear generating functions—given any finite series of chaotic measurements. If, in the case of Equation 1, a were set at 3.83 or above, all trajectory variance would appear in our error term (see Poole, 1977). Yet if a were 3.82 or below, all trajectory variance would be accountable. Herein is our second lesson. Error variance is not necessarily the result of random variation of unmeasured variables, nor is it the result of unreliability of generating processes or measurement. The chaos of the function discussed previously is merely an extension of their regularity. It strongly suggests that we must radically alter or expand our concepts of randomness to detect, if not explain, many possibly important patterns of behavior (see Chaitin, 1974; Poole, 1977).

A third lesson can be derived from the observations of convergent and divergent amplification. Patterns exhibiting the former are, of course, robust and resilient (see Holling, 1973). Occasional perturbations are quickly attenuated in classic homeostatic fashion; the pattern appears to "maintain itself." When we discover such patterns—as we sometimes do in research on perceptual constancy, motivation and drive, attitude maintenance and change, or stress coping behavior—we are likely to conclude they reflect fundamental and universal psychological mechanisms; indeed, they may.

In contrast, patterns exhibiting divergent amplification are neither robust nor resilient. Occasional, and even small, perturbations can have very large and lasting effects on the pattern, so much so that we may never detect a pattern, or never find a pattern repeated, except at the level of analogy, across people or time. We would likely describe any short-lived pattern "slices" we detected by resorting to anecdote. And we would likely explain them by involving such concepts as metaphor, critical events, and history. In fact, we would have no alternative but to describe and explain the patterns this way.

It is ironic, therefore, that the nomothetic, mechanistic, scientific-sounding accounts of convergent amplification and the idiographic, humanistic, artistic-sounding accounts of divergent amplification can arise from exactly the same pattern-generating rule. In the case of patterns generated by Equation 1, the shift from a scientific to an artistic account of its patterns will occur when $a > 3.82$. This suggests that the two styles of pattern description and explanation, for so long contrasted and viewed as antithetical, may actually be quite compatible, and that they may be unified by assuming nonlinearity of cause and effect.

Possibilities of unification aside, there can be little doubt that a considerable proportion of human activity exhibits divergent amplification, or some combination of convergent and divergent amplification. For example, though most goal directed behavior may show convergent amplification, the setting of goals may exhibit contrasting divergent properties. Thus, if I drive from campus to downtown, or work on this chapter, events that perturb my activities (traffic jams, snowstorms, telephone calls, or other interruptions) will probably stimulate compensatory behaviors intended to attenuate the effects of the events and lead me once again to converge on my goals: in this case, delivering a proposal to a government agency, and meeting the deadlines of Jaan Valsiner. These goals, however, are the result of divergent amplification. I can trace the proposal delivery to a host of seemingly small life events that caused large changes in goals: a television executive who missed an appointment for my job interview in 1967; a request of a former colleague to aid him in a local obscenity trial in 1971; a discussion by Denys Parsons about his musical dictionary on the CBC radio program "As it Happens" in 1975. Similarly, my desire to meet Jaan Valsiner's deadlines would almost surely not have arisen had I not seen an announcement of a conference at Carleton University in 1974, requested to attend, met Lloyd Strickland, obtained employment at Carleton, and spoken to Jaan while he was at Carleton visiting Lloyd.

Response inconsistency is likely to have a direct effect on chances of divergent amplification. As we increase our inconsistency, we tend to increase the diversity of its consequences. In doing so we offer ourselves more opportunities to be rearranged by new or unique events (see Dabrowski, 1964; London, 1949). Floundering can thus be seen as an adaptive response to current unsatisfactory arrangements. So too can fantasy, and flights of fancy.

METHODOLOGICAL IMPLICATIONS

Though a single divergent trajectory may show no discernible pattern, it is still possible that many of these trajectories will show similar, analogous, or otherwise related features. It may be, for example, that several divergent life trajectories of individuals begin to diverge with a single, common, historical event, or with events of similar kind. It may also be that the unique path of one individual has several features common to the paths of others for some fixed period of time (e.g., during a war, or following a natural disaster). Chance events may cause the trajectory of one individual to diverge toward the path of another; this in turn may lead to lasting friendship, and a convergence of their trajectories that will, in turn, diverge from the trajectories each of them would have followed alone.

Plotting divergent trajectories and analyzing their similarities and differences may not be science. But it is certainly an empirical endeavor, and one which addresses the same sort of historical questions and issues of concern to cosmologists, paleontologists, anthropologists, and geographers. It is also an intellectual challenge capable of sustaining and satisfying even the most gifted and curious in our midst.

The analysis of divergent phenomena is not, however, amenable to traditional laboratory procedures. To collect information on divergence and convergence we must develop research skills akin to those of the naturalist in biology, the ethnographer in anthropology, the ethnomethodolist in sociology, the cultural historian. In this regard, I think we have much to learn from the examples of Ivan and Miriam London. For the past quarter century they have done exactly what is prescribed here: gather stories of the lives of hundreds of individuals, and trace the sources of convergence and divergence within and between them. (For a partial list of their published works, see references in London & Thorngate, 1981.) Additionally, they have written extensively on research and interview methods necessary to gather relevant information (e.g., London, 1974, 1975; London & London, 1959). Their work deserves close attention.

CONCLUSION

I hope that the examples and discussion above have alerted you to some of the problems of allowing our research methods to influence where we look for patterns and how we respond to their absence. They point to a great need for clear thinking about the psychological assumptions upon which our observations and analyses rest. We can no longer accept traditional notions that observations are "signal-plus-noise," or that analyses must be sensitive only to aggregates and covariates. The study of individuals and their shifts among and between patterned and patternless behaviors can no longer be viewed as an amusing but unscientific, anecdotal pastime. For it is utterly necessary to the development of psychology as a science.

Too often we have thoughtlessly pursued research by traditional means on the assumption that we cannot develop good theories unless and until we find general patterns to which the theories can address themselves. But the central issue in psychology—indeed, in all science—is not whether we can generalize from one person or situation to another, or from a sample to a population. The central issue is whether we can generalize from our theories to specific instances. If we allow our research methods and statistics to bear the burden of generalization, we necessarily give them the status of psychological theory without demonstrating their validity.

It would be naive, however, to assume that our discipline will undergo any noticeable change in research habits (dare I say "patterns"?) by force of criticism or reason alone. To list what psychologists should not do, or to wave one's hands in the general direction of what should be done, is no way to start what is, in effect, a social movement. At very least, some concrete examples of alternative research practices suggested by the criticism or reason must be provided. In short, we who criticize tradition must sooner or later put up or shut up. This is no small challenge, but in the spirit of meeting it, Barbara Carroll and I should like to offer *Ordinal Pattern Analysis* in another chapter of this volume for your consideration.

ACKNOWLEDGMENTS

The ideas presented in this chapter were developed in conversation and correspondence with the late Ivan London, a brilliant and dedicated scholar, and extraordinary man. I should like to dedicate the chapter to his memory, and encourage its readers to attend to his works.

APPENDIXES

APPENDIX A

```
1  ' FEEDBACK SIMULATION BASED ON PAPER BY WILLIM POWERS (1978)
2  ' WRITTEN FOR RADIO SHACK COLOUR COMPUTER WITH ONE JOYSTICK
3  ' (EASILY MODIFIED FOR OTHER COMPUTERS)
4  ' "J" SHOWS THE POSITION OF THE JOYSTICK ON THE SCREEN
5  ' THE TRAJECTORY OF "S" ON THE SCREEN IS CONTROLLED BY JOYSTICK
6  ' THE TRAJECTORY OF "D" HAS NO RELATION TO THE JOYSTICK
7  ' LINE 70 ADDS SOUND TO THE "S"+JOYSTICK COMBINATION
8  '
10 CLS 'CLEAR SCREEN
20 FOR I=-179 TO 180 STEP 10
30 Y=JOYSTK(0)/2
40 X=SIN(I/57.3)*14+16.5
50 PRINTTAB(X);"D"
60 PRINTTAB(Y);"J"
70 SOUND(X+Y)*3+1
80 PRINTTAB((X+Y)/2);"S"
90 NEXT:GOTO20
```

APPENDIX B

```
1  'SIMULATION OF CHAOS FROM ARTICLE BY ROBERT MAY (1976)
2  'THE CONSTANT "A" SHOULD RANGE FROM 1.0 TO 4.0
3  'THE STARTING POINT OF VARIABLE "X" SHOULD RANGE FROM 0.0 TO 1.0
4  'LINE 60 ADDS A SOUND TO FOLLOW THE TRAJECTORY OF "#"
5  'THE GENERATING EQUATION APPEARS IN LINE 70
6  ' IT CAN BE CHANGED TO EQS. 2A AND 2B AS SHOWN IN TEXT
10 INPUT"CONSTANT (A)":A
20 INPUT"STARTING POINT (X)";X
30 CLS 'CLEAR SCREEN
40 FORI=1TO100
50 PRINT TAB(32*X);"#"
60 SOUND 254*X+1,1
70 X=A*X*(1-X)
80 NEXT:GOTO10
```

REFERENCES

Bakan, D. (1954). A generalization of Sidman's results on group and individual functions, and a criterion. *Psychological Bulletin, 51*, 63–64.

Bem, D., & Allen, A. (1974). On predicting some of the people some of the time: The search for cross-situational consistencies in behavior. *Psychological Review, 81*, 506–520.

Chaitin, G. (1974). Information-theoretic limitations of formal systems. *Journal of the Association of Computing Machinery, 21*, 403–424.

Dabrowski, K. (1964). *Positive disintegration.* Boston, MA: Little, Brown.

Dobzhansky, T., Ayala, F., Stebbins, G., & Valentine, J. (1977). *Evolution.* San Francisco: Freeman.

Einhorn, H., & Hogarth, R. (1981). Behavioral decision theory: Processes of judgment and choice. *Annual Review of Psychology, 32*, 53–88.

Feldman, S. (1966). *Cognitive consistency.* New York: Academic Press.

Fischhoff, B., Slovic, P., & Lichtenstein, S. (1980). Knowing what you want: Measuring labile values. In T. Wallston (Ed.), *Cognitive processes in choice and decision behavior* (pp. 117–141). Hillsdale, NJ: Erlbaum.

Gleick, J. (1984, June 10). Solving the mathematical riddle of chaos. *New York Times Magazine,* pp. 31–32ff.

Hogarth, R. (1982). On the surprise and delight of inconsistent responses. In R. Hogarth (Ed.), *New directions for methodology of social and behavioral science: Question framing and response consistency* (pp. 3–20). San Francisco: Jossey-Bass.

Holling, C. (1973). Resilience and stability of ecological systems. *Annual Review of Ecology and Systematics, 4*, 1–23.

Katz, F. (1974). Indeterminary in the structure of systems. *Behavioral Science, 19*, 394–403.

Kirk, R. (1968). *Experimental design: Procedures for the behavioral sciences.* Belmont, CA: Brooks/Cole.

Langmuir, I. (1943). Science, common sense and decency. *Science, 97*, 1–7.

Levins, R. (1966). The strategy of model building in population biology. *American Scientist, 54*, 421–431.

London, I. (1946). Some consequences for history and psychology of Langmuir's concept of convergence and divergence of phenomena. *Psychological Review, 53*, 170–188.

London, I. D. (1949). The developing person as a function of convergence and divergence. *Journal of Social Psychology, 29*, 167–187.

London, I. D. (1974). *The revenge of heaven:* A brief methodological account. *Psychological Reports, 34*, 1023–1030.

London, I. D. (1975). Interviewing in sinology: Observations on methods and fundamental concepts. *Psychological Reports, 36*, 683–691.

London, I. D. & London, M. (1959). A case study of the reliability of research on foreign peoples. *Psychological Reports, 5*, 36–69.

London, I., & Thorngate, W. (1981). Divergent amplification and social behavior: Some methodological considerations. *Psychological Reports, 48*, 203–228.

Luce, D. (1959). *Individual choice behavior.* New York: Wiley.

May, R. (1976). Simple mathematical models with very complicated dynamics. *Nature, 261*, 459–467.

Maynard-Smith, J. (1958). *The theory of evolution.* London: Penguin.

Mitroff, I., & Featheringham, T. (1974). On systematic problem solving and the error of the third kind. *Behavioral Science, 19*, 383–393.

Nagao, D., & Davis, J. (1980). Some implications of temporal drift in social parameters. *Journal of Experimental Social Psychology, 16*, 479–496.

Newell, A. (1973). You can't play 20 questions with nature and win. In W. Chase (Ed.), *Visual information processing* (pp. 283–308). New York: Academic Press.

Payne, J. (1976). Task complexity and contingent processing in decision making: An information search and protocol analysis. *Organizational Behavior and Human Performance, 16*, 366–387.

Pielou, E. (1969). *An introduction to mathematical ecology.* New York: Wiley.

Poole, R. (1977). Periodic, pseudoperiodic and chaotic population fluxuations. *Ecology, 58*, 210–213.

Powers, W. (1978). Quantitative analysis of purposive systems: Some spadework at the foundations of scientific psychology. *Psychological Review, 85*, 417–435.

Prigogine, I. (1980). *From being to becoming.* San Francisco: Freeman.

Robinson, W. (1950). Ecological correlations and the behavior of individuals. *American Sociological Review, 15*, 351–357.

Sidman, M. (1952). A note on functional relations obtained from group data. *Psychological Bulletin, 49*, 263–269.

Thorngate, W. (1975). Process invariance: Another red herring. *Personality and Social Psychology Bulletin, 1*, 185–188.

Thorngate, W. (1976). Possible limits on a science of social behavior. In J. Strickland, F. Abound, & K. Gergen (Eds.), *Social psychology in transition* (pp. 121–139). New York: Plenum Press.

Thorngate, W., & Maki, J. (1976). *Decision heuristics and the choice of political candidates* (Report 76-1). Edmonton: University of Alberta, Social Psychology Labs.

Tversky, A., & Kahneman, D. (1974). Judgment under uncertainty: Heuristics and biases. *Science, 185*, 1124–1131.

Wilson, K. (1973). Linear regression equations as behavior models. In J. Royce (Ed.), *Multivariate analysis and psychological theory* (pp. 45–73). New York: Academic Press.

PART II

Group versus Individual-Based Inference in Psychology: Logic and Practice

CHAPTER 4

Phenomena Lost

Issues in the Study of Development

ROBERT B. CAIRNS

In a prescient essay that appeared in the first *Handbook of Child Psychology*, Kurt Lewin (1931) outlined the need to study the dynamic relations between the individual and the environment in the concrete particularity of the total situation. Lewin challenged the then-dominant methodology of psychological inquiry on two counts; namely (a) its reliance on phenotypic aggregation for description and statistical analysis, and (b) its focus on psychological variables without explicit reference to the milieu of their operation. On the first matter, Lewin observed that sampling methods that attempt to identify the "average child" tend to obscure the dynamic processes that contribute to behavior of a specific child in a specific setting. For Lewin, clumping together children who were similar with respect to salient extrinsic or demographic characteristics—but different with respect to key psychological dynamics—was unlikely to yield generalizable psychological laws.

More than a half century has passed since Lewin's seminal contributions on method and analysis. Some of his ideas (on the need to establish ecological validity; on the embeddedness of social actions in the social context) have been incorporated into the mainstream of the field, and they are no longer identified as distinctively Lewinian. But one nuclear issue—namely, the need to understand the behavior of specific

ROBERT B. CAIRNS • Developmental Psychology Program, Department of Psychology, University of North Carolina at Chapel Hill, Chapel Hill, North Carolina 27514. This work was supported by a grant from the National Science Foundation (BNS 83-16921).

individuals in specific settings—remains to be resolved by developmental psychologists. At first blush, the problem of when or whether to pool subjects may seem to be an important though technical issue, one that should be decided by statistical and methodological journeymen. But choice of the unit of analysis—whether the person, the variable, or the group—is the cornerstone upon which all subsequent decisions of research strategy necessarily rest. An issue so central to the research enterprise is appropriately confronted by the investigators themselves.

In the present era, as in Lewin's time, the research models of general experimental psychology and psychometrics provide the criteria for the evaluation of developmental investigations. To illustrate the hegemony of statistical concerns, the most frequently cited article that has appeared in *Child Development* since 1973 has been one concerned with statistical methods (McCall & Applebaum, 1973). Likewise, statistical and research design issues play a large role in judgments of publishability for empirical reports. With rejection rates hovering between 80% to 90% in the primary developmental journals, there is scant margin for error or flexibility in standards of acceptance. To the extent that the quality of research design analysis has become equated with psychometric sophistication and complexity, increasingly higher levels of statistical expertise are required by the researcher and by members of the editorial boards who evaluate the work.

One upshot of this trend has been to move the field even further away from the Lewinian model than it had been in the 1930s. Articles published in developmental journals from 1981 to 1984, relative to those published in the same journals 50 years ago, include fewer summary descriptions of phenomena and more inferential statistics (Cairns & Valsiner, 1984). Consistent with this research trend, methodological strategies that do not conform to standard forms of statistical analysis appear with decreasing frequency.

The acceptance of the methods and procedures of general experimental psychology by developmentalists has deep historical roots. At the inauguration of the first department of psychology in an American university (at Johns Hopkins), child psychology was assigned to second-class scientific status—behind experimental psychology—precisely because of its methodological limitations (Hall, 1885). It was believed that the descriptive, historical, and correlational constraints of developmental and comparative research were inherently inferior to experimental methods. Ironically, this evaluation of developmental methods was offered by G. Stanley Hall, who is now honored as the father of American child psychology.

On the entry of the field into its second century, several writers have initiated a critical review of the methods of developmental research (e.g. Kuo, 1967; Magnusson, 1985; Nesselroade & Ford, in press; Wohlwill, 1973). In a cogent analysis of the issues, Magnusson (1985) has outlined a frame of reference for developmental research that has radical implications for how the scientific enterprise should proceed. In brief, Magnusson argues that the field should make a commitment to *understand* the processes of human behavior, and that technological goals (i.e., prediction and control) should be secondary. Moreover, the configuration of factors that operate in each individual life require that attention be given to the behavior of each person as an "integrated totality." The action patterns of every person represent a unique mosaic of biological-interactional-situational factors. To the extent that these configurations are distinctive and individual over time, the person's status on a single variable (or set of variables, independent of context) is a fallible guide for understanding. Hence studies of social development require person-oriented methods as opposed to variable-oriented research designs.

To adumbrate a primary conclusion of this chapter, major revisions are called for in the criteria that are employed to judge the adequacy of developmental research designs and analyses. Such revisions should begin with a recognition of the distinctive properties of developmental phenomena, consistent with the proposals outlined in Magnusson (1985), Wohlwill (1973), and Kuo (1967). Nondevelopmental concepts and the statistical procedures associated with them have been permitted to become dominant for the field, even though they may blur the diversity and individuality of behavioral development. The paradox of modern technology is that some of the procedures that have been introduced to refine data collection and analyses may actually retard psychological understanding. The retardation comes about partly because of the distance created between the concepts of investigators and the phenomena they wish to explain, and the elimination of developmental changes.

Following a discussion of some statistical and design issues in contemporary research, I will comment upon some procedural revisions that could help regain some of the phenomena lost.

PSYCHOLOGICAL COMPLEXITY, LEVELS OF ANALYSIS, AND STATISTICAL PARSIMONY

For any empirical investigation, the logic of the research task and the nature of the phenomenon should be the criteria against which the adequacy of a given design or statistical analysis is judged. Accordingly,

it seems reasonable to assume that because developmental phenomena
are multidetermined and complex, the procedures and statistical analyses employed to study them should be equally complex. Although this
assumption appears at first blush to be self-evident, it demands critical
scrutiny. Exactly the opposite conclusion may be reached if the principal
task for developmental research is to understand and clarify. Reliance
upon complexity in analysis to disentangle the network of multiple causes
and multiple outcomes may reflect shortcomings in other phases of the
research task. To clarify the issues, it is often the case that the simpler
the statistic the better. Parsimony in analysis may be permitted because
the major analytic solutions have already been reached in the conception
of the study, in the methods adopted, and in the forms of the data
available for analysis. When complexity of the phenomenon is permitted
to breed complexity in the analysis and interpretation, it often means
that there has been a failure in theoretical conceptions or a lack of creativity in research design.

LEVELS OF ANALYSIS

On the matter of the relation between subject sampling and psychological understanding, Alfred Binet (1903) wrote:

> American authors, who love to do things big, often publish experiments made
> on hundreds or even thousands of persons; they believe that the conclusive
> value of a study is proportional to the number of observations. That is only
> an illusion. (p. 299)

It should be noted that Binet did not have a bias against bigness
per se; nor did he oppose the use of parametric statistics when the
phenomena would be clarified by them. In his own research, Binet employed multiple levels of analysis, ranging from the intensive study of
one or two individuals over time in diverse settings, to the short-term
and experimental study of large groups of children of different ages. The
choice of research strategy was intimately related, in Binet's thinking and
work, to the nature of the question that he addressed. Although significant
advances have been made in research design and statistical analysis in
the 80 years since Binet wrote, his fundamental point on the need for
multiple levels of analysis in order to disentangle complex phenomena
should not be forgotten.

STATISTICAL CONSIDERATIONS: TWO LISREL EXAMPLES

For some issues, large numbers of subjects must be studied in conjunction with advanced statistical techniques in order to arrive at solu-

tions. By way of example, Wolfle (1985) addressed the question of whether the processes that led to success in advanced education were the same for different ethnic groups in the United States, namely, "Are black teenagers less influenced in their eventual educational attainment by social background factors than are white teenagers?" To address this problem, Wolfle used longitudinal data from the National Longitudinal Study of the High School Class of 1972, including some 7,258 persons ($N = 6,825$ white, and $N = 433$ black). The first wave of testing provided information on parental status, high school grades, high school curriculum, and scores on tests of scholastic ability. The seven-year follow-up survey indicated the level of education eventually reached by subjects. Wolfle (1985) tested specific hypotheses of equivalence and difference between the black and white samples by procedures available in the LISREL computer program (Jöreskog and Sörbom, 1985; see also Hertzog, 1985). The similarity of the structural equations derived for the black and white samples could be evaluated by comparisons of the covariance matrices computed for the two groups. The outcomes of these analyses led Wolfle (1985) to conclude that social background, high school performance, and ability factors had parallel effects for black and white samples in determining eventual levels of educational attainment.

The Wolfle (1985) analysis took advantage of a sophisticated procedure for the solution and comparison of structural linear equations. The enormous data set available required a strategy where the researcher could conceptualize both the complexity of the relations and the extent to which they were parallel in the two ethnic groups. Given the range of decisions about parameter values that must be made in the course of the solution, the LISREL analysis itself became an "experiment." The LISREL computer program also makes it possible for subsequent investigators to replicate the adequacy of the analysis, because the researcher included key sets of information in the report.* As it turns out, independent reanalyses show a reasonably close fit to the solutions originally reported (although there appear to be significant omissions in the published report, J. B. Carroll, personal communication, 29 April 1986).

But use of the LISREL statistical program does not guarantee replicable outcomes. On this score, a report that recently appeared in the *Journal of Experimental Psychology: General* illustrates the hazards of employing these statistical procedures when they are inappropriate for

---

* These are (a) the covariance matrices which contain the values that are analyzed and (b) the parameters that are constrained or permitted to vary in the solution. With this information at hand, any reader can then verify the computations and determine independently whether the investigator's solution is confirmed (and whether an alternative model may better fit the data).

the data or inappropriately executed. In a study of individual differences in short-term and long-term memory, two groups of subjects were given a variety of memory tasks and measures of physiology and attention were obtained (Geiselman, Woodward, & Beatty, 1982). On the basis of a LISREL analysis of the observed correlational matrices, the authors concluded that the two-factor memory model fitted the data better than a single-factor memory model. As in the preceding illustration, information was included in the report that permitted subsequent investigators to assess the accuracy of the solution. Unlike the preceding report, it seems impossible to duplicate certain LISREL outcomes reported by Geiselman *et al.* (1982), even when a "corrected" correlational matrix is employed. Hence one basic conclusion of the study—that the one-factor memory model fits the obtained data less well than the two-factor model—remains open to challenge.

The pitfalls in statistical application are not limited to the use of advanced algorithms. In the developmental literature, even familiar statistics, including correlations and chi-squares, have been routinely misused or misapplied. Furthermore, the failure to employ simple statistics when they are called for may handicap an investigator's understanding of the data.

The general lesson is that modern statistical procedures offer the developmental investigator useful tools for scientific analysis, but they must be interpreted in the context of the total design. Inferential statistics should be servants, not masters and controllers, in the problem-solving process. Difficulties arise when the mechanical requirements for parametric statistical analysis have been permitted to become the overriding concern in research design. Among other things, they tend to remove the investigator from a direct confrontation with the data. A primary casualty has been the low priority given to the intensive study of individuals or families in concrete settings, even when this strategy is appropriate. Either level of analysis—large scale aggregation or individual or family unit—may be productively employed, depending on the goals of the work.

Statistical Significance or Developmental Meaningfulness?

In the light of the above considerations, one cause of concern has been the direction in which the "Results" sections of research articles have evolved. It is not unusual to find published reports replete with information about inferential statistics and associated probability values, but with only modest descriptive information on mean levels of performance, or the range of individual scores. The amount of attention given to the description of the behavior of individuals, or to summaries of individual performance has been decreasing. Rather than having more complete and careful descriptions of individual phenomena in concrete settings, as Lewin (1931) proposed, we now have less. The scarcity of journal space, combined with the complexity of the information necessary for replication, allows less attention to be given to summary description.

One unhappy consequence of the omission of adequate descriptions of outcomes is that it has become increasingly difficult to determine where an effort to replicate has succeeded or failed. Elimination of the descriptive statistics, such as correlational or covariance matrices, makes it difficult to judge from research reports whether problems in replication lie in the data themselves or in the analyses conducted by the investigator. Because it has become convenient to regard inferential statistics and associated probability values as the primary research outcomes, the area has become increasingly dependent upon the meta-analysis of probabilities.

Reliance upon probability values as a judge of meaningfulness or importance has its own hazards. With enough subjects, even the most trivial findings are statistically significant at $p < .01$. Given the multidetermination of a given developmental outcome on the one hand, and the multiple outcomes produced by a given developmental event on the other, empirical support can be obtained for virtually any proposal on behavioral development that is not wholly inane. Developmental researchers are usually confronted with networks of relationships, not single antecedent–consequent linkages. This state of affairs has yielded a cornucopia of positive findings and interpretations. The abundance of "significant" antecedent–consequent linkages in contemporary developmental research has also had a negative side in that it has shifted the responsibility for understanding phenomena away from the data themselves. The findings have often become projective tests for the field, where interpretations are determined as much by presuppositions of coherence as by patterns in the empirical data.

An alternative would be to view the achievement of $p < .01$ as being the first step in empirical analysis, not the last. The research task would then become one of determining how the several factors are organized to produce the referent phenomena of development. This latter step would lead, in many instances, to the exploration of empirical levels of analysis that are person-oriented as opposed to variable-oriented (Magnusson, 1985). The problem becomes not merely to determine whether a given variable has an effect, but to determine how effects are integrated and given distinctive weight and meaning, depending on the properties of the environment and the developmental status of the individual.

It has been argued by Sameroff (1975) that the developmental model is inadequately served by an emphasis on main effects, and that the study of interaction effects is required. That proposal may be subsumed by the more radical proposition that certain developmental outcomes are sufficiently unique as to require analysis at the individual, configural level rather than at the sample, population level. Generalizations may be reached by an analysis of individuals *prior* to aggregating them into "samples" or "conditions," not after the aggregation has occurred (see Allport, 1937). That is, the unit of study becomes the individual, not the sample of individuals. General laws may then be identified by commonalities across persons or across relationships. Other processes, which permit the assumption of within-sample homogeneity and equivalence, may yield to standard parametric analyses.

PHENOMENA LOST

One enduring problem in behavioral study concerns the issue of how to translate emphemeral processes, actions and interactions, into quantifiable patterns without a loss of essential information. This matter is of special concern for developmental investigators, because the patterns–categories are themselves likely to undergo transformation as individuals mature. The challenge for the field has been to develop procedures, codes, and measurement scales that are both veridical and reliable. The hazard has been that the to-be-explained developmental phenomena may themselves become distorted, or lost, by the very operations designed to make them accessible to empirical analysis. Moreover, the loss may be not be wholly without function. On the whole, it is easier to conceptualize and analyze behaviors as static categories or structures rather than as dynamic and developing processes. The elimination of developmental change can be introduced at any stage of data transduction; namely, when the data are coded, when the outcomes are scaled, or when norms are established.

TRANSFORMATIONS AND DEVELOPMENTAL HOMOGENIZATION

It is a minor irony that many of the conceptual and statistical procedures introduced to study children of different ages have served to reduce or eliminate the impact of developmental change. For example, the statistic of choice for the study of continuity and change—the product–moment correlation—effectively eliminates real differences in performance associated with maturation. Similarly, arithmetic transformations that underlie the IQ ratio appear to have been introduced in order to get rid of age-related differences in cognitive functioning. Modern refinements of scaling have achieved the same outcome through standard scores, where same-age peers provide the reference group. These scaling techniques are not limited to the study of intelligence. They have become the strategies of choice for other domains of developmental assessment, including measures of peer acceptance, temperament, social interchange, and teacher ratings.

For students of development, the widespread acceptance of standard score transformations is curious. Despite obvious statistical convenience, this scaling convention eliminates mean differences and the possibility for tracking age-related changes in the absolute levels of performance. The problem extends beyond the formal transformation of raw data into Z scores. Measurement techniques may themselves be biased against the detection of age-related changes. Rating scales, for instance, presuppose that raters employ some reference group as the standard against which to compare the behavior of the child to be evaluated. Regardless of the identity of the rater—whether teacher, parent, peer, or the child himself or herself—the comparison group typically is comprised of the same-age, same-sex children with whom the subject is affiliated.

But the perceptual standardization introduced by raters is ordinarily just the first step. A second filter is introduced when the ratings are formally standardized by Z scores in order to remove contemporary frame-of-reference differences, including between-rater variations in mean levels of ratings, or variations in the range of ratings. So the information relevant to individuals undergoes two separate transformations: one introduced by the rater and another by the investigator. Although such double standardization may facilitate the identification of interindividual differences, it inevitably washes out absolute developmental changes.

In a more subtle fashion, developmental differences may be muted by the theoretical constructs themselves. In this regard, it is not unusual for longitudinal studies to employ a single construct to describe behaviors that differ markedly in quality and form, but that are presumed to share common functions at different age levels. These constructs include, for instance, *intelligence, aggression, attachment,* and *altruism.* Transformed scores and the nominal equivalence of disparate behaviors, when combined in a single study, promote the perception of behavioral identities and constancies, despite remarkable changes in referent behaviors. On this score, it has become conventional in the science to adopt a social consensus category in referring to the intelligence of an 18-month-old infant, an 8-year-old child, and an 80-year-old adult. Employment of the same term at different ages implies that the same unitary dimension is assessed at each stage. But it is also known that there exist marked differences in the factor structure of intelligence tests at these different ages, and there are large qualitative and quantitative differences in the performances of individuals.

The problems associated with the nominal equivalence of different behaviors by using a single, unchanging category are multiplied when social dispositions are considered, such as aggression, attachment, and altruism. These dispositions are necessarily interactional in performance and multidetermined in development. Although the use of a single non-developmental construct to classify disparate actions over time—because they appear to share a salient functional similarity—may aid in identifying continuities, there are also hazards. Such "monothetic generalization" may obscure real developmental changes in the form and function of behaviors at different age levels (Cairns, 1979; Jensen, 1967). The more general lesson is that statistical transformations are not merely issues of psychometry, and classification schemes are not simply matters of method. Such operations should be justified as much by the theoretical framework of the study as by analytic convenience.

CODING BEHAVIOR: VERIDICALITY OR RELIABILITY?

In the effort to achieve higher levels of ecological validity, developmental psychologists over the past 2 decades have explored the use of naturalistic categories that capture qualitative differences in performance. Simple quantitative measures (frequency, time) are preferred, but alternatives are now tolerated. Because of the greater risks of bias in qualitative classifications, considerable attention has been given to the formulation of codes that yield high levels of agreement among independent observers. The goal is to reduce errors of subjectivity in the primary data records. As it turns out, the selection of the behavioral classifications that are most reliable reinforces biases of selective attention and social consensus. Codes that are easiest to employ with little training are typically those that are most familiar to the observers and about which observers share a prior consensus.

Consider, for example, a behavioral coding scheme where the observer is required to make a simple judgment on whether a child's action is one of compliance or noncompliance, or whether a mother's action is one of discipline, aggression, or neglect. The judgments may appear simple—in that the judge-observer is required only to make dichotomous evaluations—but they are hardly simple in terms of the complexity of the evaluation process required of observers. If the classifications are made in real time, then on-the-spot judgments are required by observers with respect to several issues, including the motivation of the 2 members of the relationship, their intentions in acting and reacting, the developmental status of each person, their relationship to each other, their cognitive attributions, and the objective functions of the action. The entire process must be duplicated every 5 or 6 seconds, without the benefit of information about what occurs later in the sequence.

Despite these restrictions, or perhaps because of them, it is not unusual for observers of complex categories of social behavior, such as compliance or coercion, to achieve high levels of agreement. This agreement seems to have been won because the unique capabilities of the human observer are combined in a single operation as recorder, transducer, and interpreter. Human beings as computers are highly practiced in such integration and in arriving at everyday judgments consistent with social consensus. Most social judgments also presuppose that observers automatically take into account "obvious" factors, such as the child's age, sex, and context, in arriving at interpretations on the meaning of a given behavior sequence. Accordingly, reliable observations may reflect constraints on the judgment process as much as identities in the behaviors observed. The operations of measurement thus may work against the discovery of relationships that transcend the social filtering process. In this fashion, new information, or information discordant with consensus social categories and judgments, becomes lost at the first stage of recording or becomes relegated to measurement error.

How might the investigator transcend the constraints of common sense and move beyond social consensus in behavioral categorization? As Yarrow and Waxler (1979) observe, there is no substitute for an intimate working knowledge of the phnenomena to be explained prior to the formulation of the coding system. In the initial stages of behavioral analysis, the researcher should be free to employ imaginative and flexible procedures in order to arrive at a veridical recording system.

One solution would be to recognize that there is a continuous tug-of-war between reliability and veridicality in the development of codes and categories. The classification system that is most reliable is not necessarily the most veridical, and vice-versa. Rather than giving the first priority to agreement and social consensus, a higher priority might be assigned to the task of capturing the actual events, the mutual meanings, and the functions of behaviors. Another, related, solution would be to disentangle explicitly the several steps of the coding process, and to separate the recording phase from the judgment, transduction, and classification phases, and the latter three from each other. That is, the burden for classifying the intentionality and the meaning of acts need not be assigned to observers whose attention is limited to 5-second intervals; the primary observer's job may be structured to ensure that the information is recorded accurately. The task of classifying and interpreting can be reserved for secondary or tertiary steps, which are themselves public and open to verification. For most research applications, the reliability of the codes could logically become the last concern of the investigator rather than the first.

THE DISAPPEARING INVESTIGATOR

To the extent that keen observation is a source for fundamental advances in understanding, it may be misguided to assign data collection exclusively to the least informed or most naive members of a research team. Exactly the opposite research strategy should be followed, if history is a reliable guide to the achievement of progress. The essential contributions of Binet, Freud, Piaget, J. M. Baldwin, Kuo, Schneirla, and Gesell may be traced, in large measure, to instances where the investigators themselves participated in the data collection.

But in the evaluation of contemporary research, the intimate involvement of the investigator in the collection of data is usually viewed with question, or suspicion. Presumably the closer investigators are to the data to be interpreted, the more likely it is that their subjective biases will influence the outcome of the work. So assistants who are employed at the first stage of data collection should be blind to the purposes of the study and the meaning of information that is being recorded. And the professional message to students and colleagues is that intellectual responsibility and seniority is tantamount to removal from the tedium of data collection.

One difficulty with this modern definition of research roles is that persons whose responsibility it is to understand the subtle organization of behavior patterns have become insulated from direct contact with them. The distance inevitably dulls the researcher's perceptions of the critical elements of the phenomena to be explained.

Can principal investigators be introduced at the first level of observation without compromising the scientific integrity and objectivity of the study? Doubtless they can, if reasonable safeguards are adopted. For instance, objectivity in measurement can be preserved at the first stage of investigation by blinding the primary investigator with respect to subjects' identification. If this control is not feasible, as in many cases it is not, then the problem of bias can be solved by additional precautions. These may include, for example, the simultaneous involvement of an independent researcher who is unaware of the identity of the subject or his or her status on critical research variables. Or the entire experiment (or selected samples) may be recorded or videotaped for subsequent independent analysis. The precaution adopted must of course be appropriate for the research question.

A CONCLUDING COMMENT

It is difficult for many of us to imagine what a scientific approach to child development would be like in the absence of the framework of inferential statistics. That is sufficient cause to be wary. Statistical analyses and the probability values associated with them are appropriately viewed as a means to the end, not as goals in themselves. The aim is to determine whether a research outcome is a chance occurrence or an objective, stable phenomenon. Replicability can also be determined by precisely duplicating the research, or by varying nonessential parameters in subsequent work. Indeed, replication has long been the standard for evaluating studies in physiological and comparative psychology, providing a powerful model for developmental researchers. Whenever the methods of a discipline threaten to distort the phenomena they purport to clarify—because of the requirement to create an "average" view of the phenomenon by aggregating across subjects—alternative techniques should be explored and convergences determined. In the enterprise of science, no single research method or analytic tool should be permitted to become sacrosanct or an end in itself.

There is an intimate relationship between the subject matter of an area and the research methods that are appropriate. On this basis, the methods of developmental study present special problems precisely because they are addressed to dynamic and changing phenomena. The most important stuff of development has multiple antecedents and it generates multiple consequences. Moreover, true novelties and new patterns of behavioral organization arise in the course of ontogeny. These distinctive features of development would seem to require a fresh view of the basic methods of the science.

But enduring changes within the area require a consensus among knowledgable investigators with regard to the ground rules by which one judges the adequacy of research strategies. Moreover, it is not sufficient for investigators in their role as researchers to become aware of a need for greater flexibility in the evaluation of methodology and design. Editorial boards and research review committees must share a belief in the distinctiveness of development phenomena in order for new methods to evolve.

ACKNOWLEDGMENTS

I am pleased to acknowledge my indebtedness to David Magnusson and Jaan Valsiner, and to thank them for the pleasant hours that have been spent in the discussion of these issues. John B. Carroll has been generous with his time and wisdom, and I thank him for providing me with instruction on LISREL and other matters. In addition, I greatly benefited from a series of conferences on research methodology in developmental psychology sponsored by the Foundation on Child Development. O. G. Brim, Jr. and H. Sigal were instrumental in establishing these meetings, and M. Hetherington kindly invited me to participate. My colleagues in those conferences were M. Hoffman, J. Kagan, E. Maccoby, R. Parke, G. Patterson, M. Rutter, and M. Radke-Yarrow. B. D. Cairns has played a continuing role in the theoretical work discussed here.

REFERENCES

Allport, G. W. (1937). *Personality: A psychological interpretation.* New York: Holt, Rinehart & Winston.
Binet, A. (1903). *L'etude experimentale de l'intelligence.* Paris: Scheicher.
Cairns, R. B. (1979). *Social development: The origins and plasticity of social interchanges.* San Francisco: W. H. Freeman.
Cairns, R. B., & Valsiner, J. (1984). Child psychology. *Annual Review of Psychology, 35,* 553–577.
Geiselman, R. E., Woodward, J. A., & Beatty, J. (1982). Individual differences in verbal memory performance: A test of alternative information-processing models. *Journal of Experimental Psychology: General, 111,* 109–134.
Hall, G. S. (1885). The new psychology. *Andover Review, 3,* 120–135, 239–248.
Hertzog, C. (1985). Applications of confirmatory factor analysis to the study of intelligence. In D. K. Detterman (Ed.), *Current topics in human intelligence: Vol. 1. Research methodology* (pp. 59–96). Norwood, NJ: Ablex.
Jensen, D. D. (1967). Polythetic operationism and the phylogeny of learning. In W. C. Corning and S. C. Ratner (Eds.), *Chemistry of learning: Invertebrate research* (pp. 308–325). New York: Plenum Press.
Jöreskog, K. G., & Sörbom, D. (1985). *LISREL VI: Analysis of linear structural relationships by the method of maximum likelihood. User's guide.* Mooresville, IN: Scientific Software.
Kuo, Z. Y. (1967). *The dynamics of behavior development: An epigenetic view.* New York: Random House.
Lewin, K. (1931). Environmental forces in child behavior and development. In C. Murchison (Ed.), *A handbook of child psychology* (2nd ed., pp. 590–625). Worcester, MA: Clark University Press.

Magnusson, D. (1985). Implications of an interactional paradigm for research on human development. *International Journal of Behavioral Development, 8,* 115–137.

McCall, R. B., & Applebaum, M. I. (1973). Bias in analysis of repeated-measures designs: Some alternative approaches. *Child Development, 44,* 401–415.

Nesselroade, J. R., & Ford, D. H. (in press). Multivariate, replicated, single-subject designs for research on older adults: P-technique comes of age. *Research on Aging.*

Sameroff, A. J. (1975). Early influences on development: Fact or fancy? *Merrill-Palmer Quarterly, 21,* 267–294.

Wohlwill, J. F. (1073). *The study of behavioral development.* New York: Academic Press.

Wolfle, L. (1985). Post-secondary educational attainment among whites and blacks. *American Educational Research Journal, 22,* 501–525.

Yarrow, M. R., & Waxler, C. Z. (1979). Observing interaction: Confrontation with methodology. In R. B. Cairns (Ed.), *The analysis of social interactions: Methods, issues, and illustrations* (pp. 37–66). Hillsdale, NJ: Erlbaum.

CHAPTER 5

Between Groups and Individuals

Psychologists' and Laypersons' Interpretations of Correlational Findings

JAAN VALSINER

Correlation coefficients are social inventions that capture certain aspects of reality from a particular scientific perspective. Different facets of correlation coefficients are important for science. Usually the mathematical side of the bases for correlation coefficients have received the most attention. There is, however, another facet of correlations that is exceedingly important for science—the *interpretation* of correlational findings within conceptual spheres of one or another scientific discipline, and in the process of social communication of these disciplines with the lay public. Only one aspect of the interpretation of correlations—the issue of attribution of causality to different possible agents—has been given wider attention. However, even that attention has been more practical than theoretical. Numerous statistics "cookbooks" have tried to remind their users of the difficulties involved in making straightforward causal attributions on the basis of empirical correlational data. Users of these manuals may, but need not, accept such calls for caution.

Psychologists' (and laypersons') cognitive processes in interpreting correlations become explicit in the process of communication. Furthermore, certain social rules and regulations can force the person to interpret correlations in certain ways. For example, a psychologist may know

JAAN VALSINER • Development Psychology Program, Department of Psychology, University of North Carolina at Chapel Hill, Chapel Hill, North Carolina 27514.

very well that attribution of causality to either X, or Y, or Z, given a correlation r_{XY}, is a tricky issue, as his statistics teachers and handbooks have consistently pressed him to believe. However, he may, deep in his mind, believe that X causes Y in a particular content domain. Then, in one and the same research article, he may find a way to communicate both his loyalty to his statistics teachers and his personal beliefs concerning the direction of causality in the given case. While reading empirical literature, one can come across papers in which the results section echoes refusal to attribute causality, whereas in the discussion section the idea that X causes Y is put forward as a "possible speculation." Such inconsistencies within the same psychological text can alert us to the conceptual complexity of the scientific generalization process. The surplus meaning that interpreters of empirical data introduce into the process of making sense of the data keeps the process of scientific inquiry open for various new ideas that need not be in the data, but projected into them.

The analysis of interpretations given to correlational data by psychologists and laypersons is limited in this chapter to a special domain. This domain includes correlational data based on groups of subjects (samples from some population). From the perspective of the goals of the present volume, it is important to analyze the process of thinking that leads psychologists into the habit of making inferences from groups of subjects to individual subjects in a group. I will argue in this chapter that inference from interindividual (group) data to intraindividual psychological mechanisms is warranted only if groups of subjects are considered to be homogeneous classes. It is argued here that this is not the case in psychology. Theoretically, human beings (as well as all other organisms) are open systems that have developed a complex hierarchical organization, where individual uniqueness is widespread. Empirically, it is usual to observe high interindividual variability in psychological data of many kinds.

Transfer of interpretations of group-based correlational findings to understanding psychological functioning of individual subjects is further facilitated by the lack of distinction between singular and generic terms in language. Linguistic support for cognition makes it possible for laypersons and psychologists alike to carry over interpretations of correlational data from groups to individuals. Empirical investigations presented in this chapter reveal the presence of such carryover among subjects of various educational backgrounds—from first-year undergraduates to graduate students and instructors in psychology. Furthermore, fluctuation between populational and individual levels of discourse was also observed in published articles in psychology. Last but not least, similar fluctuation between levels of discourse was evident in Francis Galton's writing at the time he introduced his idea of correlation into the literature of the last century. The chapter also includes an effort to reconstruct the course of cognitive processes that leads to interpreting data obtained from groups of subjects in individual terms.

FLUCTUATION OF INTERPRETATIONS BETWEEN POPULATIONAL AND INDIVIDUAL LEVELS OF DISCOURSE

PUBLISHED SCIENTIFIC TEXTS: INTERPRETATIONS BY CHILD PSYCHOLOGISTS

In psychologists' discourse, the interpretation of particular findings is an inevitable and necessary part of social communication, both within the discipline and across its boundaries (o.g., to the lay public). Many examples from contemporary child psychology can be utilized to illustrate how psychologists, in their "natural habitat" (i.e., on the pages of a major professional journal), explicitly talk about group-based correlational data. Examples here are taken from child psychology. Table 1 includes examples from different articles, published in the journal *Child Development*.

As can be seen from Table 1, the authors of these selected interpretations shift constantly between populational (i.e., the level to which their original data belong) and individual levels of interpretation of the data. This shift between the levels can be traced by analyzing how the discourse is encoded, with the use of singular (e.g., "parent's "), intermediate (e.g., "parental"), and plural (e.g., "parents' ") verbal forms. The intermediate forms can be interpreted as pertaining to the individuals (e.g., "parental" = "parent's") and to populations (e.g., "parental" = "parents'") either simultaneously, or successively. These forms make it easy to shift between individual and populational levels of analysis, often without any conscious notice of it by the writer-researcher.

If we read the excerpts in Table 1 more carefully and literally, we can discover quite curious and paradoxical issues. For example, the last sentence of excerpt 2 in Table 1, although it is used to speculate about relationships of parents and adolescents, and is based on empirical data from an American sample of subjects (where, one can expect, strict monogamy within families is a social norm), reads as if it pertained to the cross-culturally rare case of polyandry! The sentence refers to "*the* mother," and "adolescents" who are said to be involved with "their fathers," which is assumed to irritate "*the* mother" because that involvement comes at the expense of time that these adolescents spend with *her*. Of course,

Table 1. *Illustrations of Interpretational Shifting between Populational and Individual Levels of Interpretation from Contemporary Research Articles in Child Psychology*[a]

Excerpt from the text	Analysis
1. "We also correlated total Maternal Interaction scores with 36 child and family variables.... Surprisingly, correlations significant at the .05 level were found ony for *mother's* educational aspirations for *her* child ... *father* presence vs. absence, and *father's* age. In the 1976 cohort, we correlated item and total Maternal Interaction scores with all first-grade measures pertaining to *the child.*" Source: *Child Development*, 1984, *55*, p. 242.	Most of the description, based on group (cohort) data, shifts immediately to the singular mode ("mother's," "her child," "father's," "the child"), from the initial basis of talking about variables (which, of course, were measured in the sample).
2. "Although the level of conflict with *parents* was not correlated with time spent with *peers* or time alone, *adolescents* who frequently argued with their *mothers* spent more time with their *fathers*....At least two hypotheses for this very intriguing finding could be tested. First, *the father* may serve as a kind of emotional reserve whom *adolescents* seek out for companionship and support mainly when their relationship with their *mother* is stressful....Another possibility is that frequent arguments with *the mother* are not the cause of increased *father* involvement but the result of it. *The mother* may be more easily irritated with *adolescents.* who are involved with *their fathers,* especially if that involvement comes at the expense of time with *her.*" Source: *Child Development*, 1982, *53*, p. 1,518.	The discussion starts at the populational level, as specified by plural forms ("parents," "peers," "mothers," "adolescents," "fathers") and when additional hypotheses are advanced the discourse shifts to the use of singular forms. It shifts back to a mixture of singular and plural forms in the discourse by the end of excerpt ("the mother" who is irritated by "adolescents" who spend time with "their fathers" at the expense of time with "her."
3a. "The factor *Parent's Responsiveness* to the *Infant's Needs* is predicted by the same behavior at 1 and 6 months	The intermingling of populational and individual levels of analysis starts with attaching singular-form labels to

(*continued*)

*Table 1. (**Continued**)*

Excerpt from the text	Analysis
and also by the greater *soothability* and *social responsiveness* of the *infant* at 3 and 6 months. That *the parent's responsiveness* to *the infant*, especially to crying, is associated with the emergence of social responsiveness *in the child* during the first year of life replicates the work of Ainsworth *et al.* (1978), linking *responsiveness* and the development of secure *attachments.* Clearly, *parental* responsiveness is also associated with *maternal* characteristics observable before birth. *The parent* who was later responsive to *the infant* had been assessed in a prenatal period as *someone who was* efficient yet flexible, who related well to *her* ·peers, who *was* not too anxious, and who gave little evidence of approaching others with a critical suspicious attitude." Source: *Child Development*, 1983, *54,* p. 206.	factors that were found empirically in the group data ("Parent's Responsiveness," "Infant's Needs"). The discourse then shifts to singular mode ("the" Infant, parent, child), and proceeds in an intermediate (neither clearly singular, nor plural) mode ("parent" and "parental," "maternal"), before shifting again to the singular mode ("the parent," "the infant," "someone who was," "her").
3b. "The *parent's* tendency to stimulate cognitive experience at 12 months is predicted by that *same parent behavior* and the *infant's alertness* and *social responsiveness* at 1 month. The emerging interactional character of *parent* stimulation and *infant* alertness is highlighted by these findings. *Parents* who stimulated their *infants* at 12 months also tended to encourage new experience in a context of positive feeling at 6 months. Finally, *the mother's prebirth Adaptation-Competence* becomes a significant predictor in the regression analyses performed both 1 and 3 months." Source: *Child Development*, 1983, *54,* p. 206.	The second excerpt from the same text starts at the individual level ("parent's," "that same parent behavior," "infant's"), gets then connected with intermediate (variable) reference ("parent stimulation," "infant alertness"), and then proceeds to the populational level as exemplified by the use of plural forms ("parents," "infants"). Finally, the discourse shifts back to singular form-variable blend ("the" mother's "prebirth Adaptation-Competence").

[a]All emphases (in italic) are added.

this very literal reading is employed here only as a heuristic device to underline the paradoxicalness of the phenomenon of shifting levels of conceptualization in the process of discourse.

The present analysis of psychologists' messages that involve interpretation of correlational data is based on the assumption that the use of language by the authors can tell us a story about what is going on in their minds. Immediate explanations for the phenomenon of mixing populational and individuals levels in discourse are obviously available. For example, we may want to attribute it to the sloppy writing habits of the particular authors. That explanation does not preclude its parallel in the cognitive sphere—sloppy thinking habits. My task in this chapter is not to evaluate individual persons' thinking or writing from a normative perspective, but to try to trace what cognitive conditions allow such texts to be generated. Inference errors or sloppy writing habits are real psychological phenomena that require explanation on their own. Furthermore, the number of individual persons who can be observed to show such looseness in their use of language is not relevant for the present discussion. Even only if one person were to describe correlational data alternating between populational and individual levels of discourse, that would be sufficient to verify that the phenomenon exists. Its existence— even if only in few cases—requires its analysis.

HISTORICAL BACKGROUND: GALTON'S INTERPRETATION

Correlation is a concept with a relatively short cultural history. It is a concept that has its roots in the pluralistic traditions of occidental philosophy over the past four centuries (Dixon & Nesselroade, 1983). However, its actual invention is intimately related to Francis Galton's theoretical and practical efforts to understand principles of heredity, as those are explicated in different aspects of offsprings' similarity to their parents.

Galton (1888, p. 135) provided an explicit definition of correlation in the following way:

> "Co-relation or correlation of structure" is a phrase much used in biology, and not least in that branch of it which refers to heredity, and the idea is even more frequently present than the phrase; but I am not aware of any previous attempts to define it clearly, to trace its mode of action in detail, or to show how to measure its degree.
>
> Two variable organs are said to be co-related when the variation of the one is accompanied on the average by more or less variation of the other, and in the same direction. Thus the length of the arm is said to be co-related with that of the leg, because a person with a long arm has usually a long leg, and conversely. If the co-relation be close, then a person with a very long arm

would *usually* have *a very long leg;* if it be moderately close, then the length of *his* leg would *usually* be *only long, not very long;* and if there were no correlation at all then the length of *his* leg would *on the average* be mediocre. (emphasis added)

This quote—the very first explicit description of correlation and its interpretation by Galton himself—illustrates the long history of mixing population inference with references to individual cases. Galton's explanation starts at the population level (variation of one organ size within a sample is accompanied, *on the average,* by more or less variation in the size of the other organ within that sample). In the next sentence, the shift occurs—Galton brings in the reference to "the arm" and "the leg" (the definite article here serves as a vehicle for shifting from variation within a sample, i.e., between *different* legs and arms), but relates that these two limbs by reference to "a person" who has "a long arm" (individual level of analysis), who "*usually*" would also have "a long leg." Here, the term *usually* pertains to the populational level of analysis (because the intraindividual use of the term in this context would result in the nonsensical idea that the same person's leg length varies depending on circumstances, so that sometimes the person has a short leg, sometimes a long one, the latter being more "usual" for him than the former). However, immediately after referring to the populational level of analysis, Galton shifts back to an intermediate mixing of levels ("a long leg" in the context of the particular individual person refers to the concrete leg, but in the populational context would refer to one specimen in the class of legs, which, relative to others, is described as "long"). Similar ambiguity in the level of description is evident in the following sentence. Again, reference is made to "a person with a very long arm" who "*usually*" has "a very long leg," and even—if correlation is "moderately close" then "*his*" leg length would "*usually*" (sic!) be "only long, not very long." Finally, we can again observe the connection between the individual level of analysis ("his leg") and the populational level ("on the average").

Galton seemed not at all worried about such instant and free shifts between the two levels of analysis. Neither has that problem occupied the minds of the majority of psychologists ever since, who have used correlational techniques in their empirical studies. Because similar verbal forms in English (e.g., "this leg" vs. "leg" in general) can be used to denote both particular and generic concepts, the problem of mixing interpretation levels could have easily gone unnoticed. It is an interesting historical question to try to reconstruct the reasons why Galton fluctuated so easily between the two levels. Perhaps an answer to that question can be found in the interplay of two separate lines of thinking that Galton followed in his work, and tried to integrate.

The first of these was Quetelet's emphasis on the average man (Quetelet, 1842), which was directly related to the belief in the law of error. It provided Galton with a dilemma. On the one hand, he believed in the generality of the normal distribution of phenomena. On the other hand, as far as human societies were concerned, Galton disliked the idea that the ultimate ideal for a society is its average person. In his system of eugenics, Galton suggested that social reforms should be introduced that would have prevented the British society from approaching (regressing to) the democratic ideal—the average man (Galton, 1904).

The second direction in Galton's thought was related to Darwin's evolutionary theory. Galton's thinking was directly influenced by Darwin's ideas, which set up the issues of variability within populations as theoretically important for understanding changes across generations in a species. Galton's own attempts to develop further Darwin's theory of pangenesis (Galton, 1876) provided the theoretical framework in which the invention of statistical methods of regression and correlation was a necessary next goal to attain. While working toward that goal, Galton introduced some new concepts that played a significant role in his thinking about co-relations and their measurement—the concepts of the mid-parent and nepotal center. These concepts may be considered to constitute the vehicle by which Galton overcame the difficulties of connecting the populational and individual levels of analysis.

Let us look at these auxiliary concepts more thoroughly. The concept of mid-parent occupied a central position in Galton's analysis of the heredity of stature. He defined mid-parent as "an ideal progenitor, whose stature is the average of that of the father on the one hand and of that of the mother on the other, after her stature had been transmuted into its male equivalent by the multiplication of the factor of 1.08" (Galton, 1888, p. 143). Defined in this way, the mid-parent is an average of the two concrete, individually specific, parents of the particular children. On the other hand, it is an average that is construed to be "ideal progenitor" of these children. In the concept of mid-parent, Galton brings the populational and individual levels of analysis as close together as possible. The mid-parent represents both the individual parents in their concreteness, and the minimal possible population of the parents of the given set of children.

The concept of nepotal center captures a similar fusion of the populational and individual aspects of analysis. According to Galton (1886, p. 50):

> Each family of nephews affords a series of statures that are distributed above and below the common mean of them. They are deviations from a central family value, or, as we may phrase it, from a nepotal centre, and it will be

found as we proceed . . . that these deviations are in conformity with the law
of error.

Nepotal center is thus another specific average—that for the family—
which can be treated both as a group (populational level of analysis) and
as a representation of concrete individuals (individual level of analysis).
The concepts of mid-parent and nepotal center may have helped Galton
to overlook differences between the two levels of analysis, because through
these concepts it becomes remarkably easy to fluctuate between thinking
and talking about concrete parents of particular children and parents of
many children in general. Likewise, the law of error can, through these
concepts, be made cognitively applicable to concrete families (e.g., stature
of members in one family can be expected to distribute around the
family's nepotal center), and the children of a concrete family can be
expected to tend to regress towards the average characteristics of that
family, and not of the population as a whole.

This short venture into Galton's thinking illustrates the ease with
which the populational and individual levels of analysis of psychological
and anthropological phenomena can become intermingled in scientists'
thinking.

ELICITED INTERPRETATIONS: SUBJECTS WITH PROFESSIONAL SOCIALIZATION

The examples of interpretation of a group-based correlational data
that were presented in Table 1 constituted a selected set of a class that
the present author encounted in his occasional reading of the particular
journal. If the phenomenon of individual-populational shift is a wider
one, it should be evident in the everyday discourse of psychologists.

Two small groups of subjects from two widely different universities—
one in the United States and the other in Sweden—served as informants
in a study of interpretations of correlational data. All subjects were psychology graduate students working towards completion of their Ph.D.
degrees, or psychology faculty. The subjects were given, in written form,
a description of a correlational finding:

> In a study of 256 mother–infant dyads (152 with male, 104 with female infants),
> correlational relationships between mothers' and children's behavior were
> sought. A Pearson product-moment correlation coefficient $r = +0.75$ (signif-
> icant at $p < 0.001$ level) was found between two variables: (a) the rate of moth-
> ers' utterances, while talking to their infants; (b) the rate of infants' vocali-
> zations.

The description of the correlational finding was purposefully constructed using plural forms (mother–infant dyads, mothers, children,

infants). In this respect, the story that specified the correlational data was set up to lead the subjects toward keeping their interpretations at the populational level, with the aim of revealing any shift between levels that would occur, despite that lead. The subjects were asked to provide their interpretations, in two versions:

> QUESTION 1: How would you interpret this finding, if you were to provide your interpretation of it to another professional (a psychologist)?

> QUESTION 2: How would you interpret this finding, while explaining it to a layperson (for example, to one of the mothers who participated in this study)?

The reason for asking subjects to provide two interpretations, rather than one, is grounded in the nature of human communication. When communication occurs, explicit messages that are sent by communicators and received by recipients are always embedded in the framework of some shared implicit knowledge. Psychologists, through the similarity in their professional socialization, can be expected to share some basic knowledge about the concept of correlation. They also share socialized knowledge about how to interpret correlations. This shared knowledge can be expected to be absent if a psychologist interacts with laypersons whose knowledge base is socialized differently. Smedslund (1963) has demonstrated that at least one professional group of adults, whose training obviously differs from that of psychologists—nurses—did not share psychologists' concept of correlation when confronted with inference tasks from 2×2 contingency tables. The second question asked in this study was aimed at creating an imaginary situation, where the subject (the encoder of the interpretative message) is unlikely to assume that the individual layperson has similar understanding of correlation as the psychologist does. Besides a concrete individual recipient of the message, particularly if she is defined in the example as a mother who participated in the study from which the data were obtained, will be perceived as interested in the concrete, applicable outcomes of the study. If that is the case, a subject was expected to be under some pressure by the task to make a shift from the populational to the individual level of explanation that would fit the hypothetical mother's perceived perspective towards the issues involved.

Twenty-two subjects (11 males, 11 females, age range 21–41 years) gave their interpretation of the correlational finding. Seven subjects were given the task in Sweden, in Swedish, and their answers were analyzed with the help of their translation back into English. Only 3 subjects used plural verbal form to explain the data in their answers to both questions. Others utilized different ways of writing about the data—some (4) used singular forms to answer both questions, others (4) switched from the plural in answering one question to the singular in the response to the other, or vice versa. Finally, five subjects showed a consistent shift between the plural and singular *within* the particular answers to both of the questions. Examples of answers that exemplify shifts between plural and singular forms in their responses are presented in Table 2.

As we found in the examples from published articles (Table 1), the singular-plural fluctuation is observed in the subjects' answers.

The findings from this study provided further evidence about the presence of populational-individual level shifts in professional discourse about group-based correlational data. The majority of the subjects were in the process of acquiring their professional socialization in different universities in two countries. However, the tendency to diverge from the populational level of interpretation of correlational relationships in favor of the individual level of explanation may be present more widely in a population than just in a particular professional sample. Smedslund (1963, p. 169) reports that some of the nurses in his sample expressed a *particularistic* concept of relationship. The particularistic understanding of correlation was revealed in nurses' thinking about individual cases and imputing a logical relationship (e.g., symptom A and symptom B are present in patient X—therefore, in case of X, a relationship between the symptoms exists). Smedslund's findings suggest that persons who have not learned the statistical concept of correlation can operate on the basis of their own, commonsense, concepts of relationship that need not have a common ground with the quantified populational basis of the statistical concept.

Thinking about relationships in human cognition can take different forms, some of which are applicable at the level of a populational analysis of phenomena, others at an individual level. The statistical concept of relationship applies at the former level. In contrast, at the individual level of analysis, relationships can be conceptualized in structural-qualitative terms. Consider the example of a relationship between two parts of an individual organism—for example, between the head and the body. The adequacy of a statement—"the head is related to the body"—can be verified in qualitative-structural terms, by examining an individual case, and finding that the head is related to (connected with) the body with the help of a structure that we tend to call neck. This slightly absurd example illustrates the adequacy of the nurses' concept of relationship in Smedslund's (1963) study. A statistical approach to conceptualizing the relationship of head and body can proceed in two directions, both of which give the investigator information about the group (sample from a population) of cases studied. For example, an investigator may examine a sample of ($N = 1000$) living human beings, counting the frequency of

Table 2. Examples of Psychology Graduate Students' and Faculty's Interpretations of Correlational Data

Subject	Question 1: Message to another psychologist	Question 2: Message to a layperson
12	A positive relationship was found between rate of mothers' utterances to *their* infants and the rate of infants'vocalizations. (plural)	The rate of mothers' speaking to *their* babies making sounds are positively related . . . (plural) . . . as *a mother's* talking increases, *the baby's* vocalizations increase . . . (singular). This does not show that *the mother's* amount of talking causes how much *the baby* vocalizes or that *the baby's* vocalizations causes how much *the mother* talks. (singular)
18	There is a high correlation between the rate of mothers' utterances to *their* infants and the rate of infants' vocalization. (plural) . . . We cannot say whether mothers' speech elicits vocalizing from the *children* or vice versa, or whether perhaps both mothers' and infants' rate of vocalization are correlated through some third variable. (plural)	The study found a straightforward relationship between the rate of mothers' speech to *their* infants' "speech"—their infants' vocalizations. (plural) . . . We don't have enough information to say what causes the relation. It may be that when mothers talk alot, *their* infants learn to talk alot, or perhaps a very vocal *infant* inspires *the mother* to respond with lots of speech; or maybe talkative mothers tend to have talkative babies because of inheritance. (plural–singular–plural shift)
19	There is a strong correlation (+ 0.75) between rate of mothers' utterances and the rate of *their* infants vocalization. (plural)	*You* can get *your baby* to talk fast by talking fast to *him*. (singular)
20	There is a high positive correlation between mothers utterances and vocalizations by *the child*. The more *the mother* vocalizes the more *the child* does. (singular)	The more frequently a mother speaks to *her child*, the more *the child* tries to talk back, and, conversely the less a mother talks to *her baby*, the less *the child* will attempt to talk, and *the infant* will be a quieter baby. (plural–singular shift)

how often heads of individuals are linked to their bodies. Or—in the second direction—the relationship can be calculated for the sample through examining correlations of specially created measures of the head and body (e.g., their lengths). On both these two occasions, the resulting correlation coefficient describes some relation within the *sample* studied. It may be applicable to individual cases in the population only under rare circumstances.

There exists, however, some temptation for investigators to use populational-level data to make inferences to theoretical issues that pertain to individuals. In psychology, the prevailing dominance of statistical methodology seems to have fed into that temptation. A reverse inferential temptation exists in anthropology, where individual informants' accounts often are interpreted as full descriptions of their culture.

Sometimes, laypersons may be more sensitive to the issue of difference between populational and individual levels of analysis than statistically socialized psychologists can. For example, the controversy in cognitive psychology about the fact that some subjects regularly refuse to take populational base rates of phenomena into account when solving problems that concentrate on some aspects of an individual phenomenon illustrates the laypersons' awarenesss of the issue (see Cohen, 1981; Kahneman & Tversky, 1972; Levi, 1983). Refusal to base one's decisions about an individual case on populational base rates may be a cognitive adaptation to the nature of the problem-solving task—we may understand that some problems are formulated in terms of populations (groups, samples), and others pertain to individuals as self-organized systems. Consequently, we select the perspective with which we approach the particular problem.

In order to look further into the possibility that human cognitive processes provide their user—an individual person—with the certain margin of freedom of interpretation of a particular thinking task, we will have to turn to the issue of how subjects who lack the psychologists' socialized understanding of correlations put forward interpretations to explain given empirical correlational data.

LAYPERSONS' RANKING OF INTERPRETATIONAL OPTIONS

The term *layperson* in the present context is synonymous with an American undergraduate student. Two samples of students participated in this part of the study. Sample 1 included 95 subjects who participated in the study for credit in their introductory psychology course. This sample included 61 females and 34 males, all within the age range of 18

to 22 years (average age for the group was 19.1 years), and within the first
3 years of their college education: 63 were freshman (first-year students),
21 sophomores (second year in college), and 11 juniors (third year in
college). Sample 2 included 35 subjects from a population of undergraduate students who were slightly more advanced in their studies (no
freshmen were in that sample, 10 were sophomores, 16, juniors, and 9
seniors, that is, fourth-year students). Sample 2 consisted of students
who had elected to participate in an introductory course on child development, and who had taken introductory psychology courses earlier
in their studies. The sample included 30 females and 5 males, all in the
age range of 19 to 23 years (average age was 19.9 years).

In group testing conditions, subjects were given a questionnaire that
asked them to reveal their knowledge (or lack of it) about correlation
coefficients, and presented them with the task of interpretation of correlational data. Included in the questionnaire were the questions: "What
is a correlation coefficient?" "What is the difference between positive and
negative coefficients of correlation?" "Where did you learn about correlation coefficients?" The subjects were also asked to list all mathematics
and statistics classes they had taken.

The main purpose of the study was to present the subjects with a
task of interpretation of correlational data. The following description of
the data was given to subjects in both samples:

> In a study of mother–child dyadic interaction, a correlational relationship
> between mothers' and children's behaviors was revealed. A Pearson product-
> moment correlation coefficient, $r = +0.75$ ($p < 0.01$) was found between the
> following variables: (a) the rate of mothers' speech, directed to their infants
> and (b) the rate of infants' vocalization. This empirical result can be written
> about in different ways. Below you will find 4 different sentences that char-
> acterize the correlational finding. Your task is to rank-order these sentences,
> from the one that seems to fit the extended description of the results of the
> study best, to the one that fits that description least well.

The 4 options to be rank-ordered were the following:

- Mothers whose rate of speech, directed towards infants, was higher,
  tended to have infants who vocalized more often.
- An infant who vocalized more often, tended to have a mother whose
  rate of speech, directed towards the infant, was higher.
- Infants who vocalized more often tended to have mothers whose
  rate of speech, directed towards infants, was higher.
- A mother whose rate of speech, directed towards her infant, was
  higher, tended to have an infant who vocalized more often.

After the subjects had finished their ranking, they had to answer the
openended question: "Explain the rationale of your ranking."

As expected, only the minority of subjects in Sample 1 demonstrated an adequate understanding of correlation coefficients (13 out of 95). In Sample 2, the proportion of subjects with adequate understanding of correlation was higher—21 out of 35. Adequacy of understanding of the correlation concept was evaluated from answers to the questions about the correlation coefficient, and about the difference between positive and negative correlation coefficients. Only if a subject was aware of a correlation as being a relationship between two variables, and described (in writing or with the help of a graph) the difference between positive and negative correlation coefficients, was he or she considered to have adequate understanding of the concept.

Results on how the two samples of subjects ranked the interpretation options are presented in Table 3.

First, the term *preference pattern* used in Table 3 needs clarification. The analysis of data in Table 3 took place separately for the two contrasts embedded in the ranking task. The two contrasts were separated by the implicit causality encoded into the ranked options. This was accomplished through including statements where mother(s) were mentioned to have infant(s) [Contrast 1—the case of implicit direction of causality from mother(s) to infant(s)], and reversed versions where infant(s) were

Table 3. Frequencies of Occurrence of Different Preference Patterns in Laypersons' Interpretations of the Correlational Finding

A. Sample 1 (N = 91)

		Contrast 2 [Infant(s) → Mother(s)]	
		Plural (Is → Ms)	Singular (I → M)
Contrast 1	Plural (Ms → Is)	33 (36.3%)	20 (22.0%)
	Singular (M → I)	10 (11.0%)	28 (30.7%)

B. Sample 2 (N = 34)

		Contrast 2 [Infant(s) → Mother(s)]	
		Plural (Is → Ms)	Singular (I → M)
Contrast 1	Plural (Ms → Is)	20 (58.9%)	2 (5.9%)
	Singular (M → I)	1 (2.9%)	11 (32.3%)

mentioned to have mother(s) [Contrast 2—with implicit direction of causality from infant(s) to mother(s)]. *Within* each of the two contrasts, an individual subject's assigned ranks were compared separately. For example, if a subject ranked the plural form in Contrast 1 (mothers—infants) higher than the singular form (mother–infant), then the subject was classified as preferring the populational level of discourse for Contrast 1. The actual ranks given were not important in recognition of that preference. For example, different subjects that gave ranks 1 and 2, 1 and 3, 1 and 4, 2 and 3, 2 and 4, and 3 and 4 to the plural and singular options (respectively) of Contrast 1 were all considered to prefer the populational level of analysis to the individual level, and classified as such. A similar procedure was performed with individual subjects' rankings of Contrast 2 plural and singular forms. The data in Table 3 are given in the form of bivariate classification of subjects, based on the two contrasts.

Let us consider some theoretically expected distributions for data in Table 3 that would serve as the background for examining the actual data. If interpretation at the population level of the correlational finding of a relationship between speaking and vocalizing by mothers and infants were the rule, 100% of the cases should occur in the plural–plural cells in the Table 3 (A and B). It should not be forgotten that the description of the correlational finding that was given to the subjects was purposefully worded in plural forms, with an expectation that if subjects were tempted to switch the interpretation to the individual level, the plural-form leading information would eliminate the switching of levels. Likewise, if the rule for interpretations dictated strictly the individual level, all the responses in Table 3 should have been in the singular–singular cells of both samples.

As is evident from Table 3, different individual subjects had different ways of interpreting the data—some of them (36.3% in Sample 1 and 58.9% in Sample 2) did indeed keep their interpretations at the populational level, but others (30.7% in Sample 1 and 32.3% in Sample 2) demonstrated preference for interpretations at the individual level in the case of both contrasts. Even more interesting are the cases—especially numerous in Sample 1—where the preferences for levels of interpretation varied from one contrast to the other. Thus, 22% of the subjects in Sample 1 demonstrated preference for the mixture of levels: mothers . . . have infants . . . and an infant . . . has a mother. Likewise, 11% of subjects in Sample 1 demonstrated preference for a pattern that included a reverse switching of levels: a mother . . . has an infant . . . and infants . . . have mothers. The latter interaction of implicit causal direction and level of interpretation is in need of further study. For the purposes of the present study, it was found that in total 72.7% of the subjects in Sample 1, and 41.1% of subjects in Sample 2, did *not* show consistent preference for the population level of interpretation, despite the lead towards the use of plural verbal forms for interpretation that was embedded in the description of the data.

A closer look into the thinking processes of individual subjects within the samples can be obtained from an analysis of the explanations of the particular subject's rationale for one or another ranking pattern.

The task of introspecting on one's own rationale while ranking the interpretation options was not an easy one for subjects in Sample 1 The majority of the subjects left the question unanswered, some subjects provided repetitions of the task description, others confessed that they had guessed. Some subjects referred to their intuitive understanding of the interpretations as sentences in the language (e.g., "made more sense," "were more clearly written than the others," "was more easily understood"). Selected samples of individual subjects' ranking and rationale descriptions are presented in Table 4.

Examples in Table 4 were selected by two criteria: (a) the particular subject demonstrated inadequate understanding of correlation; (b) the answer to the rationale question had to explicate his or her reasoning process in a more extended form than simple references to acceptability of sentences by language norms, guessing, or reference to "more sense." Cases that satisfied these criteria are presented in Table 4. These examples of laypersons' introspection efforts reveal different ways in which subjects attempted to explain the issue. Some of the subjects (Nos. 37 and 64) gave reasons why the interpretation that was preferred had to remain at the population level, whereas others switched directly over to the individual level in explaining their rationale. The transition to the individual level in explaining a subject's ranking preferences was interdependent with attempts to construct some causal explanations to the correlation data, that would fit with the given subject's understanding of the issue of how a mother's relationship with her infant is organized, and functions (e.g., Subjects No. 15, 23, 74, and 79). The interdependence of efforts at causal explanation and levels of interpretation revealed some interesting cases of inconsistencies between a subject's ranking and explanation. For example, Subject No. 20 ranked the plural 'mothers-infants' higher than the singular 'mother-infant,' and then proceeded to attribute causality in the explanation: "*the* mother would cause the behavior rather than *the* infant." Similar shifts from preference for populational interpretation in the ranking and individual-level explanation is evident in Subjects No. 23, 74, and 85 in Table 4.

In summary, empirical evidence about interpretation of group-based correlations by psychologists, psychology graduate students, and undergraduate students as laypersons illustrates that some individuals would

Table 4. *Examples of Explanations for Particular Patterns of Ranking Interpretation Options in Task 1 by Individual Subjects without Adequate Knowledge about Correlation (From Sample 1)*

Subject number	Actual ranking	Explanation
10	1. Mother–infant 2. Mothers–infants 3. Infant–mother 4. Infants–mothers	"1 - sounded plausible and had 'her child' in it. 2 - didn't seem like her child. 3 - made the child seem responsible, not the mother. 4 - incomprehensible."
15	1. Infants–mothers 2. Infant–mother 3. Mother–infant 4. Mothers–infants	"If an infant is talked to and not shunned, most of the time it will talk more. It will not be as timid."
20	1. Mothers–infants 2. Mother–infant 3. Infant–mother 4. Infants–mothers	"It makes more sense that the mother would cause the behavior rather than the infant, but either way they say about the same thing."
23	1. Mothers–infants 2. Mother–infant 3. Infants–mothers 4. Infant–mother	"The infants need to have examples set first so they can model them."
37	1. Mothers–infants 2. Mother–infant 3. Infant–mother 4. Infants–mothers	"Plural mothers tend to speak to larger population and it is clearer to put the cause then the effect, rather than the effect and then cause."
64	1. Mothers–infants 2. Infants–mothers 3. Mother–infant 4. Infant–mother	"Plural is more general—not with every infant/mother does this happen."
74	1. Mothers–infants 2. Infant–mother 3. Mother–infant 4. Infants–mothers	"A mother dominates her child—the mother effects the child; the child does effect the mother."
79	1. Infant–mother 2. Infants–mothers 3. Mother–infant 4. Mothers–infants	"It seemed logical that an infant would learn to speak faster if it was talked to."
85	1. Mothers–infants 2. Mother–infant 3. Infants–mothers 4. Infant–mother	" + .75 refers to if a mom's speech, as increased, the infants vocalization increased."

shift their interpretation of correlational data from the populational level of analysis (where the data belong) to the individual level and vice versa. The particular number of individuals in the subset of people referred to here as "some individuals" may vary from one sample to another, and can easily depend on the particular education they have had and on the everyday thinking necessary to regulate their activities (professional or otherwise). The cognitive conditions under which this shift from the populational to individual level of interpretations can (but need not) take place is the next issue that needs to be covered in this chapter.

THE COGNITIVE BASIS OF THINKING ABOUT INDIVIDUAL SUBJECTS

THINKING OF CORRELATION: SYNCHRONIC AND TEMPORAL CONCEPTUALIZATIONS

A change in interpretation of correlational findings from populational individual levels requires a shift of perspective from which the relationship is thought about. At the populational level, a sample-based correlation coefficient can be conceptualized in a *synchronic* way. Individual subjects in the sample serve as elements of the whole sample. Even if these individual subjects are tested at different times, the data that are entered for calculation of correlational relationships are treated as if these were collected at the same time, that is, synchronically. The relationship of variables that are correlated is in this case a populational finding—it characterizes the *sample* of subjects on the basis of which it was established. If proper statistical safeguards are used, the correlational relationship may be generalized to some population at large. In this case, the generalized relationship of the correlated variables is a characteristic of a synchronic state of the population.

Generalization of a group-based empirical correlational relationships to abstract synchronic relationships of variables in a population is the most appropriate route for thinking about such correlational data. However, as the data in this chapter reveal, many persons who interpret group-based correlations habitually make inferences from the sample to individual subjects. In their thinking, they often translate the correlation from a synchronic to a *temporal* relationship. The process of interpretation may start from a synchronic conceptualization (e.g., "mothers who talk more to their infants have infants who vocalize more"), and subsequently proceed to think of the relationship as if it applied to ongoing (temporal) relationship within an individual case. In the terms of the example, the statement about mothers and infants is rephrased in the singular, which implies the assumption that repeated occurences of the variables ("a mother's talking" and "her infant's vocalizing") within the individual case over time are related to the same degree as the sample-based correlation indicates. In other words, a relationship of variables in a synchronically analyzed sample of subjects is interpreted in terms of the repeated co-occurrence of variables over time within an individual case. If a correlation coefficient $r = +.75$ is found in a sample of 256 mother–infant dyads, then the interpretation of that correlational relationship can move into considering it equivalent to a relationship, based on repeated measurement of the varibles 256 times in the case of an individual mother–infant dyad.

Furthermore, the translation of populational-synchronic relationships into individual-temporal ones can go even a step further. A temporal interpretation of a relationship in a case may be interpreted as if it illustrated an experimental case, where variables are thought of as manipulable. A high correlation coefficient between talking and vocalizing could be interpreted, for example, by the statement, "if *the* mother speaks *faster* to *her* infant, then *the* infant will vocalize more often." In this interpretation, experimental manipulation of a variable (modification of her rate of speech, by the mother) is implied. It is also assumed that once that experimental manipulation occurs, the correlation coefficient allows us to predict what would happen with *the* infant's vocalization rate. The former is conceptualized as an independent variable, the latter as a dependent variable, in this cognitive translation of observational data into an experimental model, which is assumed to be applicable to individual cases.

Somehow, the professional socialization of psychologists has bypassed this issue of cognitive transformation of empirical relationships from populational to individual levels of interpretation, and within the latter, from observational to experimental ways of thinking about them. The problem of such transformations seems to have been paid attention to mostly as a practical obstacle in empirical research, rather than as a theoretical issue. An excerpt from the introductory part of the chapter on correlation from a widely used textbook of psychological statistics (McNemar, 1969, p. 122) illustrates that well:

> One of the chief tasks of a science is the analysis of the interrelationships of the variables with which it deals. In the physical sciences, and frequently in the biological sciences, the interrelations can be determined by noting *how much of a change* in one variable is associated with a *change in another*. The physicist studying the relationship between temperature and pressure exerted by a gas *can vary* the former *at will* so as to determine the pressure at different temperatures. In the social sciences, and sometimes in the biological sciences,

the variables studied are apt to be characteristics of individuals (plant or animal), thus to study relationships *the experimenter is compelled to make measurements of several individuals.* (emphasis added)

This quote is noteworthy because it attempts to establish interpretational equivalence between correlational studies of groups of subjects, and experimental studies where direct effects of investigator's varying of some variable ("at will") can be related to changes in another. Many variables in social sciences cannot be varied by a scientist simply because they are unalterable characteristics of individuals; therefore investigators study the interindividual variability of these characteristics within a sample from a population, with the ultimate aim of arriving at interpetations that would tell us something about the phenomena, *as if* these could be controlled by experimental manipulations ("the experimenter *is compelled* to make measurements of several individuals"). In other words, it is implicitly assumed that, theoretically, we can treat interindividual variation of some characteristic in a sample as if it represented the (experimentally unattainable) variation within individuals in the sample, under different stimulus conditions (values of the independent variable). For example, it may be implicitly assumed that a *difference* in some quantifiable variables (e.g., leg length and arm length) *between* specimens (e.g., John, who has long legs and arms, and Jim who has short ones) in a sample is an adequate model for the impossible situation where (by an experimental miracle!) Jim's arms are made to grow to be similar in length to those of John's, and then his legs would follow suit to reach John's leg length. This assumption is similar in character to the attribution of a probability of occurrence of some outcome, calculated on the basis of a sample from a population, to each and every individual specimen in the population. Gordon W. Allport's criticism of that application (cf. chapter by Franck, this volume) is directed exactly at that issue—if, in a sample of delinquents, 80% of those come from broken homes, it does not necessarily follow that *this* particular delinquent from a broken home has an 80% chance of becoming a recidivist.

Finally, thinking about correlational relationships can shift from a quantitative to a qualitative (structural) realm. If that happens, the original framework of interindividual quantitative dissimilarities that served as the basis of establishing the corelation of the variables is replaced by the notion of qualitative-structural interdependence of the phenomena in question. From that perspective, the nurses studied by Smedslund (1963) who diagnosed a relationship between symptoms A and B in a single case of a patient, if A and B co-occurred, were operating within the qualitative-structural mindset that is applicable to individual cases without any necessity for using a population frame of reference. Or—if *my* body has developed cancer in one of its organs, it is quite irrelevant to console myself (and my relatives) with the populational information that that type of cancer is extremely rare in a population. The relationship between my body and the particular cancerous organ in my body is structural, and no quantitative measurement of correlational relationships between some parameters of bodies and the particular cancerous organs in a sample can clarify the structural relationship. Instead, it can lead to further confusion in the thinking of everybody concerned.

Let us return to the example that has been used all through this chapter. Interpretations of the correlation between talking and vocalizing in mother–infant dyads can take the form of, for example, the belief that talking to an infant is good for an infant's vocalizations. in this case, no quantitative aspects of the original data are retained, and the interpretation has moved into a qualitative-structural realm.

SEPARATION OF QUALITY AND QUANTITY

The application of quanitification in contemporary psychology has become a very widespread tradition, despite its obvious inadequacies in case of many psychological phenomena (Kvale, 1983). This tradition has introduced a paradoxical situation into psychology, where qualitative aspects of phenomena are translated into quantitative data, the latter are analyzed, and their interpretation often moves back to generalized statements about qualitative aspects of the phenomena. In other words, qualitative and quantitative aspects of phenomena are separated from each other at different stages of the research process. Very often, psychologists consider their subjects to be qualitatively similar in respect to a particular phenomenon, but quantitatively dissimilar. The latter assumption is relevant for many practical applications of statistics to the data, which have previously been constructed in some quantitative form through abstracting some aspects from the original phenomena. Calculation of Pearson's *r* is a good example. It can be calculated if quantitatively measured variability is present in the sample. The assumption of qualitative similarity of subjects is important at the more abstract level of analysis—it allows generalizations from some subjects to all subjects. If that assumption is rejected, then any generalization (from samples of subjects to other subjects, i.e., populations, or from one individual subject to some other person) would be entirely suspect. All this has a certain Orwellian character—all persons are equal (in quality), but some of them are more equal than others.

Separation of quantitative and qualitative facets of psychological phenomena from one another effectively limits the adequacy of psychological explanations of reality. This separation promotes a basically static world view—any changes observed in psychological phenomena are conceptualized as those of degree, and not of kind. This precludes thinking about development of qualitatively new psychological phenomena. In other words, if psychological phenomena develop into a new qualitative state, psychological research on these phenomena that is based on the assumption of qualitative sameness of the phenomena will be conceptually blind to the emergence of qualitative novelties. Given the assumption of unchanging quality, the only way that development can be conceptualized is embedded in the world of thinking about quantitative changes in some psychological variables. The latter are oftentimes artificial constructs that are made up in the process of research activities, which naturally leads to doubts about their adequacy to the real world.

HOMOGENEITY AND HETEROGENEITY OF CLASSES OF PHYSICAL AND NATURAL KINDS

Psychologists' continuing fascination with the scientific rigor of some of the so-called hard sciences has led them to mimic the thinking of classical physicists (cf. Brandt, 1973). If imitation of other scientists' thinking is part of one's professional role, it is remarkably easy to take over from classical physics some implicit assumptions, which may be perfectly adequate in physics, but not necessarily so in psychology. Perhaps the most central implicit assumption that often surfaces in psychology is that psychological phenomena are not qualitatively different from physical phenomena. That implicit assumption legitimizes psychologists' efforts to resemble classical physicists in the way of thinking and research efforts. Here, too, it is possible to trace the presence of the particular way of considering two large classes of phenomena—physical and psychological—similar in quality, and concentrate on the quantitative aspects of the latter in ways similar to the manner in which the former have been studied.

Incidentally, it was James Clerk Maxwell who, long ago, cautioned about the assumption of similarity of physical and biological phenomena. Maxwell (1875, p. 330) emphasized *qualitative* difference between classes of phenomena in physics and biology:

> We have been thus led by our study of visible things to a theory that they are made up of a finite number of parts or molecules, each of which has a definite mass, and possesses other properties. The molecules of the same substance are all exactly alike, but different from those of other substances. There is not a regular gradation in the mass of molecules from that of hydrogen ... to that

of bismuth; but they all fall into a limited number of classes of species, the individuals of each species being exactly similar to each other, and no intermediate links are found to connect one species with another by a uniform gradation.

We are here reminded of certain speculations concerning the relations between the species of living things. We find that in these also the individuals are naturally grouped into species, and that intermediate links between species are wanting. *But in each species variations occur, and there is a perpetual generation and destruction of the individuals of which the species consist.* Hence it is possible to frame a theory to account for the present state of things by means of generation, variation, and discriminative destruction.

In the case of molecules, however, each individual is permanent; there is no generation or destruction, and no variation, or rather no difference, between individuals of each species. (emphasis added)

Maxwell's statement about the difference of biological species and classes of molecules is interesting because it comes from the side of the hard sciences that psychologists have so often held as their scientific ideal. Maxwell's warning against considering physical and biological classes of phenomena similar in their qualitative nature was a suggestion that could have been at the foundation of psychology. Ironically, it was not. Psychology emerged on the basis of a selective set of ideas that were only partially borrowed from other sciences, and relied on cultural "folk models" of the given society.

The basic assumptions about the objects of a particular science are seldom taken from the scientists' immediate experience. Rather, these assumptions originate in the scientists' general world views, which result from their cultural rather than professional socialization. Assumptions about the nature of individual specimens in a class are fundamental points of departure for any empirical study. Classes of phenomena can be assumed to be either homogeneous or heterogeneous (or inhomogeneous—cf. Elsasser, 1966, 1970, 1981). A homogeneous class is one in which any two members of it are fully or nearly alike; in a fully homogeneous (also labeled "congruence class" by Elsasser, 1970, p. 138) class any two members are completely indistinguishable and substitutable for each other. In contrast, an inhomogeneous (heterogeneous) class is one in which a member of the class is not generally like another member (Elsasser, 1966, p. 32).

A decision made by a scientist to think of his phenomena as either a homogeneous or heterogeneous class is a step that determines what kind of empirical findings are possible for him to obtain in the research process. In some sciences, homogeneity of classes of phenomena is both feasible (from the standpoint of the nature of the phenomena) and theoretically necessary. For example, the mathematics of quantum mechanics cannot be formulated without the axiom that all electrons are rigorously alike and interchangeable (Elsasser, 1970, p. 139). Physicists in general tend to treat their phenomena, and with reasonable justification, as homogeneous classes. If the classes in practical research turn out to be not fully homogeneous (not congruence classes), but if it is still reasonable to conceptualize the classes as homogeneous theoretically, then a physicist can get rid of inhomogeneity be averaging out small irregularities in the observed phenomena. It is in these cases where the law of error has reasonable applicability—given that the scientist's conceptualization of a given class as homogeneous in principle fits the actual reality well. In the case of homogeneous classes, the average specimen of the given class can be considered to be an adequate representation of each and every specimen in the class. In the case of congruence (fully homogeneous) classes, every specimen of the class is a typical representative of the whole class.

Classes of biological and psychological phenomena, perhaps with the exception of those where elementary constituents of complex organisms are studied, include high variability. Traditionally, biologists and psychologists have proceeded to deal with that variability within classes in different ways. Biologists, who have followed the lead of evolutionary theory in explaining the functions and causes of variability at the level of species (i.e., populations consisting of different individuals), have tended toward the use of populational thinking in their discourse. However, even with the help of evolutionary traditions in biology, understanding of the differences between populations and individuals in scientific discourse has not been assimilated fully among biologists (Mayr, 1972) and continues to puzzle many of them (Ghiselin, 1974; Hull, 1976).

In psychology, in the overwhelming majority of cases researchers have proceeded to consider classes of their phenomena under study to be homogeneous in principle. The empirically evident inhomogeneities are "smoothed out" by statistical analyses of the data. Psychologists' publication practices in 1980s very often assign low relevance to reporting within-class (within-sample) variability in their data. The information on variability within samples is often not available in published articles, in the so-called prestigious and nonprestigious journals alike. Within a research domain, statistical methods have been developed that make possible the elimination of inhomogeneity from the sample of empirical articles on a given topic (see Rosenthal, 1984). Inhomogeneity in research findings is eliminated through quantitative estimation of the direction in which the majority in the data is pointing. In published articles, the typical responses of the *majority* of subjects are generalized to refer to *all* subjects. These methods of establishing majority trends are similar to the social ideals of democratic governance, or the "tyranny of the majority over minority" as de Tocqueville (1835/1956) candidly remarked more than a century ago. Between published papers, analysis of the majority of effects in quantitative accumulation of findings performs a similar role. Psychologists translate their phenomena, which both theoretically and empirically form heterogeneous classes, into homogeneous classes, and operate with the latter in their discourse. Such practice is based on the assumptions that phenomena in a class are qualitatively similar and only quantitatively variable. This practice fits well with our commonsense thinking about the world, but obscures some important issues in the nature of biological and psychological phenomena.

THINKING AND LANGUAGE: LINGUISTIC PREQUISITES FOR DATA INTERPRETATION

Language provides the vehicle for our communication in ways that enable the encoding of ideas into messages and the setting of constraints on our efforts of that encoding. In the present context, the ways in which the English language directs an investigator's interpretation of the data should be understood, so that the range of linguistic freedom of interpretation is clarified. In psycholinguistics, the issue of how linguistic organization of languages and cognitive practices of the language bearers are related has been extensively investigated. The important result in that field has been the understanding that there exists no one-to-one relationships between language and cognition (Bertalanffy, 1955; Tulviste, 1981). Instead, language use is tied to the content of people's actions, to their goals involved in communication about these actions, and to the world at large. The widely popular and oversimplified interpretation of the Sapir–Whorf hypothesis—that our language determines our thinking—should not be accepted at its face value. Whorf himself stated that

> the forms of a person's thoughts are controlled by inexorable laws of pattern of which he is unconscious. These patterns are the unperceived intricate systematizations of his own language—shown readily enough by a candid comparison and contrast with other languages. (Whorf, 1956, p. 252)

This statement can be interpreted differently, depending on the meaning that the term *controlled* is given. If controlled is held to mean "strictly determined," then thinking would be fully at the mercy of language rules, and lack any independence from them. On the other hand, if controlled is considered to mean "guided" or "directed," then the role of language rules in thinking is that of *canalizers*—circumstances that guide the thinking process in one or another direction, but do not determine the outcomes of the process in a strict manner. In this respect, a person is afforded a range of freedom of thought, constrained (at its outer boundaries) by the linguistic structure of the language used by the person. Bloom (1981, p. 20) has suggested that a language, which may allow a certain way of labeling a specific mode of experience, cannot determine whether its speakers will think about the experience using the label. It can only encourage the speakers to use the given label to denote the experience, but cannot guarantee that they will do so.

In the context of the present analysis of the interpretation of correlations, it is essential to analyze the ways in which the system of the English language affords intermittent use of populational and individual levels of reference in discourse. The lack of explicit linguistic distinction between reference to concrete, specific objects (e.g., "the child"—one, previously specified, concrete child, out of a class of children), and generic terms specifying the class of objects (e.g., "the child"—meaning the generic concept of child, true for all concrete children in a class) is suggested here as the basic condition that makes switching between individual and populational levels of discourse possible. Bloom (1981, pp. 34–35) has described this lack of distinction:

> If an English speaker speaks of the "the kangaroo" while standing next to a large marsupial or after just having discussed his friend's pet kangaroo, his use of "the" will be interpreted as entailing reference to the particular relevant kangaroo. But if the same speaker talks of "the kangaroo" in the absence of any actual kangaroo or previous mutual familarity with one, "the" will no longer be interpreted as entailing reference to any particular kangaroo, but will rather be interpreted as a signal of the generic kangaroo—as a signal to the listener to direct his attention to a theoretical entity extracted from the world of actual kangaroos.

In the empirical examples presented earlier in this chapter where switching between populational and individual levels of discourse was very often present, the change from one level to the other might therefore not indicate a change from talking about population to a particular individual, but to the generic individual. Because the difference between reference to particular individuals (the particular mother, the specific infant) and generic individuals (the mother, the infant) is linguistically unmarked, it is easy for a person to start from populational discourse, then switch to talking about the generic case (abstracting theoretical properties of the case from the group of specimens). From the generic concept, it is easy for the thinker (language user) to proceed to talking about *specific* cases—as the term that is used to characterize the generic case is not different in English from occasions when it is used to denote specific instances. This characteristic of the English language illustrates how an encoding of the world in terms of homogeneous classes (where the generic term for the class is the same as the concrete term used to denote its members) is facilitated by the language system. In languages other than English, this is not necessarily the case. For example, in the Chinese language such a lack of distinction in the form of particular (individual) instances in a class and the generic term usable for every specimen in the class, is not present (Bloom, 1981, p. 35–36). Therefore it can be complicated (although by all means possible) for a native speaker of Chinese to proceed from the individual level of discourse on to abstractions, because the similarity of the generic term to terms used to refer to individuals is not present to help him.

Another characteristic of the English language can facilitate the transfer of interpretation of group-based correlations to the intraindividual sphere of an individual (or generic) person. In English, plurality is applied to objects in two ways—as real or imaginary plurals (Whorf, 1956, p. 139). The real plurals refer to objectively perceptible aggregates of objects, for example, "10 children" can be objectively perceived as 10 in one group, present at the same time. On the other hand, we talk about objects that, in principle, cannot be perceived simultaneously in a group, for example, "10 days." In this case we operate with imaginary plurals—we can certainly wait for 10 consecutive 24-hour periods, counting them as they pass by, and at the end say that we have waited "10 days"—which is an imaginary, mentally constructed, group. Objects–events that can be counted over time give rise to imaginary plurals (e.g., "10 steps forward," "10 strokes on a bell"). The English language does not distinguish these two, qualitatively different, conditions of establishing plurality, and treats them as similar. The number 10 in either case ("10 children were playing in the park" vs. "Mary gave birth to 10 children") denotes plurality, independently of the conditions (simultaneous or successive) under which the particular case of plurality occurred. Given this characteristic of English, it is not surprising that psychologists often transpose interpretations of correlational data based on interindividual variation, to denote a supposedly intraindividual relationship between the variables studied. Or—in other terms—if a correlation between X and Y is found in a sample of subjects, a psychologist is assisted by the lack of distinction between real and imaginary plurals in the language to think of the relationship as if it would also be true in an individual, if values of X and Y could be varied experimentally in a single-case study.

These examples illustrate the relevance of the linguistic canalization of thinking. Its inevitable presence creates the cognitive context in which persons solve their problems. In the present content domain, native speakers of Indo-European languages may be aided in their interpretation of correlations by their language structure. Language aids cognition by scaffolding the thinking process. The process of making inferences from correlational data is dependent on language, but is not stricly determined by it.

THE PROCESS OF INTERFERENCE FROM GROUP-BASED CORRELATIONS: A COGNITIVE RECONSTRUCTION

Thus far, I have outlined different aspects of interpretation of group-based correlation coefficients. Now it is possible to synthesize these different aspects into a possible model of the thinking process that can be assumed to generate the empirical phenomena that were described in the first part of this chapter, and that can be observed to occur with remarkable persistence in psychologists' discourse. It is important to stress at this point that psychologists' shifts of interpretations between individual and populational levels of analysis are not erroneous. Instead, such shifts follow from the implicit general assumptions that guide commonsense thinking of both laypersons and psychologists.

The process of inference can be conceptualized to include the following steps:

1. *Axiomatic Basis.* It is *a priori* assumed that:
    1.1. In terms of quality of the phenomena studied, all individuals in a sample from a population form a homogeneous class.
    1.2. In terms of quantitative aspects of the phenomena, clearly observable interindividual variability exists within the sample.

    Comment: In Step 1, the stage is set for quantitative transformation of the phenomena into the data, and for further statistical analysis of the latter. Separation of the qualitative and quantitative aspects of the phenomena is accomplished at this step. This separation serves to reconcile the belief (supported by language structure and common sense) that members in a class are similar, with our experience and perception that they are different from one another.

2. *Operation of Measurement.* Relationships of constructed variables that represent the phenomena are measured on the basis of quantitative heterogeneity within the sample of subjects using correlational techniques.

3. *Extension of the Meaning of the Relationship Concept.* The meaning of the concept of relationship can become extended in the mind of the investigator, including in addition to the original meaning

of statistical correlational relationship also the qualitative-structural notion, and individual-experimental notion of the concept.

Comment: At Step 3, the distinction between quality and quantity that was the starting point (at Step 1) becomes blurred. This fusion of meanings makes it possible to attempt generalizations starting from the quantitative-correlational finding.

4. *Reduction of Heterogeneity to Homogeneity in Interpretation.* Because the phenomena studied were assumed to be homogeneous in quality, and because the concept of relationship has been extended in its meaning, particular interpretations that make intermittent use of populational and individual levels of analysis, or these even only at the individual level, can be generated.

Comment: At this step, interpretations move into the realm of quality, where all individuals are assumed to be similar to one another. This assumption is facilitated by the semantics of Indo-European languages, and by common sense. Given this, it is natural for an investigator to consider populational-level data as adequate grounds for making individual-level inferences.

5. *Reverse Transformation of Quality into Quantity.* (optional) After intermittent use of populational and individual levels of interpretation, an investigator may explain some interpretation through thinking about a quantitative relationship in an individual subject. For example, after interpreting a correlation coefficient between mothers' talking to infants and infants' vocalizing in a sample of mother–infant dyads, and after extending the relationship from its quantitative basis to a relationship in quality (where it is assumed to apply to every mother–infant dyad), then the investigator may explain that "if a mother begins to talk more to her infant, the infant will start to vocalize more."

In other words, interpretation of group-based corelational data is characterized by implicit shifting of thought from quantitative to qualitative aspects, and possibly followed by a reverse shift to the quantitative aspect in an individual. All this is possible because qualitative and quantitative aspects of the phenomena are axiomatically separated from each other in the thinking process. The qualitative aspects are used to establish the status of a homogeneous class of the sample (and population) in respect to the phenomena studied. Separately from that, the aspects of quantity are used to cope with the observable heterogeneity in the population in respect to these phenomena. The observable heterogeneity is used to arrive at a relationship between the phenomena, after which the heterogeneity becomes an obstacle for further generalization of the findings, and is eliminated from further thinking by switching to the qualitative level of discourse. After all, psychologists are primarily interested in generalizations (the ideal models of phenomena) that could be applicable to the individual person *in general*. Any information that tends to undermine the principal attainability of that goal through traditional means borrowed from classical physics is potentially dangerous for the integrity of the world view of psychologists, and can easily be disgarded. The ultimate example of the elimination of observable variability in psychology is the process of averaging—which serves as a means of preserving the belief among psychologists in the homogeneous-class status of their phenomena. Even in those (not too frequent) cases where psychologists bother to present data on observed variability within the samples they study (by including standard deviations, or coefficients of variance in their published papers) the information about variability is often not used as a finding on its own, but rather as some aspect of "error."

The case of corelational inference is more complex. Here the observed heterogeneity is not eliminated immediately, but only after it has been exploited to arrive at an empirical result that is then given meaning within an interpretation framework that assumes homogeneity of the class.

THE BASIS FOR EXTENDED INTERPRETATIONS: COMMON SENSE AND IMPLICIT CAUSALITY

All persons who attempt to interpret a correlation coefficient are well socialized in their culture, and share the commonsense explanatory repertoire used within the culture. The commonsense experience with psychological phenomena is primarily individual and only secondarily based on sampling of other people from a population. Therefore, it is not surprising that the rules of everyday, commonsense explanations of relationships between different psychological characteristics take the form of implications coded in singular forms and applicable to particular instances as everyday life experiences demand it. For example, a mother participating in a study of mother–infant dyadic interaction may be interested in the ways how her talking to her infant relates to the language development of her child. A psychologist talking to that mother shares the mother's commonsense knowledge that the object of interest is the particular child, and can modify his explanation of the correlational result according to the mother's expectations. Certain cultural beliefs about causality (e.g., the parent forms the child, not the child the parent) can also be shared by the psychologist and the layperson. These beliefs can constrain the person's freedom of construction of plausible commonsense explanations of the correlational data. Likewise, one of the explicit tasks that psychology students are expected to learn in the course of their professional schooling is the unacceptability of using causal explanations when interpreting correlational data. This is an example of a limit on explanation that is introduced simultaneously with the students' learning about correlations. However, many students remain faithful to the commonsense belief system while becoming schooled as psychologists—and in their explanations of correlational findings it is possible to observe blending of the professional knowledge (statements about impossibility of making causal inference) and their commonsense background (actual making of that inference, in an implicit form, anyway). If the system of commonsense prescribes thinking about an individual case, any pertinent information will make sense only if it is translated into terminology that is adequate for the individual case. For example, if a person of one sex wants to marry another person of the opposite sex, all the information used in making up his or her decision, even if it were to come from a countrywide sociological survey of marriage relationships in many couples, gets translated to a frame of reference that pertains to *that* particular would-be husband or wife. For certain other tasks, common sense may provide expectations that pertinent information be translated from individual cases to the level of populations. For example, if Martians existed and only some individual representatives of their population visited the planet Earth, then people on Earth would be expected to make generalizations about Martians as a whole population on the basis of a limited number of individual specimens. Again, if, in respect to the particular issue (variable) that is the target of generalization, the assumption that it forms a homogeneous class is adequate, such inference is warranted. If, however, it is not—then inference may lead to very curious and inadequate generalizations.

In science, there are many occasions where the inference is by necessity, or by social convention, made from individual specimens toward generalizations about the class. For example, our knowledge of protohominids is, by necessity, based on a limited number of skeleton specimens that archaeologists have been lucky to discover. Or, cultural anthropologists who live in a foreign culture establish working relationships with a limited number of representatives (informants) of the culture they want to learn about, and make inferences to the culture as a whole (populational level) on the basis of information obtained from these individuals.

To summarize, a person's interpretation of any object—be this a group-based correlation coefficient between certain variables or general ways of life in a far-off unknown society—makes use of the available information that is transformed into a state in which it could be used for the purposes of the particular task. The psychologist's task has often been to arrive at general knowledge about individuals. In line with that, correlational data based on interindividual variability of specimens in a class gets translated into knowledge that is believed to be applicable to individual cases. A similarly unwarranted extension of empirical results in the opposite direction—from individuals to populations—often takes place in cultural anthropology, where the task for the scientist is set up in that direction. Laypersons in a culture may use these differently oriented generalizations intermittently, depending on their particular goals— whether they want to marry a certain particular individual, or to create a commonsense picture of the customs and habits of a far-off country in general, once they have only been exposed to a single (or few) representative(s) from that country. Such task specificity, and the extension of the thinking process beyond the information given are certainly no news for psychologists of a cognitive persuasion (cf. Bruner, 1973). In fact, interpretational activity that goes beyond what is immediately available for the person, must by necessity, alter the framework from which the starting data are obtained in order to solve problems in domains where direct information cannot be obtained. The inclination of many (but not all) persons to interpret correlational data on mothers' and infants' vocalizations in dyadic interaction in terms of an individual relationship between a mother and her infant may perhaps tell us what people— psychologists and laypersons alike—try to conceptualize. Any empirical data serve only as auxiliary material in solving that cognitive problem, rather than as an absolute cornerstone of the thinking process.

CONCLUSION: PSYCHOLOGISTS' GOALS AND INFERENCE FROM GROUPS TO INDIVIDUALS

In this chapter, two lines of argument have been pursued in parallel. One of them has been an empirical demonstration that inference from group data to individual psychological mechanisms, although unwarranted, takes place quite often when correlational data are interpreted. It was also observed that cases of such inference are present among people with different socialization histories in psychology and statistical methodology—from college entrants to Ph.D.-level psychologists.

The second line of argument in this chapter has been theoretical. It was suggested that the inference transfer from the populational- to individual-level analysis of psychological phenomena need not be explained away by considering such instances errors in thinking. Instead, the shifting between populational and individual levels of explanation was seen to follow directly from the axiomatic assumption that all individuals are similar in their quality, and differ only in respect to the quantitative (and thus measurable) aspects of that quality, as it expresses itself in cases of concrete individuals. Or—more generally—human beings tend to operate at first in terms of *quantitatively heterogeneous* classes when a correlation coefficient is calculated, and then switch their thinking to the *qualitatively homogeneous* view of the given class of phenomena as they proceed with an interpretation of the correlation.

Such cognitive duality—fluctuation between axioms of heterogeneity and homogeneity of classes—is not surprising if we consider what goals psychologists (and laypersons, as well) have when they think and talk about psychological phenomena. From its very beginning, psychology as a discipline has been concerned with explaining its phenomena in ways that pertain to an abstract, typical, individual person. The interest in such abstract individual case has its counterpart in laypersons' cognitive efforts to make sense of people around them—often by making up indigenous classes of people that are construed as homogeneous (e.g., good vs. bad people, friends vs. enemies, etc.). When an individual categorizes other persons around him into such qualitatively homogeneous, commonsense classes, then one's thinking about everyday problems that come up in one's life is assisted by the knowledge of the existence of these sets of good and bad people, friends and enemies, in those cases in which the person has no need to go beyond the general classification and its implications. Thus, from the knowledge that "X is an enemy," the person can construct his view of X on the basis of features common to the whole class of enemies. The person then does not go to the trouble of finding out more about X, as he believes that the characteristics of the class enemy are immediately applicable to X, who is a member of that class. Needless to say, such a way of thinking saves the layperson some mental (and perhaps physical) energy, even as it can lead to huge distortions in his understanding of reality. Psychologists and laypersons seem to share this cognitive short cut when they have difficulties understanding and explaining some phenomenon; they resort to classifying it, to making up homogeneous classes.

However, there are limitations that constrain us from very extensive homogenization of classes. Extended experience with individual members of a class can lead us to observe heterogeneity within the class. In order to preserve our belief in the homogeneity of the class, and in the immutability of the static nature of our world, we can resort to the use of some aspects of statistical methodology in order to reinstate homogeneity of a class. The belief in the law of error that has been transferred to psychology from the Platonic-theological world view of 18th century philsophy, with the help of 19th century astronomers (cf. Gower, 1982; Sheynin, 1984), has provided a comfortable basis for psychologists' use of commonsense ways of thinking. It has preserved among psychologists a nostalgic longing for a static world of psychological laws that could be attained if only the psychologists would work diligently, amassing large amounts of data. That, eventually, will bring them close to the truth about psychological phenomena, in its assumed static and immutable existence, similar to that of rocks or planets.

Psychology, from the time of its separation from philosophy, has existed in a curious position between the ideals of hard science and the commonsense roots of the discipline. The relationship (in the qualitative-structural sense of the term!) between common sense and scientific methodology in psychology has been one of mutual support for each other. Commonsense questions in psychology are quite comfortably answered by the use of methodologies that have been borrowed from the science of 19th century physics, and the scientific status of the latter provides support for formulating new commonsense questions for psychology to answer. Undoubtedly there have been periodic changes in the fashions for one or another kind of questions, as well as in methods for answering them, that psychologists have followed during one decade or another (cf. Buss, 1978). However, these fashions have not moved psychology away from circular interdependence of the discipline and common sense in western industrialized cultures. For laypersons and psychologists alike, the mixing of populational and individual levels of interpretation of group-based correlational data is a cognitive necessity that is deeply engrained in the hidden assumptions that constitute common sense concerning the homogeneity of classes and the relationship between the quality and quantity of phenomena.

But why, and how, should our thinking proceed in other directions?—this can indeed be a legitimate question at this point. Is it not the case that common sense is a sufficiently rich and adequate knowledge base for the thinking of psychologists? Undoubtedly common sense is rich, and sufficient for our thinking about everyday problems in our lives. In addition to that, it is highly complex and controversial—it contains multiple models of thinking about psychological phenomena, and about causality related to those phenomena (Hargreaves, 1980). Besides, for an applied psychologist—such as the hypothetical one talking to a mother about the study on mother–infant dyadic speech—it may be simply irrelevant to be cautious about inferences from a sample and applied to an individual case. A particular inference may be deeply rooted in common sense (shared by both the psychologist and the subjects) as a true or plausible explanation. As such, it may easily become inferred from sample-based correlational study, if the data afford it. Furthermore, the particular idea may be adequate indeed—only independently of any efforts to prove it through a study of a sample. For example, why should one refuse to accept the commonsense idea that when a mother begins to talk more frequently to *her* infant, the latter has an increased exposure to the adult's speech, which in its turn may lead to more frequent vocalizing by the child, and be instrumental in the child's learning of language in the long run? This systemic explanation of the mother–infant dyadic organization may be adequate for some, many, or perhaps even all individual mother–infant dyads. However, a correlational study that is based on a sample of dyads can neither prove nor disprove that systemic model of the dyad. Instead, it is practically inconsequential—because inference from groups to individuals in the case of heterogeneous classes of phenomena is indeterminate. In contrast, an experimental study of a single mother–infant dyad, where the changes in infant vocalization rate in conjunction with variation in the mother's speech rate are investigated, can provide a direct empirical proof or disproof of the model—within the individual dyad and in the case of that particular dyad.

The assumptions of homogeneity versus heterogeneity of classes are cognitive tools that investigators use in their thinking. A psychologist—implicitly or explicitly—starts from certain assumptions about the phenomena. Once these assumptions are in place, all empirical research by that psychologist is affected by them. The acceptance of the assumption that psychological phenomena can be adequately treated to form homogeneous classes, even if only in respect of their qualitative aspects, leads directly to the dismissal of individual specimens of a class as such from extra consideration, and to satisfaction with group averages, specially selected prototypical specimens, or correlation coefficients that are assumed to characterize relationships between variables in general. Likewise, the acceptance of the assumption that psychological phenomena are open systems, capable of maintaining themselves in a steady state, or of undergoing growth—all thanks to their interdependence with their environment—leads the psychologist to the understanding of the fundamental relevance of the organization of individual specimens in a heterogeneous class.

Finally, how can one make warranted inferences from populations to individuals in the case of sample-based correlation coefficients? Again, if we assume that our phenomena form homogeneous classes, the traditional inference from groups to individuals is acceptable, *relative to the assumption.* However, if the assumption of heterogeneity of classes is accepted, then it is in principle impossible to infer anything about individuals in a sample, from an investigation of the sample. In that case, a sample of subjects can provide adequate information about the functioning of that sample as an individual organism of some kind. In that case, individual subjects in the sample are not treated as individual specimens of a class, but as parts of the phenomenon. For example, in discussions about species in evolutionary biology, it is feasible to think of species as individuals (Ghiselin, 1974), where particular organisms that belong to the same species are thought of as parts of the individual (= species), which are related with one another and result in the species' developmental history in the process of evolution. Here the particular specimens in a species are thought of not as members of a homogeneous class, but as parts of an individual organism. A study of relationships between these parts can thus lead us to unwarranted inferences about the species (population) *as* an individual. Returning to the concrete example used in this chapter—a study of a sample of mother–infant dyads—it can be argued that when the interpretation of the correlation finding is maintained at the populational level of explanation, it can help us to understand how that sample functions as an individual. Of course, a haphazardly assembled sample, as was given in the interpretation task, lacks the reality that would make inferences pertaining to such samples scientifically interesting. On the other hand, the purposeful selection of samples that have the reality of being individuals (e.g., *school class* as a collective individual) can warrant inference about that individual, based on correlational data obtained in a between-subjects analysis of relationships between variables. In this case, a group-based correlation coefficient would allow inference, based on the group of subjects (students in that individual school class), that would give us information about *that* class. Following this line of thinking, correlational techniques used to characterize relationships between variables within the same individual subject afford inference about that individual subject. The adequacy of inference from correlational data depends on how the investigator conceptualizes the system that the data are supposed to characterize. That concept of system is set up in the psychologist's thinking. Very often, a psychologist's thinking about that issue is loose, and the system is set up in a way where only unwarranted inference from the data is possible. The use of correlational techniques on these occasions cannot lead to new knowledge, and at its worst will give fortuitious support to commonsense explanatory myths, the scientific value of which is often negligible.

ACKNOWLEDGMENTS

The author is grateful to Peter A. Ornstein, Thomas Wallsten, and Robert B. Cairns for their reading and comments on the preliminary draft of this chapter, and useful suggestions for further elaboration of ideas discussed in it. The help of Tommy Gärling and Anita Svensson—Gärling in some aspects of data collection—is gratefully acknowledged.

REFERENCES

Bertalanffy, L. von. (1955). An essay on the relativity of categories. *Philosophy of Science, 22* (4), 243–263.

Bloom, A. (1981). *The linguistic shaping of thought.* Hillsdale, NJ: Erlbaum.

Brandt, L. W. (1973). The physics of the physicist and the physics of the psychologist. *International Journal of Psychology, 8(1),* 61–72.

Bruner, J. (1973). Beyond the information given. In J. Bruner, *Beyond the information given: Studies in the psychology of knowing* (pp. 218–238). New York: W. W. Norton.

Buss, A. R. (1978). The structure of psychological revolutions. *Journal of the History of the Behavioral Sciences, 14,* 57–64.

Cohen, L. J. (1981). Can human irrationality be experimentally demonstrated? *Behavioral and Brain Sciences, 4,* 317–331.

Dixon, R. A., & Nesselroade, J. R. (1983). Pluralism and correlation analysis in developmental psychology: Historical commonalties. In R. M. Lerner (Ed.), *Developmental psychology: Historical and philosophical perspectives* (pp. 113–145). Hillsdale, NJ: Erlbaum.

Elsasser, W. M. (1966). *Atom and organism: A new approach to theoretical biology.* Princeton, NJ: Princeton University Press.

Elsasser, W. M. (1970). The role of individuality in biological theory. In C. H. Waddington (Ed.), *Towards a theoretical biology* (Vol. 3, pp. 137–166). Chicago, IL: Aldine.

Elsasser, W. M. (1981). Principles of a new biological theory: A summary. *Journal of Theoretical Biology, 89,* 131–150.

Galton, F., (1876). A theory of heredity. *Journal of the Anthropological Institute, 5,* 329–348.

Galton, F. (1886). Family likeness in stature. *Proceedings of the Royal Society of London, 40,* 42–73.

Galton, F. (1888). Co-relations and their measurement, chiefly from anthropometric data. *Proceedings of the Royal Society of London, 45,* 135–145.

Galton, F. (1904). Eugenics: Its definition, scope and aims. *Nature, 70* (1804), 82.

Ghiselin, M. T. (1974). A radical solution to the species problem. *Systematic Zoology, 23,* 536–544.

Gower, B. (1982). Astronomy and probability: Forbes versus Michell on the distribution of the stars. *Annals of Science, 39,* 145–160.

Hargraeves, D. H. (1980). Common-sense models of action. In A. Chapman & D. M. Jones (Eds.), *Models of man* (pp. 215–225). Leicester: British Psychological Society.

Hull, D. L. (1976). Are species really individuals? *Systematic Zoology, 25,* 174–191.

Kahneman, D., & Tversky, A. (1972). Subjective probability: A judgement of representativeness. *Cognitive Psychology, 3,* 430–454.

Kvale, S. (1983). The quantification of knowledge in education: On resistance toward qualitative evaluation and research. In B. Bain (Ed.), *The sociogenesis of language and human conduct* (pp. 433–447). New York: Plenum Press.

Levi, I. (1983). Who commits the base rate fallacy? *Behavioral and Brain Sciences, 3*, 502–506.

Maxwell, J. C. (1875). *Theory of heat* (4th ed.). London: Longmans, Green, & Co.

Mayr, E. (1972). The nature of the Darwinian revolution. *Science, 176*, 981–989.

McNemar, Q. (1969). *Psychological statistics* (4th ed). New York: Wiley.

Quetelet, L. A. J. (1842). *A treatise on man and the development of his faculties.* Edinburgh: W. & R. Chambers.

Rosenthal, R. (1984). *Meta-analytic procedures for social research.* Beverly Hills, CA: Sage.

Sheynin, O. B. (1984). On the history of the statistical method in astronomy. *Archive for History of Exact Sciences, 29* (2), 151–199.

Smedslund, J. (1963). The concept of correlation in adults. *Scandinavian Journal of Psychology, 4*, 165–173.

Tocqueville, A. de. (1956). *Democracy in America.* New York: New American Library. (Original work published 1835)

Tulviste, P. (1981), Mõtlemise keelelisest ja tegevuslikust relatiivsusest (On the linguistic and activity-bound relativity of thinking). *Keel ja Kirjandus, 24*, 6 329–336 (in Estonian).

Whorf, B. L. (1956). *Language, thought, and reality.* Cambridge, MA: M.I.T. Press.

CHAPTER 6

The Individual Subject in Behavior Analysis Research

F. CHARLES MACE and THOMAS R. KRATOCHWILL

Behavior analysis is a small but growing subset of the diverse field of scientific psychology. Its roots are in the experimental analysis of infrahuman behavior conducted in controlled laboratory settings (Skinner, 1938). B. F. Skinner, the progenitor of behavior analysis, sought simply to apply the methods of the natural sciences (e.g., physics and biology) to the study of behavior in the hope of discovering lawful relations between environmental events and the behavior of organisms (Ferster & Skinner, 1957; Skinner, 1938, 1953). Principles of behavior were established not by comparing different groups under various experimental conditions, but rather through careful examination of the responses of individual animals during the presence and absence of specific independent variables (Sidman, 1960). In this chapter, we discuss the development of the various methodologies used in behavior analysis research. The role of the individual subject in achieving experimental validity is also examined along with implications for selecting single-case designs to meet various experimental objectives.

To understand the logic of behavior analysis research methods it is useful to apprehend the nature of the discipline. Several behaviorists have attempted to delineate the distinguishing characteristics of behavior analysis. Some have done so by differentiating between applied and basic

F. CHARLES MACE • School Psychology Program, School of Education, Lehigh University, Bethlehem, Pennsylvania 18015. THOMAS R. KRATOCHWILL • Department of Educational Psychology, University of Wisconsin, Madison, Wisconsin 53706.

research camps (Baer, 1981; Baer, Wolf, & Risley, 1968; Epling & Pierce, 1983; Michael, 1980; Pierce & Epling, 1980; Wolf, 1978). Others have discussed how the technical shift of the field threatens advances in the science of behavior (Dietz, 1982; Hayes, Rincover, & Solnick, 1980).

Baer *et al.* (1968) and Pennypacker (1981) have defined the field by analyzing the term *behavior analysis.* Baer *et al.* (1968) refer to *behavior* as a physical event produced by an organism that results in observable change in the environment. Thus, behavior is what individuals do and is detectable through observation. Behavior is further considered a product of the interaction of an organism's biological functions and environmental contingencies for responding (Pennypacker, 1981). *Analysis,* on the other hand, is the tool used to understand the processes responsible for a behavioral act. The interaction of an organism with its environment is often a complex process that the analytic endeavor illuminates by dissecting the behavior–environment interface into component parts or elements. The objective of the analysis of behavior is to describe the relationship between the observed response(s) and environmental variables that correlate with their occurrence (Pennypacker, 1981, p. 159). When specific responses are shown to covary with the introduction and withdrawal of an independent variable(s),* a functional relation is established. The analysis is not completed, however, until all the variability in a given behavior is accounted for by environmental factors, a goal seldom realized.

The subject matter of behavior analysis and the approach to understanding it naturally influenced the adoption and evaluation of the field's experimental methods. Some historical perspectives on the role measurement played in the process are presented next, followed by a discussion of the implications measurement has for experimentation and design selection.

APPROACHES TO MEASUREMENT

Measurement is the hallmark of any science. It is the activity that permits scientists to quantify their observations in a systematic manner and, in so doing, distinguish their endeavor from philosophy. Johnston

---

* The term *independent variable* is used in the broad sense to refer to a set of stimulus conditions that are manipulated by the experimenter either by varying its level or by introducing and withdrawing the treatment condition.

and Pennypacker (1980)* have traced the roots of scientific measurement. They describe the development of two distinct approaches to measurement that emerged from a common origin. Whereas most divisions of scientific psychology adopted methodologies born of vaganotic approaches to measurement, research in behavior analysis has its history in idemnotic measurement strategies. Differences between these two measurement approaches (elaborated in the following) are linked to differences between behavior analysis and most other areas of psychology, particularly with respect to the role of the individual subject in research.

Vaganotic Measurement

The emergence of calculus-based probability theory had a profound effect on measurement practices beginning in the early 19th century. Noting that measures of physical phenomena varied over time, mathematicians Legendre, Euler, and Gauss in 1806 used probability theory to formulate the least squares method of defining the "best" measure of a natural event given a set of variable data. Measures that deviated from the "best" estimate of the phenomenon were considered "measurement errors" that were shown to be normally distributed around the central tendency of the group of scores. It was a short leap to consider these measurement "errors" as deviations from the "true" value of the property being assessed. This idea was bolstered by the development of the law of large numbers that demonstrated that measurement error decreased as the number of measurement samples increased.

In the early 1800s, a Belgian statistician, Quetelet, applied these statistical procedures to the measurement of human characteristics (Quetelet, 1835). He found that measuring large numbers of individuals on a particular characteristic resulted in a normal distribution of scores that approximated those derived from probability theory used to estimate measurement error. Quetelet reasoned that the central tendency of a distribution of, for example, heights of adult males, would represent nature's ideal for that characteristic. Comparing an individual to the group average could then be interpreted as the extent to which that person deviated from nature's tendency. During the late 19th and early 20th centuries, Galton, Binet, and Cattell extended these techniques to develop tests of mental ability. Utilizing the law of normal error, Galton

* The following discussion of vaganotic and idemnotic measurement strategies is based largely on Johnston and Pennypacker's (1980) formulation of these concepts and their historical developments. Readers are referred to this volume for a complete presentation of these issues.

established a theoretical normal curve of mental ability divided into a 14-step interval scale, thus defining the construct of intelligence and creating a new scale to measure it (Galton, 1889). On the heels of Galton's work, Binet and Cattell devised and published standardized tests of mental ability. Items for these tests were selected because they produced variable performance in the population. Variability among persons taking these tests resulted in a normal curve of intelligence scores that paralleled the theoretical distributions used by Legendre, Quetelet, and Galton.

These developments set the stage for the propogation of a wide variety of psychological and educational tests. Underlying these tests of traits and abilities is the vaganotic approach to measurement. *Vaganotic* measurement involves the development of scales and units of measurement based on the variability of a set of observations among individuals. College entrance examinations are examples of vaganotic measurement. Test items from various academic subjects are pooled and administered to what becomes a norm group of college applicants. The mean and standard deviation of this group are determined and assigned arbitrary values on a contrived equal interval scale. For example, the means and standard deviation of the Scholastic Aptitude Test (SAT) are 500 and 100, respectively. These particular values are arbitrary and could have assumed many other values (e.g., T-scores: 50, 10; or z-scores: 0, 1).

There are three important consequences of vaganotic measurement (Johnston & Pennypacker, 1980). First, this strategy provides a mechanism for defining phenomena into existence. In the previous example, the construct "scholastic aptitude" was created and defined as the range of possible scores on the SAT. Yet "scholastic aptitude" does not exist in the physical world independent of a particular test of it. It is this freedom from physical referents that some believe has led to the proliferation of psychological and educational tests, many of which purport to measure the same construct. A second consequence is that scales and units of measurement are created when a phenomenon is defined by the variability in a set of observations among individuals. Using the SAT example, the measurement scale selected ranges from 200 to 800. Other tests of the same construct, however, employ different units and scales (e.g., American College Testing). Because one unit of "aptitude" on the SAT is not equal to a unit on the ACT, comparisons across measurement scales are not possible without reference to the mean and standard deviation of the norm group. Likewise, conversions from one scale to the other are not possible in the same way inches may be converted to centimeters. Third, a score obtained via vaganotic measurement derives its meaning from its relative position in the reference group. For example, a score of 600 on the SAT is interpreted according to the distribution of scores for a particular administration of the test. On one occasion an SAT of 600 may fall at the 85th percentile and on another administration be equivalent to the 75th percentile. In neither case however, may the score be interpreted as 600 units of the construct being measured (i.e., scholastic aptitude). This is in contrast to the natural sciences where measures correspond to and derive their meaning from the amount of the physical property being measured.

Adoption of vaganotic measurement strategies did much to define the subject matter of many fields of psychological study (Boring, 1950). As noted earlier, vaganotic measurement and scaling permitted the creation of numerous psychological constructs and instruments to measure them. Many of these constructs were personality, aptitude, and achievement variables that were, for the most part, considered relatively static characteristics or traits of the individual. The bulk of psychological inquiry focused on identifying the correlates of these constructs and independent variables that altered them.

IDEMNOTIC MEASUREMENT

Prior to the development and application of probability theory, scientific measurement consisted of counting and timing parameters of physical events. As measurement techniques developed, scales and units were established that were both standard and absolute in the sense that they were anchored to the amount of the physical property being measured. Johnston and Pennypacker (1980) have coined the term *idemnotic* to refer to measurement strategies that employ standard and absolute units that exist independent of variability in the phenomenon.

The development of the natural sciences has been linked closely to advances in idemnotic measurement. Early approaches to measurement established units and scales for determining quantities of mass, distance, and time. As sciences formed and evolved, the dimensions of mass, distance, and time were algebraically combined to yield novel units and scales for measuring newfound phenomena. To illustrate, in physics contemporary light wave theory is premised on basic concepts of light wave measurement. Light may be represented as a wave profile that is analogous to vibrating one end of a stretched rope. Waves traveling along the vibrated rope will have important characteristics that, when representing light waves, distinguish one type of light from another. These characteristics include (a) the displacement of the rope from its normal position (i.e., wave height, ξ), (b) the distance between successive wave crests (i.e., wave length, λ), and (c) the time between wave crests (i.e., γ). Two important measures used to classify light waves may be derived from these basic wave dimensions. The *temporal frequency* of a wave profile is the number of vibrations per unit of time, or $1/\gamma$. Similarly, the *spatial frequency* of a wave series is represented by $1/\lambda$, or the number of waves per unit of length. In this case, combining standard measures of distance and time resulted in a measurement scale used to classify known (e.g., visible light) and discovered forms of light (e.g., infrared light). Unlike vaganotic strategies, however, light wave measures correspond to physical characteristics of the subject matter and exist independently of variability in a set of wave measurements.

Idemnotic strategies were likewise dominant in the evolution of behavior analysis (Johnston & Pennypacker, 1980). Advances in the measurement of behavior made by Ebbinghaus, Thorndike, Watson, Pavlov, and Hull, among others, laid the groundwork for a contemporary science of behavior. Ebbinghaus (1885) was among the first to analyze experimentally variables that affect human learning and memory. A typical experimental procedure consisted of having subjects memorize lists of nonsense syllables. Memory was measured as the number of nonsense syllables repeated correctly and the number of trials required to learn a list. Using this procedure Ebbinghaus showed that memory was a function of specific parameters such as the length of the list and the time interval between learning and recall. Thorndike (1898) employed similar idemnotic strategies to measure the construct of intelligence in animals. His experiments involved presenting cats with problem-solving tasks, in the course of which conditions that affected the number of unsuccessful problem-solving attempts and solution time were studied. Watson (1919), the first avowed behaviorist, criticized his predecessors' reliance on unobservable constructs to account for behavior. Instead he favored the objective measurement of stimuli and responses, including study of the physiological functioning of an organism. He went so far as to conjecture that human thought processes could eventually be understood by measuring physiological correlates of thought, such as the occurrence, duration, and amplitude of tongue movements. Along similar lines, Pavlov (1927) suggested ways of measurement of reflex activity in his work on respondent conditioning. Devices were developed to collect animal saliva and divide it into drops of equal size. Finally, Hull proposed a series of theories to account for learning. Although Hull ascribed learned behavior to an array of intervening constructs (e.g., habit strength, drive, and incentive motivation), he employed standard and absolute units in the measurement of independent variables, such as the number of prior reinforcements and the weight of reward, along with dependent variables, including reaction latency and the number of nonreinforced responses to extinction (Hull, 1952).

The tradition of idemnotic measurement in early behavioral research apparently influenced Skinner's conceptualization of the subject matter of a science of behavior as well as the strategies he employed to measure it (Skinner, 1938, p. 58). For Skinner, and his descendants in behavior analysis, the subject matter was defined as the observable actions of a single organism over time. Because behavior shares "many of the characteristics of matter in motion, the same principles of measurement are applicable" (Johnston & Pennypacker, 1980, p. 73). That is, standard units and scales used in the physical sciences to measure mass, distance, and time were deemed appropriate for the study of behavior and were widely adopted by the field.

EXPERIMENTATION AND VALIDITY

Fundamental differences between vaganotic and idemnotic strategies have tended to promote different methods for achieving experimental validity. A major objective of experimental research is to determine whether the independent variable is causally or "functionally" related to the dependent variable (Skinner, 1953). Valid judgments regarding this relationship, however, are threatened by numerous factors extraneous to the variables under study that could account for the results of a given experiment and limit the generality of its findings. The degree to which these factors may be discounted represents the extent to which valid conclusion may be drawn. Yet because of the complexity of psychological research, complete control of all extraneous variables is unlikely (Cook & Campbell, 1970; Kiesler, 1981). The goal is to minimize potential validity threats and thereby strengthen the believability of the experimental effect.

The notion of experimental validity was elucidated by Campbell and Stanley (1963), later refined by Cook and Campbell (1979). In these works, two types of validity are considered central in evaluating experimental research.* Internally valid experiments are those that control factors that may suggest alternate explanations for the results of a study. Thus, the effect of the independent variable on the dependent is isolated and confounding factors are either eliminated or controlled. External validity, on the other hand, refers to the generality of the relationship across different subject or situational variables. We hope to show that vaganotic and

---

* A brief summary of internal and external validity is provided here. Readers are referred to Kazdin (1980) and Cook and Campbell (1979) for detailed treatments of experimental validity, including, in the latter reference, statistical conclusion validity and construct validity.

idemnotic approaches to measurement promote different strategies for establishing internal and external validity that have important implications for the role individual subjects play in experimental research.

INTERNAL VALIDITY

The preeminent question in all experimental research is, "Is the variability in the dependent measure a function of the variable manipulated during the experiment?" Without affirmation of the relationship between the independent and dependent variables, questions concerning the generality of the findings are moot (Campbell & Stanley, 1963; Cook & Campbell, 1979; Kerlinger, 1973). Yet the task of isolating this functional relation can be formidable due to the myriad of alternate explanations for the results that must be dispelled in order to draw valid conclusions. These factors are called threats to internal validity and have been catalogued in detail by Cook and Campbell (1979).

To illustrate the concept of internal validity, consider a behavior analyst who is investigating the effects of different schedules of reinforcement on a preschool child's rate of learning correct pronunciation of new words. The experimenter trains a single observer to record correct and incorrect pronunciations according to a set of rules. Data collected under three schedules of reinforcement reveal initial differences between the schedules that become less apparent over the course of the study. One possible threat to the internal validity of this study is instrumentation. That is, it is possible that changes in records of pronunciation may have been due to inconsistent measurement procedures during the experiment. Factors such as observer drift and bias (Kazdin, 1977) and changes in response definitions (Kent, O'Leary, Diament, & Dietz, 1974) may be responsible for the lack of schedule differences at the end of the study. The experimenter could have eliminated this threat by making frequent accuracy checks on the observer (Cone, 1981; Johnston & Pennypacker, 1980) and employing design elements that reduce the plausability of instrumentation threats (Barlow, Hays, & Nelson, 1984; Cook & Campbell, 1979; Kazdin, 1980, 1982). Which experimental designs are selected, however, may be influenced by the measurement strategies employed.

Vaganotic Approaches

Johnston and Pennypacker (1980) argued that the measurement practices adopted by a field of study have considerable influence on their experimental methods. As noted earlier, vaganotic strategies have spawned numerous measures of ability, personality, and traits (Nunnally, 1967, 1970, 1972) that are widely used as dependent measures in psychological research (Kerlinger, 1973). For several reasons, such tests are not always useful for repeated observations of subject performance. Rather, vaganotic measures are best suited for outcome assessment taken at limited points in time during the experiment (e.g., posttest only; pre- and posttest; pretest, posttest, and follow-up). From a conceptual standpoint, ongoing assessment is unnecessary because the subject matter purportedly being measured is a characteristic of the individual that is relatively stable across time and environmental conditions. Thus, in the vaganotic tradition, a "one-shot" approach to measurement is justified because, without exposure to a specific independent variable, differences across measurement occasions are considered a defect of the instrument rather than a reflection of variability in the phenomenon being measured (Nunnally, 1967). Psychometric and experimental considerations also militate against repeated measurement with vaganotic devices. Multiple exposures to a given instrument can result in testing effects that can be confused with those due to an independent variable (Campbell & Stanley, 1963). If alternate forms of a test are used to control testing effects, differences between the forms can inflate measurement error and decrease the power of statistical tests (Kirk, 1982; Nunnally, 1972). A final obstacle to repeated measurement is the practical problem of administering often lengthy ability or personality tests to subjects on a regular basis. Time constraints and subject cooperation are among the practical problems encountered with ongoing measurement using vaganotic devices (Goldfried & Kent, 1972).

In addition to infrequent assessment, vaganotic strategies also promote measurement of large groups of individuals. Inferential statistics born of the vaganotic tradition, such as analysis of variance (Fisher, 1925), are often used to evaluate the effects of different treatment levels of an independent variable on grouped data. An example is taken from the use of one-way analysis of variance design (Kirk, 1982). In this design, a large group of randomly selected subjects are distributed across different treatment groups in a random fashion. Due to randomization the groups are considered equivalent prior to the administration of the experimental treatment. If statistical analyses subsequent to that treatment reveal differences among group means that overshadow variability within the groups, disparate effects are attributed to the levels of the independent variable(s).

Randomization to achieve equivalent groups is the key to controlling many threats to internal validity using vaganotic approaches (Campbell & Stanley, 1963; Cook & Campbell, 1979). Competing hypotheses that may account for group differences are ruled out because it is presumed that threats such as history, maturation, testing and instrumentation, will operate equally across individuals and—consequently—groups. Given this assumption, it follows that differences among groups following exposure to the experimental procedures are the result of some properties of the treatment conditions because, apart from their differential treatment during the experiment, groups were "equivalent" at the start.

The vaganotic system of measurement and experimentation has several implications for the individual subject's role in achieving internal validity that we are now ready to discuss. First, the practice of limited assessment restricts the contribution of every subject to the experimental endeavor. Nearly all branches of scientific psychology recognize that psychological processes occur at the level of the individual (Boring, 1950). The process is a continuous one that is manifested in variable behavior over time (Johnston & Pennypacker, 1980). Group approaches to research, however, reduce this continuous process to a few measurements that are intended to represent an individual's performance on the dependent measure. As a result, the product captures only a small sample of this process, thus limiting the role of the single subject in research.

Second, the objective of vaganotic-based experiments is to discover between-group differences that are statistically significant. This is done by comparing the performance of groups under different experimental conditions. Group performance is a composite of individual measures that are represented by various descriptive statistics, such as the mean and standard deviation. Here again, the contribution of the individual subject is limited to (typically) a single score in a summary measure. The greater the number of subjects in a group, the smaller is the impact of the individual, and the less that is known about how individuals respond to the treatment variables in question (Kratochwill & Levin, 1979; Michael, 1974).

Third, the logic underlying the control of most threats to internal validity using group designs rests on the assumption of "equivalent" groups prior to treatment. Equivalence is assumed to be a natural by-product of randomization and is sometimes verified by showing that group means do not differ significantly on a pretest. However, the notion of equivalence as it relates to control of internal validity threats appears to be flawed on two counts. To begin with, nonsignificant group differences prior to administration of the independent variable do not necessarily denote the absence of sizable differences among individuals. The possibility exists that slight changes following treatment may combine with preexperimental differences to yield significant effects without adequate control for pretreatment differences among subjects. A second flaw concerns the assumption that internal validity threats, if present, will operate on all subjects to produce comparable effects. The problem again seems to be the failure to recognize that behavior change occurs at the individual level. Not all subjects will be exposed to all validity threats operating in an experiment, nor can it be assumed that individual responses to various threats will be equivalent. If indeed the change process occurs at the level of each subject, then control procedures should also be exercised with each individual (Sidman, 1960).

Finally, with vaganotic approaches inferences regarding the effects of an independent variable are made on the basis of between-group differences illuminated by statistical analyses. Conclusions are made about the relationship between variables irrespective of the responses of individuals to the experimental procedures (Hersen & Barlow, 1976; Michael, 1974). For example, Kendrick, Craig, Lawson, and Davidson (1982) compared the effectiveness of a cognitive-behavior therapy with a behavioral rehearsal procedure and a waiting-list control condition in the treatment of a debilitating "music performance anxiety." Fifty-three pianists distributed across the three conditions were assessed on six vaganotic and two idemnotic measures related to cognitive, physiological, and behavioral expressions of performance anxiety. Multivariate analyses of variance (MANOVAs) conducted on posttest and follow-up measures led the researchers to conclude "both the cognitively-based treatment program and the behavioral-rehearsal program proved effective in reducing musical performance anxiety, in contrast to a waiting list control condition" (p. 359). This conclusion was offered despite rather large standard deviations across virtually all groups and dependent measures, suggesting that there were individual differences in subjects' response to treatment. Such conclusions overlook the failure of the treatment to effect change in some clients and prematurely terminate the investigation without analyzing the reasons for mixed effects. As a result, general statements regarding the effects of the independent variables are not warranted (Kiesler, 1981) and scientific advances may be more apparent than real (Johnston & Pennypacker, 1980).

It should be emphasized that vaganotic measures do not, by their very nature, preclude repeated measurement and the analysis of individual data. However, vaganotic measurement is consistently associated with between-group experimentation and the attendant limitations noted above. Due to these factors, vaganotic measures are rarely employed in behavior analysis research.

Idemnotic Approaches

Adoption of idemnotic measurement strategies beckons researchers to employ a very different set of experimental methods that have been the hallmark of behavior analysis research (Johnson & Pennypacker, 1980;

Sidman, 1960; Skinner, 1938, 1953). Central to the development of these methods has been the recognition that behavior is a function of an individual's interaction with its environment. From this assumption it follows that the unit of measurement and experimental analysis is the single response or response class of the individual organism as it experiences various independent variables over time.

The previously presented perspective on the appropriate subject matter for a science of behavior has important implications for experimental design to control threats to internal validity. Because the unit of analysis is the single response measure for each subject, controls should be in place at that level of analysis. It is at the level of the individual response that internal validity threats may operate to confound the results of an experiment. Consider for instance a study attempting to isolate the effects of differential reinforcement of other behavior (DRO) on the "psychotic" talk of five adolescents in a private school. Following a period of baseline observations to establish pretreatment patterns of psychotic talk, a DRO procedure is administered individually to each youth. Unbeknown to the researcher, different threats to internal validity are present and may affect the language of three of the five subjects. One of the students experienced an historical event that could have a supressive effect on psychotic speech; a second reacted to continued observation of his behavior; and the third subject, selected for the study because of recent escalation in psychotic language, returned to normal speech patterns during the first week of treatment.

Rival explanations for improved language in these three subjects must be controlled before valid conclusions about the effects of the DRO procedure are made. Validity threats in this hypothetical example are unique to three subjects. Control procedures employed for one subject will not eliminate threats to the others. For example, the historical threat experienced by the first subject will, if operative, only affect the psychotic speech of that individual. In order to discount history as a plausible account for changes in speech patterns, the researcher must demonstrate that the independent variable is responsible for observed changes in psychotic talk for that subject. This is achieved by replicating the independent variable within each subject and tracking its effect on the dependent measure as the treatment is alternately introduced and withdrawn (Sidman, 1960). Similar strategies must be employed for the other subjects in order to control internal validity threats at the level of the single-response or response class. These assumptions regarding the unit of analysis and level at which threats to internal validity need to be controlled have important implications for the selection of experimental designs, which are discussed in a later section.

EXTERNAL VALIDITY

With the establishment of the effects of definable features of the environment on behavior by studying the organism over time and under different conditions comes the question of the generality of the relationship. Traditionally, the extent to which the results of a given investigation hold true outside the original experimental situation is referred to as the external validity of the study (Campbell & Stanley, 1963; Cook & Campbell, 1979). Concern centers on the generalizability of findings to different times, settings, behaviors, and subjects (Drabman, Hammer, & Rosenbaum, 1979). As with internal validity, vaganotic and idemnotic-based approaches to experimentation adopt different strategies for achieving external validity.

Vaganotic Approaches

Experimental design in the vaganotic tradition generally involves the statistical comparison of two or more groups of individuals that are exposed to one or more experimental conditions. Comparisons that are statistically significant yield results that are considered to have broad or limited generality depending on the randomization procedures employed. For example, to establish the external validity of a between-group experiment across the population of subject variables, the researcher using vaganotic methods would, theoretically, obtain a random sample of subjects from the population to which generalizations are important. By random sample we mean that every subject in the population of interest has an equal probability of being selected. The goal is to have a sample of individuals that are representative of all elements of the population and is free of selection biases (Kerlinger, 1973). A major criterion for determining whether a sample is representative is the degree to which its elements possess prevalent characteristics of the population (Stilson, 1966). If experimental effects are evident with representative samples, the rules of vaganotic research permit inferences to be made to the entire population of subjects from which the random samples were drawn.

A similar tack is taken in drawing inferences about the generality of relationships across different levels of independent variables. In group research, independent variables or factors are considered to be fixed or random. Fixed factors are independent variables whose levels in a given experiment have been determined either arbitrarily or systematically (Kirk, 1982). Specific levels may be chosen because of their importance to the topic under study or due to the inaccessibility of all possible levels of an independent variable. In a hypothetical study examining the effects of varying dosage levels of a drug (Dilantin) on the control of seizures, an experimenter may treat the independent variable medication as a fixed-effects factor. Dosage levels may be selected by the experimenter on a systematic basis (e.g., 50 mg, 100 mg, 150 mg, 200 mg) in order to answer research questions pertaining to those particular dosages. By contrast, random factors are independent variables whose levels in an experiment have been randomly selected from a population of levels that the variable could assume (Kirk, 1982). Returning to the medication example, Dilantin also could have been a random factor if the dosage levels included in the study were randomly selected from the population of dosages within a specified range (e.g., 0–300 mg).

Different conclusions regarding the generality of findings are drawn when fixed versus random factors are involved in group research (Hays, 1981). Given that subjects were randomly selected and assigned to experimental conditions, results pertaining to a fixed factor would generalize only to those levels of the independent variable contained in the original experiment. That is, the effects illuminated by experimental manipulation would, according to some probability value, be representative of the population of subjects exposed to the fixed levels of the independent variable. Effects due to random factors, on the other hand, may be generalized to all possible levels that the treatment variable could assume. When coupled with random selection of subjects, effects attributed to random factors are considered to have broad generality across the population of subjects and levels of the independent variable (Kerlinger, 1973; Kirk, 1982).

This approach to establishing external validity influences the role that individual subjects play in group research. The key to drawing general conclusions from vaganotic studies is randomization to obtain representative samples of subjects and levels of independent variables. Two factors that affect the representativeness of samples are the sample size in relation to the size of the population and the heterogeneity of subjects (Cook & Campbell, 1979; Kerlinger, 1973). In general, the larger the proportion of sample subjects to population elements, the more likely it is that the sample will exemplify the characteristics of the population. In psychological research, many populations of interest are quite large, which often necessitates obtaining large samples in order to achieve a representative subset of the population. When this occurs the contribution of the individual's data to the conclusions of a study diminish as sample size increases.

Because randomization is central to establishing the external validity of group investigation, several considerations should be mentioned regarding the adequacy of this procedure. First, true random selection from a population is more a theoretical notion than it is a viable option in most psychological research (Kerlinger, 1973). Access to all elements of a population to which generalizations are made is seldom possible. Thus, in order to satisfy the requirements for randomization, the population is typically narrowly defined (Kazdin, 1980). For example, drawing a random sample of all hyperactive children attending a given elementary school may be a plausible option. Drawing inferences beyond this narrow population, however, would require replication studies conducted with other similarly narrow groups. Although this strategy may reveal the reliability of the central tendency, it will shed little, if any, light on the generality on the phenomenon at the individual level (Sidman, 1960). Second, without knowledge of the characteristics of the population, the researcher never knows whether the sample at hand is indeed representative of the parent group. All random sampling assures that the long-run average of several samples will reflect most elements in the population. Firm generalizations are usually not justified until comparable effects are obtained with other population members. A third consideration has to do with heterogeneous responses of subjects exposed to the same independent variable. Because significant effects based on group averages are possible despite the absence of behavior change in some individuals, the parameters of generalization are unknown and precise relationships among variables are difficult to establish (Bergin, 1966; Chassan, 1979; Sidman, 1960).

Idemnotic Approaches

The nature of behavior analysis research requires a much different approach to external validity than those described for the group or vaganotic approach. From the behavior analytic perspective, internal and external validity are separate issues that are addressed with different procedures and often at different times of the experiment. The influence of idemnotic measurement on methods of establishing internal validity extends logically to strategies for determining the generality of experimental effects (Johnson & Pennypacker, 1980). Given success in ruling out threats to internal validity, the next step in the experimental process is to determine the boundaries of generality for the functional relation detected with an individual subject.

The strategy used to establish the generality of functional relations is *replication*. Sidman (1960) refers to two types of replication that pertain to different classes or dimensions of generalization. *Direct replication* is the repetition of an experiment with the same subject or with different subjects. Repetition of an experiment involves duplication of all conditions in the original study, including variables such as the experimenter, setting, length of experimental phases, and procedures involved in administering the independent variable. The purpose of direct replication with the same subject is to increase the reliability of the effect for that subject in cases where the researcher has reason to believe that application of the experimental technique may have been faulty. More common, however, is direction replication across different subjects. When intersubject replication is performed, the researcher is establishing the generality of the phenomenon across subjects. During initial efforts it is often prudent to replicate procedures with relatively homogeneous subjects, especially in applied research, so as to minimize the number of variables that may be responsible for a failure to produce comparable effects with all subjects (Barlow & Hersen, 1984; Hersen & Barlow, 1976). A lack of consistent effects across subjects may then be traced to a limited number of variables that, in turn, can be analyzed in future replication series (e.g., the length of time prior to treatment that a subject engaged in a problem behavior).

In addition to concern for external validity across subjects, behavioral science is also charged with determining the generality or robustness of an experimental treatment as parameters of the experiment vary. This is achieved through *systematic replication* of research across a range of situations (Sidman, 1960). Systematic replication involves the repetition of an experiment while varying a certain number of its components. When data patterns persist under these varied conditions, conclusions can be extended to these areas. A failure to reproduce the original findings, however, suggests that these varying aspects of the study are critical to the functional relation.

Several authors have provided conceptual frameworks for establishing the generality of effects via systematic replication. Among the frameworks developed are the Generalization Map (Drabman *et al.*, 1979), the Behavioral Assessment Grid (Cone, 1978), stimulus and response generalization (e.g., Kendall, 1981; Stokes & Baer, 1977) and clinical trials (Hersen & Barlow, 1976; Kazdin, 1984). We will present a framework for generalization proposed by Sidman (1960) and later modified by Johnston (1979) and Johnston and Pennypacker (1980).

Johnston (1979) and Johnston and Pennypacker (1980) have defined seven dimensions of generality of a functional relation.* Figure 1 illustrates that a functional relation can be evaluated in terms of the generality *of* variables, methods and processes that are evident *across* different

---

* Interested readers are referred to Johnston (1979) and Johnston and Pennypacker (1980) for detailed accounts of dimensions of generality.

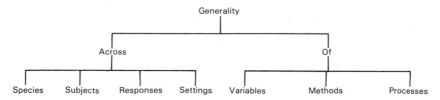

Figure 1. Dimensions of generality. From "On the Relation between Generalization and Generality" J. Johnston, 1979, *The Behavior Analyst*, 2, p. 4. Copyright 1979 by the Society for Advancement of Behavior Analysis. Reprinted by permission.

species, subjects, responses, and settings. *Generality of variables* refers to the extent to which an independent variable produces a defined effect under a wide range of experimental conditions. A robust variable of this kind in behavioral psychology is intermittent reinforcement. That reinforcement administered on a noncontinuous but systematic basis yields consistent patterns of behavior has been well documented in a variety of laboratory and applied settings. *Generality of methods* concerns the reliable effects of procedures or techniques, such as overcorrection. Procedural variations of overcorrection have proven effective in suppressing a spectrum of deviant responses across subject and setting variables (Miltenberger & Fuqua, 1981). *Process generality* pertains to the robust effects of a behavioral process involving two or more variables. An example of a behavioral process having broad generality is positive reinforcement.

The generality of variables, methods, and processes is determined via replication studies across four dimensions. *Species generality* refers to demonstrations that a phenomenon, such as discrimination learning, holds true as procedures are replicated up the phylogenic scale. *Subject generality* concerns the representativeness of an experimental effect across individuals within a species. This dimension of generality is analogous to Sidman's (1960) notion of direct replication discussed earlier. *Generality across responses* occurs when replications of an experiment yield comparable findings when applied to different behaviors within a subject or to behaviors of similar or dissimilar topographies across subjects. Finally, *setting generality* concerns the degree to which effects hold up when aspects of the experimental setting are varied systematically. Shaping is a behavioral procedure that has proven useful in response acquisition studies conducted in countless laboratory, school, hospital, and institutional settings.

The point we wish to emphasize is that there are several dimensions of generality that are important to establishing a scientific basis for psychological phenomena. Delimiting the parameters of generality for a functional relation is a deliberate process that begins with the intense study of an individual subject. The question of whether a variable, method, or process produces a real effect is answered, in most cases, the first time a subject's response is shown to vary as a function of a manipulated variable. Each repetition of the experiment either across subjects or situations provides evidence for or against the generality of the relation. The idemnotic approach to measurement and experimentation compels us to maintain our investigation at the level of the individual subject (Johnston & Pennypacker, 1980). It is through direct and systematic replications with single subjects that scientists are able to control threats to internal validity unique to a given case and at the same time specify the parameters of generality for a functional relation.

DESIGN SELECTION: IMPLICATIONS OF IDEMNOTIC MEASUREMENT

The case has been made that the unit of analysis in behavioral research (i.e., the single response or response class of the individual organism) guides us to employ experimental methods that control internal validity threats at that level. Likewise we have noted that concerns for the external validity of a study may not be satisfactorily addressed until an experiment has first been shown to be internally valid. Experimental designs common to behavior analysis research meet these requirements to varying degrees. In this section we present a brief overview of major single-subject designs and discuss their adequacy for establishing the existence and generality of experimental effects. Numerous volumes have been devoted to single-case research design and we refer readers to these for in-depth coverage of design issues (Barlow et al., 1984; Barlow & Hersen, 1984; Kazdin, 1982; Kratochwill, 1978; Johnston & Pennypacker, 1980; Tawney & Gast, 1984).

Hayes (1981, 1983) and Barlow et al., (1984) have presented an organizational scheme for single-case designs that recognizes that the numerous design variations developed in the past three decades are comprised of a group of core elements. These core design elements are the building blocks from which all single-subject designs are derived. Designs may be classified according to three types: within, between, and combined series. We will discuss a major design from each of these categories.

Within-Series Designs

The within-series type of design is characterized by examining repeated measures of a single subject's behavior across different phases of an experiment. The logic of the within-series design is in comparing response measures at the point of phase change for abrupt differences in level, trend, or variability. The A-B-A-B design is exemplary of the within-series strategy and is depicted in Figure 2. An experiment commences with a baseline phase (A) in which the single response or response class of one subject is monitored under a specified set of conditions that do not include the independent variable in question. Once a stable pattern of responding is evident, the independent variable is introduced while holding all other conditions constant. Behavior change is monitored throughout this intervention phase (B) until the effects of the experimental treatment are apparent. Next, to expose threats to internal validity, the independent variable is withdrawn in a second A phase. If changes in the subject's performance observed during the first B phase is a function of the treatment variable, the subject's behavior is expected to return to baseline levels when treatment is removed. Finally, replication of the intervention in a second B phase that is followed by a concommitant change in behavior provides convincing evidence of a functional relation between the independent and dependent variables.

The within-series A-B-A-B design satisfies the requirements for establishing a functional relation in the most rigorous sense of the term.

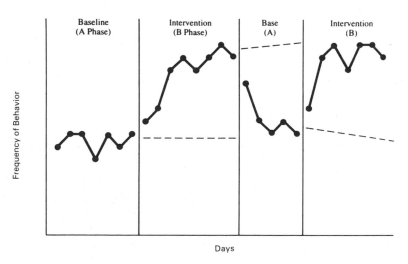

Days

Figure 2. Hypothetical data depicting the A-B-A-B within-series withdrawal design. The solid lines connecting the data points represent the subject's performance during each phase. The dashed lines indicate the predicted level of performance based on the preceding pattern of responding. From *Single-Case Research Designs: Methods for Clinical and Applied Setting* (p. 11) by A.E. Kazdin, 1982, Copyright 1982 by Oxford University Press. Reprinted by permission.

Intrasubject replication of baseline and treatment conditions for a single response measure provides excellent control for competing explanations for behavior change. Performing manipulations at the unit of analysis, the behavior of a single subject, makes possible firm conclusions regarding the effects of the independent variable because factors threatening internal validity are exposed at the level in which they may operate.

A single application of a within-series design, although capable of confirming a functional relation, does little to establish its generality. Concerns for the external validity of an experiment are addressed via replication series that target the various dimensions of generality discussed by Johnston (1979). Stevenson and Fantuzzo (1984) employed within-series replications to extend the applicability of a comprehensive self-management program for students showing poor academic achievement and inappropriate classroom behavior. Separate A-B-A-B designs were used to evaluate the self-management procedures for a single youth across settings (school and home) and responses (accurate math problems and disruptive behavior). The design revealed that the treatment procedures were functionally related to desired changes in both classes of subject behavior and were effective when used under teacher and parent supervision.

Between-Series Designs

Between-series strategies compare two or more data series across time for a single subject. The alternating treatments design (ATD) is representative of this approach (Barlow & Hayes, 1979). In this configuration, data on a single response measure are collected for one subject during exposure to separate experimental conditions for equal periods of time. Figure 3 illustrates the ATD with a set of hypothetical data in which two treatments are compared. The design typically begins with a baseline phase in which multiple sessions are conducted daily. The number of sessions per day corresponds to the number of treatment conditions being contrasted. Because all sessions are conducted under no-treatment conditions there should be no difference between data series plotted across the two time periods (see Figure 3). In the second phase, separate treatments or interventions are alternated among the different time periods each day. The sequence of times of treatment exposure should be determined randomly or by counterbalancing. If the two interventions produce different effects on the subject's behavior, comparison of the two data series should reveal differences in level, trend, or variability as is the case in Figure 3. Typically, applications of the ATD in applied research have included a third phase in which the most effective treatment is

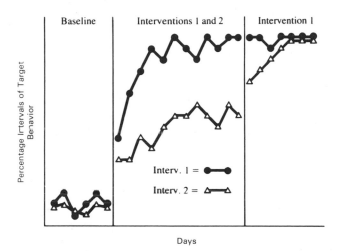

Figure 3. Hypothetical data depicting the alternative treatments design and the between-series design element. During baseline, data are plotted across two different times in which no treatment is in place. The second phase involves the application of two different treatments across the time periods in counter-balanced or randomized order. In the final phase the more effective intervention is replicated across all time periods. From *Single-Case Research Designs: Methods for Clinical and Applied Setting* (p. 127) by A.E. Kazdin, 1982. Copyright 1982 by Oxford University Press. Reprinted by permission.

administered during all time periods (Kazdin, 1982). This strategy facilitates replication of an independent variable during time periods in which less effective procedures were previously operating.

The ATD was conceived as a strategy for comparing multiple treatments as an alternative to traditional between-group vaganotic methods (Barlow & Hayes, 1979). This is achieved during the second phase of the design in which diverging data series provide evidence for the differential effectiveness of the contrasted treatments. Rival explanations for between series differences, such as history, maturation, testing, and so forth are discounted because it is unlikely that these threats would alternate along with the treatments to affect one condition more than another. As a result, the design provides good control for factors that could frustrate a valid comparison of independent variables.

A shortcoming of the design, however, is its failure to establish functional relations between the treatment conditions and the subject's dependent measure. Inspection of the design configuration in Figure 3 reveals that only an A-B element is used to compare either treatment with baseline conditions. Threats to internal validity that could account for behavior change following baseline are left uncontrolled. In other words, that the subject's response was altered at all cannot be attributed to either treatment due to the numerous rival hypotheses that are not dispelled using the A-B element. However, differences between the two data series in the second phase can be accounted for by the differential effectiveness of the interventions. A simple solution to this problem is to include a separate data series comprised of baseline observations only. This condition would be maintained throughout the second phase permitting both treatment–treatment and baseline–treatment contrasts. Again, because of the rapid alternation of experimental conditions, internal validity threats are controlled and, using this design variation, functional relations are illuminated (see Ollendick, Shapiro, & Barrett, 1981, for an illustration of this design). As with the within-series elements, establishing the generality of results gleaned from a between-series design requires replication studies across species, subjects, responses, and settings.

Combined-Series Designs

Combined-series designs utilize within- and between-series comparisons to draw inferences regarding treatment effects. A familiar example of these combined elements is the multiple baseline design (MBD). Typically, the MBD employs a within-series A-B element that is replicated across two or more independent data series. These data series can represent different subjects, different behaviors of a single subject, or the same behavior of a single subject in different settings. Figure 4 depicts hypothetical data for the MBD across three behaviors. The design purports to expose common threats to internal validity by staggering the introduction of the independent variable at different points in time across the different data series. According to the logic of the design, if changes are observed at the first A-B shift (within-series comparison) and the remaining baselines remain unaffected (between-series comparisons), alternative explanations for behavior change, such as history and maturation, are less likely. Confidence in the treatment effect is said to be enhanced each time replication of the independent variable across data series produces the anticipated pattern of performance.

The implications noted earlier that idemnotic measurement has for experimentation have bearing on an evaluation of the MBD. The point was made that our measurement and experimentation practices should take place at the appropriate unit of analysis for a science of behavior (Johnston & Pennypacker, 1980). It is at the level of the individual's response or response class that environmental events operate to control behavior, including threats to internal validity. The logic of the MBD

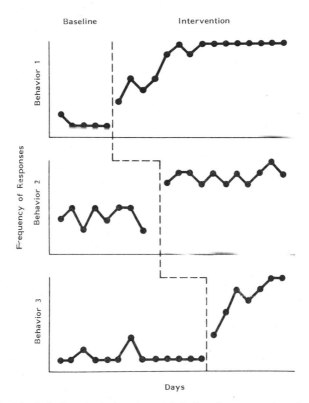

Figure 4. Hypothetical data depicting the muliple baseline or combined-series design. Following a baseline period, the same intervention is introduced in staggered fashion across three responses. From *Single-Case Research Designs: Methods for Clinical and Applied Setting* (p. 180) by A.E. Kazdin, 1982. Copyright 1982 by Oxford University Press. Reprinted by permission.

would seem to contradict these basic assumptions. The design is predicated on behavior change occurring within a data series only following the introduction of treatment. Untreated baselines serve as controls in the sense that the absence of change in these data series purportedly provides evidence that something other than the independent variable could not have been responsible for the observed change. However, this logic assumes that threats to internal validity operate equally across data series (i.e., subjects, responses, and settings). Yet such an assumption is at odds with a strict behavior analytic view of behavior. For example, much has been made of the need for independent data series when employing the MBD (Hayes, 1981; Kazdin & Kopel, 1975). This requirement is necessary to avoid generalized change in behavior across data series when the first data series is treated (as might be the case if the behaviors were members of the same response class). The problem is that by definition independent behaviors are those that are controlled by different environmental variables. It follows then that environmental change contiguous to or contingent on one behavior may influence that particular response, but have no effect on other independent behaviors. Such is often the case with threats to internal validity. It is not hard to imagine an extraneous event (i.e., history) affecting one behavior at the point of phase change and not other behaviors. When this occurs in the context of an MBD, the untreated data series do not serve as adequate controls because they may have either not been exposed to the threat or were unaffected by it.

In view of the flaw in this combined-series design, it may be more appropriate to consider the MBD as a series of A-B elements that are replicated across subjects, responses, or settings. As A-B designs, however, they leave uncontrolled various internal validity threats and consequently do not permit firm conclusions regarding the effects of an experimental variable. When combined with other design elements, the MBD can be strengthened. For example, inclusion of a within-series A-B-A-B element in one of the data series may yield evidence for a functional relation for one response. The remaining data series, although comprised of A-B elements, may provide some indication of the generality of the relation.

SUMMARY AND CONCLUSIONS

In this chapter we have provided an overview of the conceptual framework of behavior analysis research methods. Traditionally, the field behavior analysis has been defined as a subdiscipline of the field of scientific study in which behavior is perceived as a physical event produced by an organism that results in some observable change in the environment. Behavior is considered a product of the interaction of the organism's biological functions and various environmental contingencies operating at the time. The analysis of behavior refers to a set of tools used by researchers in the discipline to understand the processes responsible for the behavioral act.

Traditionally, scientific psychology has adopted a vaganotic approach to measurement wherein various scales and units of measurement are used to assess the variability in a set of observations. In contrast, behavior analysis has adopted (in theory and often in practice) idemnotic measurement, which employs standard and absolute units independent of the variability in some observed phenomena. Selection of the idemnotic measurement strategy has important implications for validity in experimental research. In contrast to vaganotic measurement approaches, idemnotic measurement is used in the context of continuous measurement of behavior over time. Moreover, idemnotic-based experiments focus on the individual unit of analysis to determine specific environmental events that may be responsible for behavior. Based on the assumption that the individual's behavior is a function of the individual's interaction with the environment, the unit of measurement and experimental analysis is the single response or response class of an individual organism as that organism is measured over time. Because the unit of analysis is the individual, a number of single-case designs have been used to establish the internal validity of the experiment. We have reviewed these designs, noting that they are uniquely suited to establishing functional relations between the measures selected for observation and the independent variables being examined. With the exception of multiple baseline designs, each of these designs are well suited to establishing the functional relationship between the individual and the environment. In the case of establishing the external validity or generalizability of research findings, idemnotic measurement requires that the investigator include replication as a formal strategy in research activity. In this regard direct and systematic replication alternatives were reviewed, and several dimensions of generality as presented by Johnston and Pennypacker (1980) were discussed.

Our discussion of the different types of measurement and design considerations of behavior analysis research suggests that many researchers who embrace the behavior analytic model only approximate the more pure form of the methodology discussed here. This is reflected in the focus on groups of subjects within intersubject replication designs, as well as the application of multiple baseline designs to multiple subjects, as noted earlier. In addition, as behavior analysis has expanded to the natural environment, a variety of research questions now extend to units that are conceptualized well beyond the individual level of analysis. Whether or not this is the best way to establish a science of behavior is readily subject to debate. Nevertheless, as part of the dissemination process, it is likely that we will continue to see researchers embrace the basic tenets of behavior analytic research but employ variations of its methodology to expand on the scope of the field.

ACKNOWLEDGMENTS

The authors express their appreciation to Brian Iwata for his helpful comments on the manuscript.

REFERENCES

Baer, D. M. (1981). A flight from behavior analysis. *The Behavior Analyst, 4*, 85–92.

Baer, D. M., Wolf, M. M., & Risley, T. R. (1968). Some current dimensions of applied behavior analysis. *Journal of Applied Behavior Analysis, 1*, 91–97.

Barlow, D. H., & Hayes, S. C. (1979). Alternating treatments design: One strategy for comparing the effects of two treatments in a single subject. *Journal of Applied Behavior Analysis, 12*, 199–210.

Barlow, D. H., & Hersen, M. (1984). *Single case experimental design: Strategies for behavior change.* New York: Pergamon Press.

Barlow, D. H., Hayes, S. C., & Nelson, R. O. (1984). *The scientist practitioner: Research and accountability in clinical and educational settings.* New York: Pergamon Press.

Bergin, A. E. (1966). Some implications of psychotherapy research for therapeutic practice. *Journal of Abnormal Psychology, 71*, 235–246.

Boring, E. G. (1950). *A history of experimental psychology* (2nd ed.). New York: Appleton-Century-Crofts.

Campbell, D. T., & Stanley, J. C. (1963). *Experimental and quasi-experimental designs for research.* Chicago, IL: Rand McNally.

Chassan, J. B. (1979). *Research design in clinical psychology and psychiatry* (2nd ed.). New York: Irvington.

Cone, J. D. (1978). The behavioral assessment grid (BAG): A conceptual framework and a taxonomy. *Behavior Therapy, 9*, 882–888.

Cone, J. D. (1981). Psychometric considerations. In M. Hersen & A. Bellack (Eds.), *Behavioral assessment: A practical handbook* (pp. 38–68). New York: Pergamon Press.

Cook, T. D., & Campbell, D. T. (Eds.). (1979). *Quasi-experimentation: Design and analysis issues for field settings.* Chicago, IL: Rand McNally.

Dietz, S. M. (1982). Defining applied behavior analysis: An historical analogy. *The Behavior Analyst, 5*, 53–64.

Drabman, R. S., Hammer, D. C., & Rosenbaum, M. S. (1979). Assessing generalization in behavior modification with children: The generalization map. *Behavioral Assessment, 1*, 203–219.

Ebbinghaus, H. (1885). *Memory.* New York: Teachers College.

Epling, W. F., & Pierce, W. D. (1983). Applied behavior analysis: New directions from the laboratory. *The Behavior Analyst, 6*, 27–38.

Ferster, C. B., & Skinner, B. F. (1957). *Schedules of reinforcement.* New York: Appleton-Century-Crofts.

Fisher, R. A. (1925). *Statistical methods for research workers.* London: Oliver & Boyd.

Galton, F. (1889). *Natural inheritance.* London: McMillan.

Goldfried, M. R., & Kent, R. N. (1972). Traditional versus behavioral assessment: A comparison of methodological and theoretical assumptions. *Psychological Bulletin, 77*, 409–420.

Hayes, S. C. (1981). Single case experimental design and empirical practice. *Journal of Consulting and Clinical Psychology, 49*, 193–211.

Hayes, S. C. (1983). The role of the individual case in the production and consumption of clinical knowledge. In M. Hersen, A. E. Kazdin, & A. Bellack (Eds.), *The clinical psychology handbook* (pp. 181–196). New York: Pergamon Press.

Hayes, S. C., Rincover, A., & Solnick, J. (1980). The technical drift of applied behavior analysis. *Journal of Applied Behavior Analysis, 13*, 275–285.

Hays, W. L. (1981). *Statistics.* New York: Holt, Rinehart, & Winston.

Hersen, M., & Barlow, D. H. (1976). *Single-case experimental designs: Strategies for studying behavior change.* New York: Pergamon Press.

Hull, C. L. (1952). *A behavior system: An introduction to behavior theory concerning the individual organism.* New Haven, CT: Yale University Press.

Johnston, J. M. (1979). On the relation between generalization and generality. *The Behavior Analyst, 2,* 1–6.

Johnston, J. M., & Pennypacker, H. S. (1980). *Strategies and tactics of human behavioral research.* Hillsdale, NJ: Erlbaum.

Kazdin, A. E. (1977). Artifact, bias, and complexity of assessment: The ABC's of reliability. *Journal of Applied Behavior Analysis, 10,* 141–150.

Kazdin, A. E. (1980). *Research design in clinical psychology.* New York: Harper & Row.

Kazdin, A. E. (1982). *Single-case research designs: Methods for clinical and applied setting.* New York: Oxford University Press.

Kazdin, A. E. (1984). Therapy analogues and clinical trials in psychotherapy research. In M. Hersen, L. Michelson, & A. Bellack (Eds.), *Issues in psychotherapy research* (pp. 227–250). New York: Plenum Press.

Kazdin, A. E., & Kopel, S. A. (1975). On resolving ambiguities in the multiple-baseline design: Problems and recommendations. *Behavior Therapy, 6,* 601–608.

Kendall, P. C. (1981). Assessing generalization and the single-subject strategies. *Behavior Modification, 5,* 307–319.

Kondriolk, M. J., Craig, K. D., Lawson, D. M., & Davidson, P. O. (1982). Cognitive and behavioral therapy for music performance anxiety. *Journal of Consulting and Clinical Psychology, 50,* 353–362.

Kent, R. N., O'Leary, K. D., Diament, C., & Dietz, A. (1974). Expectation biases in observational evaluation of therapeutic change. *Journal of Consulting and Clinical Psychology, 42,* 774–780.

Kerlinger, F. N. (1973). *Foundations of behavioral research* (2nd ed.). New York: Holt, Rinehart & Winston.

Kiesler, D. J. (1981). Empirical clinical psychology: Myth or reality? *Journal of Consulting and Clinical Psychology, 49,* 212–215.

Kirk, R. E. (1982). *Experimental design: Procedures for the behavioral sciences.* Belmont, CA: Brooks/Cole.

Kratochwill, T. R. (Ed.). (1978). *Single-subject research: Strategies for evaluating change.* New York: Academic Press.

Kratochwill, T. R., & Levin, J. R. (1979). What time-series designs may have to offer educational researchers. *Contemporary Educational Psychology, 3,* 273–329.

Michael, J. (1974). Statistical inference for individual organism research: Mixed blessing or curse? *Journal of Applied Analysis, 7,* 647–653.

Michael, J. L. (1980). Flight from behavior analysis. *The Behavior Analyst, 3,* 1–24.

Miltenberger, R. G., & Fuqua, R. W. (1981). Overcorrection: A review and critical analysis. *The Behavior Analyst, 4,* 123–142.

Nunnally, J. C. (1967). *Psychometric theory.* New York: McGraw-Hill.

Nunnally, J. C. (1970). *Introduction to psychological measurement.* New York: McGraw-Hill.

Nunnally, J. C. (1972). *Educational measurement and evaluation.* New York: McGraw-Hill.

Ollendick, T. H., Shapiro, E. S., & Barrett, R. P. (1981). Reducing stereotypic behaviors: An analysis of treatment procedures utilizing an alternating treatments design. *Behavior Therapy, 12,* 570–577.

Pavlov, I. P. (1927). *Conditioned reflexes* (G. Aurup, Trans.). London: Oxford University Press.

Pennypacker, H. S. (1981). On behavior analysis. *The Behavior Analyst, 4,* 159–161.

Pierce, W. D., & Epling, W. F. (1980). What happened to analysis in applied behavior analysis? *The Behavior Analyst, 3,* 1–10.

Quetelet, A. (1835). *Sur l'homme et le développement des ses facultés.* Paris: Bachelier.

Sidman, M. (1960). *Tactics of scientific research*. New York: Basic Books.

Skinner, B. F. (1938). *The behavior of organisms*. New York: Appleton-Century-Crofts.

Skinner, B. F. (1953). *Science and human behavior*. New York: Macmillan.

Stevenson, H. C., & Fantuzzo, J. R. (1984). Application of the "Generalization Map" to a self-control intervention with school aged children. *Journal of Applied Behavior Analysis, 17*, 203–212.

Stilson, D. (1966). *Probability and statistics in psychological research and theory*. San Francisco: Holden-Day.

Stokes, T. F., & Baer, D. M. (1977). An implicit technology of generalization. *Journal of Applied Behavior Analysis, 10*, 349–367.

Tawney, J. W., & Gast, D. L. (1984). *Single subject research in special education*. Columbus, OH: Merrill.

Thorndike, E. L. (1898). *Animal intelligence*. New York: Macmillan.

Watson, J. B. (1919). *Psychology from the standpoint of a behaviorist*. Philadelphia, PA: Lippincott.

Wolf, M. M. (1978). Social validity: The case for subjective measurement or how applied behavior analysis is finding its heart. *Journal of Applied Behavior Analysis, 11*, 203–214.

CHAPTER 7

The Time Domain in Individual Subject Research

EVELYN B. THOMAN

TEMPORAL REGULARITIES IN BEHAVIOR

"A most important function of time in psychology is as a matrix for imbedding the behavioral stream" (Galanter, 1984, p. 193). The temporal ordering of behaviors is of importance from the very earliest days of life, and this is apparent in the behavior of the single individual and in the behaviors of interacting individuals, such as a mother and her infant. Temporal order is considered to be expressed in two major ways: (a) in consistencies of behavioral sequences, either in an individual or between individuals; and (b) in the regularity of recurrence of specific behaviors. The first form of ordering is studied generally for the purpose of making cause–effect inferences; study of the latter form is typically motivated by an interest in the rhythmicity of behaviors such as sleep, activity, and even patterns of social interaction. In both instances, sequential analysis is the strategy for study.

In this chapter, we propose another approach to study of the temporal domain, one which uses nonsequential analytic strategies to assess the temporal organization of recurring behavior patterns. It will be suggested that, in fact, for some behaviors, nonsequential organization occurs developmentally before sequential order emerges. Although the notion

EVELYN B. THOMAN • Department of Biobehavioral Sciences, University of Connecticut, Storrs, Connecticut 06268.

of studying temporal organization nonsequentially may seem paradoxical, our approach is guided by a theoretical perspective, namely general systems theory (GST), which incorporates time as fundamental to process analysis. GST emphasizes ongoing feedback over time and the limits of linear-sequential models for understanding biological and behavioral phenomena. Formal systems notions, and our efforts at application, are particularly relevant for the present volume because of its focus on the individual subject. That is, within this framework, an understanding of the organismic significance of temporal patterning of behaviors can be understood only from investigation of those patterns within individual subjects.

In what follows, some major aspects of general systems theory will be presented and their relevance for developmental study and analysis at the level of the individual will be discussed. Implications of GST for the development of temporal order in behaviors will be considered, and the alternative of nonsequential investigation of temporal organization will be proposed. These notions will be illustrated with analyses of the sleep and waking states in infants observed in the home environment.

GENERAL SYSTEMS THEORY AS A FRAMEWORK FOR STUDY OF THE TEMPORAL DOMAIN

General systems theory is concerned with problems of organized complexity (Bertalanffy, 1968), which is precisely the issue for developmental research. Bertalanffy (1933) proposed that any theory of development and life in general must be a systems theory, applicable to social relationships as well as other levels of biological organization. Systems notions have been applied at a range of levels from aggregates of inanimate particles to aggregates of living organisms. Although there are many approaches to systems theory, we can consider briefly some general characteristics of systems as proposed primarily by Weiss (1969, 1971) and Bertalanffy (1933, 1968) and their implications for study of temporal organization of behaviors during development.

Weiss (1969) defines a biological system as a

> rather circumscribed complex of relatively bounded phenomena which, within these bounds, retains a relatively stationary pattern of structure in spite of sequential configuration in time, in spite of a high degree of variability in the details of distribution and interrelations among its constituent units of lower order. Not only does the system maintain its configuration and integral operation in an essentially constant environment, but it responds to alterations of the environment by an adaptive redirection of its component processes in such a manner as to counter the external change in the direction of optimum preservation of its system integrity. (Weiss, 1969, p. 12)

More simply put, "a system is a whole or a unit composed of hierarchically organized and functionally highly interdependent subunits that may themselves be systems" (Fowler, 1975, p. 26).

By these descriptions of the nature of systems, it is clear that the overall organization of any system has characteristics that are not apparent in the behavior of any of its components. Thus, its functioning cannot be deduced by all possible information on the nature of the component parts. In addition, the distinction is made between a machine as a system and a living system. In the former, the behavior of the parts determines the operation of the whole, whereas in a living system, the structure of the whole determines the operation of the parts (Weiss, 1969).

It is not simply that a living system is complex—a machine may also reach a high degree of complexity—but also that, in contrast to the infinite number of possible interactions and combinations among its constituent units that could take place in a mere complex, in the living system only a limited selection from the "grab bag" of opportunities for effective processes is being realized at any one moment. The selection can be understood solely by its bearing on the concerted harmonious performance of a task by the complex as a whole. This is the feature that distinguishes a living system from a dead body, or a functional process from a mere list of parts involved. Weiss concludes that the *ordered complexes* that we designate as a living system are products of experience with nature.

The maintenance of equilibrium is an important product of the interactive processes and ongoing feedback. That is,

> there are patterned processes which owe their typical configuration not to a prearranged, absolutely stereotyped, mosaic of single-tract component performances, but on the contrary, to the fact that the component activities have many degrees of freedom, but submit to the ordering restraints exerted upon them by the integral activity of the "whole" in its pattern systems dynamics. (Weiss, 1969, p. 9)

Of major interest to systems theorists are the functions of the brain, as a hierarchically organized system, with multiple expressions, including overt behavior.

> The function of the central nervous system, with its memory, communications, computational, and learning capabilities, is to provide algorithmic content capable of mediating the stability of internal chains, so that a satisfactory pattern of behavior emerges. (Iberall & McCulloch, 1969, p. 290)

The results of this regulation is a behavioral patterning whose richness derives from this complexity at the level of the central nervous system. That is, the patterning of behaviors is an emergent process from the organized processes of the central nervous system.

In conclusion, and most important for our interests in behavior, it is generally accepted that three principles are expressed in any living system: (a) a system involves simultaneous interaction of complex variables; (b) there is ongoing and mutual feedback within the system; and (c) the system functions to maintain its own equilibrium (Bunge, 1979; Weiss, 1971; Steve Chorover, personal communication). Each of these criteria refers to network dynamics, with major implications for the study of behavior.

It is clear from the principles specified for this theoretical perspective that to focus on any single factor (or behavior) is insufficient, but rather, one must be concerned with the interplay of numerous interacting factors; and also that the organized and organizing interactions in the network are temporally integrated. Process occurs over time and must be studied over time. A single time-frame, "snapshot" view of behavior cannot reveal the systems dynamics the developmentalist is interested in. Thus, temporal organization of the behaviors that compose a system becomes a central issue for understanding the integration of the behavior components of a definable whole and its role in the survival or adaptation of the organism.

GENERAL SYSTEMS THEORY AS A FRAMEWORK FOR STUDY OF THE INDIVIDUAL

General systems theory is concerned with complex dynamic processes whose integration occurs over time. An understanding of the principles of integration for any type of system requires (a) investigation within a time frame, and (b) a focus on the functioning of individuals. It is extremely difficult to do a general systems analysis using group data (Denenberg, 1977), given that the process of interest occurs only within the individual. This notion has long been reflected in learning studies, where it has been found that a learning curve must be determined for each individual subject, as data summed over the performances of a group of subjects may suggest a smooth function where acquisition of asymptotic performance may, in fact, be achieved by each individual as a sudden change on a single trial. Likewise in the study of biorhythms, it is essential that the functions be derived from the data of individual subjects. On the other hand, in developmental research, changes with age have too often been assessed from group averages, a form of analysis that may produce a trajectory of performance or physical changes that does not meaningfully represent the process in *any* individual (Wohlwill, 1973).

The study of single individuals has fallen into some disrepute as being "unscientific," primarily in reaction to its extensive use in the past for case studies by clinicians. There are several consequences of the too-ready dismissal of the case study. The first is the possibility of overlooking the case study as an important part of the scientific process, namely as a means of generating hypotheses. Freud used the history of his individual patients to derive a theory of psychosocial development that continues to have a pervasive influence on many areas of psychology, including experimental research with animals. And Piaget based his theory of cognitive development on observations of three individual subjects, his own children.

There also has been a general failure among critics of single-subject research to discriminate between the case study as a subjective, evaluative description and the study of an individual that is empirical, analytical, and process oriented. Case studies can be a rich source of hypotheses, and their value in this regard has been markedly underestimated except in the clinical literature. In addition, however, the study of individuals can test hypotheses, and thus serve as a basis for deriving general principles. This latter use of the single subject includes learning studies, biological-rhythm studies, psychophysiological studies, experimental studies where it is possible for an individual to serve as his or her own control for an intervention, or repeated intervention, and ethological study, where species-specific behaviors are of interest. In these instances, study of each individual may test a hypothesis, and such studies may serve as a basis for deriving general principles.

The single individual has of necessity to be the research unit if we are concerned with understanding the dynamics of behavioral mechanisms and brain–behavior relationships. The processes governing these relationships occur within each individual, and

> statistical averaging over a group is more likely to obfuscate than to elucidate the inherent nature of such processes. In the final analysis, if we wish to understand the individual, we can do so only by studying individuals, not by studying them as statistical actuarial averages. (Denenberg, 1982, p. 20)

From this perspective, an obvious issue that must be considered is that of generalization. Clearly, the serious problem when working with single cases is whether the results obtained have any degree of generality or whether they represent a unique constellation of behaviors of the individual studied. Denenberg (1979) has provided guidelines for addressing this issue. The problem is to separate, within any one person, the unique features of that person's behaviors from those features that may be shared with others in the population, as meaningful generalizations can only be obtained from commonality, not from uniqueness. As Denenberg points out, it is reasonable to argue that finding any other person with the same pattern of results should suffice to indicate that the results from one individual are not unique. In such a case, one can treat the results statistically by specifying the null hypothesis, which states that the particular pattern of results obtained do not exist in the population, that is, that the proportion of people in the population who would exhibit this pattern of results is zero. This hypothesis can be tested (Walker & Lev, 1953, pp. 433–455 and Chart VI). It is thus only necessary to find one or two additional cases to reject the null hypothesis that the population proportion is zero. Then, if one finds the same pattern occurring in only one additional case out of 50 or less, this is sufficient to reject the null hypothesis at the .05 level. In these instances, one can reject the null hypothesis. "If one knew the actual value of the proportion in the population, this would define the degree of generality of the phenomenon" (Denenberg, 1977, p. 46). In order to answer this question, it is necessary to carry out a study using a large number of cases to obtain a reliable estimate of the population proportion with a relatively small standard error. However, group studies using means and differences in means are primarily based only on rejection of zero value or differences on the average as representing population characteristics.

Denenberg describes several forms of invariances that may be found across individuals. One form of invariance is that expressed in constants that may be common across individuals. As an example, in an analysis of mother–infant interaction data from extensive observations of six infants (Thoman, Denenberg, Becker, & Freese, 1974), we found that they could be characterized as follows: If an infant cried and was picked up in less than 90 seconds, the baby stopped crying very quickly; however, if the baby cried for more than 90 seconds before being picked up, it took a much longer time before crying stopped. This 90-second threshold value was found in five of the six babies (the sixth infant was always picked up in less than 90 seconds, and thus there was no information concerning more prolonged crying). This statistical parameter, then, clearly generalizes beyond any one infant to a larger population.

Denenberg (1979) explains that it is not necessary for all of the subjects to have the same value as was the case in the study just described, although such a finding makes a conclusion more powerful, but that each infant should have a constant (that is, a measure of central tendency that remains very much the same for that infant from one set of observations to another).

Another kind of constant was suggested by our findings (Thoman *et al.*, 1974) of an apparent constancy in the amount of physical contact between a mother and her baby during the early weeks of life. The suggestion was based on data from one mother–infant pair in which the baby was held much less than the other infants in the sample. However, the mother of this infant spent a great deal of time patting and caressing the baby, when compared with the group of mother–infant dyads. Such a finding suggests the hypothesis that the amount of mother–infant contact, as defined by touching, talking to, or looking at the baby, may be a constant (Denenberg, 1979). This hypothesis can be tested from data on other subjects to determine whether such a generalization holds across subjects and over time. If this were found to be true for a significant number of infants, the generality of the hypothesis would be established. Thus, having found a pattern that describes one infant, or mother–infant pair, it is possible to determine whether this same pattern, as an expression of a common principle, is present in other infants or mother–infant dyads. If so, the generality of the phenomenon can be established in this way.

These examples illustrate that even though analysis is carried out at the level of the individual, it is necessary that a number of subjects have been studied in the same manner. Nevertheless, the idea is that each subject is a separate study or experiment. Consistent with systems theory, it is also requisite that extensive data be obtained over time for each subject. Thus one can find invariances in organizational characteristics or in process principles that characterize individuals within a specified population.

IMPLICATIONS OF APPLYING SYSTEMS THEORY NOTIONS TO THE ANALYSIS OF INFANT BEHAVIOR

One of the criticisms of general systems theory notions is the difficulty of translating them into guidelines for empirical research. Ideally, one would derive mathematical models that would be isomorphic with observed characteristics and relationships among the behaviors of interest. However, even today, there are few models for biological phenomena, and "mathematicians have not yet designed strategies appropriate for *complex* or hierarchical systems" (Abraham, 1984, p. 2). Denenberg (1980, 1981) has derived a model involving three hypothetical brain processes from the general systems equation of Weiss (1969) and shown that this model has heuristic value for the study of brain laterality. However, behavior is clearly a virgin field from this perspective.

We have taken a simpler, nonmathematical approach to the application of GST to our behavioral research, mainly by applying the notions conceptually, and thus using the characteristics of systems and the principles of their functioning as guidelines to our behavioral models. This approach influences our choice of behaviors studied, the research questions we ask, the design of procedures used, the kinds of data analyses, and the way in which we interpret our findings. Based on this theoretical perspective, we have been led to address the study of infant behavior in some untraditional ways.

For example, if one considers the infant as an integral component of the mother–infant dyadic system (where "mother" refers more generally to the infant's caregiver, and in most instances this person is, in fact, the infant's mother), it is clearly necessary to take into account that the infant's functioning as a participant in a dyadic relationship with (a) complex interactions, (b) ongoing mutual feedback, and (c) constraints on the infant's behavioral organization—the principles of living systems as set forth by Weiss (1969, 1971) and Bertalanffy (1933, 1968). The implications for study are profound. For example, in this framework, one would not address temporal organization of behavior by examining sequences of mother and infant behaviors as indexes of causal relations. That is, except in a very limited, or local, sense one would not consider specific behaviors of a mother as *causing* specific behaviors of the infant. Rather, such episodes are part of a larger flow of interchange and feedback between a mother and her infant, and they have relatively little significance in isolation. This is not to minimize the importance of the sequential nature of the interchange, but rather to indicate the limitations of a stimulus–response interpretation.

Whether one measures parameters of the behaviors of both mother and infant or of either individually, it is necessary to assume that one is making a systems statement. Where the focus is primarily on the infant, the relationship, as context, much in some way be taken into account.

NATURALISTIC STUDY OF INFANTS

An important characteristic of systems is their hierarchical organization, with functioning at each level in the hierarchy being the result of systems dynamics. In order to understand the dynamics at any level, it is important to have some understanding of the level above and below the one of interest. From this perspective, the infant can be viewed as a subsystem of the mother–infant relationship. The social milieu constitutes the level above that of the infant; whereas biological factors constitute the level below. The infant's behaviors are an emergent process from the biological dynamics, and the infant's social exchange is an emergent process from the dynamics of the infant behaviors in interaction with those of the mother. It is impossible to think of an infant's behaviors occurring independent of the higher and lower levels in the hierarchy.

Thus it is impossible to study the infant in vacuo; the infant is always studied in some setting selected in accordance with the objectives of one's research. And the conditions for observation, the context, will affect the nature of the behaviors observed. Likewise, the infant's changing endogenous status will affect any behaviors observed.

For some purposes, where an isolated process, such as perception or learning, is of interest, a common context is required to control as many extraneous variables as possible. In these circumstances, the laboratory setting is most appropriate. And inferences are drawn with the limitations of these conditions taken into consideration; that is, it is possible to determine what an infant subject *can* do from observations in the laboratory. However, laboratory research does not provide information on what the infant *does* do in the context of his or her real world. Thus, for questions of organismic functioning aimed at understanding principles of adaptation, and factors that may affect adaptative behaviors as part of the developmental process, general systems notions point to the natural, or typical, setting for the occurrence of the behaviors of interest as the appropriate context for such study. That is, observing the infant in the natural setting, as a component of a dyadic system, provides the kind of data needed for generalizations about the infant as an organism designed biologically to adapt to a social milieu.

Although we argue for the importance of observing infants in their species-appropriate environment, specific research objectives may dictate the nature and extent of the natural circumstances selected; for example, during specific types of interactions with the mother, when the infants are in their crib and alone, or, as in the case of our studies, during the full range of circumstances that occur in the typical home. Clearly, the natural setting provides the context that is ecologically valid (Gibbs, 1979) for investigating the behaviors of an infant as a functioning systems component.

In the section that follows, I will illustrate these points from analyses of behaviors of infants during the early weeks of life. More specifically, I will focus on the temporal organization of their behavioral states, as this characteristic of infants expresses the infant's endogenous status and, through interactive processes, reflects the ongoing flow of environmental circumstances.

NATURALISTIC STUDY OF THE STATES OF INFANTS

Any serious description of an infant's behavior must take into account the infant's state and the situation in which the behavior occurs, as already described. *State* is an ubiquitous behavioral characteristic throughout life and across species, expressing the functioning of the central nervous system and reflecting the ongoing interchange, primarily of a social nature, with the environment. The infant's state is both mediator and modulator of other behaviors, and thus acts as interface between the environment and central neural processes. The functions of states are multiple. In the social interchange, for example, the infant's state may serve as a cue to the mother as to the timing and nature of her intervention (crying as a state behavior may elicit picking up and holding the baby; and even if it does not elicit an overt behavior, the crying communicates to the mother—few mothers can ignore such a message), whereas at the same time state serves to mediate the impact of her ministrations (e.g., a baby perceives and responds to stimuli differently when alert and when crying), and a change in state may accompany or follow her ministrations. Clearly, an infant's states are an integral part of the communication system, or "the dance" (Gunther, 1961), of any mother and infant from the time of the infant's birth (Thoman, 1979, 1981; Thoman, Freese, & Becker, 1977).

We have devoted more than a decade to the Connecticut Longitudinal Project, a naturalistic study of mothers and infants from the time of the infant's birth. Because sleeping and waking states are ubiquitous, a major objective of this research is to investigate the state process as an index of central nervous system (CNS) status of the infant, as an indicator of behavioral organization in the infant, and as a functional behavior in the mother–infant relationship.

A brief description of our procedures for observing infants during early infancy will be useful at this point as background for data that will be presented. Infants were observed in the home in circumstances that were as natural as possible; there were no interventions or manipulations of the home circumstances for the purposes of the study. The observations to be described were made when the infants were 2, 3, 4, and 5 weeks old, with a 7-hour observation made once a week at each of these ages. Throughout each observation, behaviors of both mother and baby were recorded every 10 seconds. The categories of behaviors recorded include the proximity of mother and infant, position and location of the infant, and forms of attention to each other, including looking, touching, and vocalizing. Throughout the day, in addition to the other behaviors recorded, the infant's states of sleep and wakefulness were recorded. The state categories included are as follows: alert, nonalert wakefulness, fussing, crying, drowse, daze, sleep–wake transition, active sleep, quiet sleep, and active–quiet transition sleep (Thoman, 1975, 1982).

Two observers were involved in each observation: one recorded for 3 1/2 hours, and when the next observer arrived, both observers recorded for a 15-minute period (for ongoing assessment of reliability). The first observer then left the home, and the second observer recorded for another 3 1/2 hours. Six months' training is required for an observer to achieve reliability for these recordings, to be able to sustain the required focused attention, and to acquire the skills associated with observing and recording unobtrusively in the home.

It is important to note that despite the fact that a number of observers were involved in the observations of each mother–infant pair, and despite the fact that the circumstances of the observations were widely variable across weeks and over subjects, as might be expected in any home situation, we have found that all of the measures of the states of infants and most measures of the behaviors of the mothers and infants, show significant and individual differences over the four weekly observations (Thoman, 1982; Thoman, Becker & Freese, 1978; Thoman, Acebo, Dreyer, Becker, & Freese, 1979).

From the statistic for individual differences, it is possible to calculate reliability of measurement for each of these behaviors. Thus the behaviors recorded reliably describe individual infants. Assessment of reliability of measurement of variables used for description or prediction is rarely a part of research using naturalistic observations. However, as a prerequisite for analysis of data within individuals, its importance cannot be overestimated.

Temporal Structure of the States of Infants: Sequential Analysis

A major approach to characterizing infants' state organization has been to examine the data sequentially for evidence of periodicity, and numerous studies have reported periodicity, or rhythmicity, in the sleep states of infants. However Kraemer, Hole, and Anders (1984) point out that these findings have generally not been the result of statistical tests for rhythmicity; that is, consistent definitions and rigorous methods for state time series have not been used. The usual statistic has been the mean amount of time between recurrence of episodes of a state. However, it is typical that the variability is greatly relative to the mean recurrence time; in such a case, the mean is not very meaningful. Simple recurrence of a behavior does not necessarily mean that it occurs in a cyclic manner. "Behavioral scientists . . . have used time series terminology loosely to describe any seemingly recurrent pattern of behavioral states" (p. 4). This criticism applies equally to studies of other rhythms in infant behaviors or mother–infant interaction.

Notable exceptions to the general claim for rhythmicity in infant state are found in the studies of Harper and collaborators (Harper *et al.*,

1981; Harper, Frostig, & Taube, 1983). They applied Fourier analysis to state sequences of sleeping infants, and conclude that "the newborn seems to be characterized by rather irregular state patterning." They find that "the organization of states into a stable sequence requires 3 to 4 months of developmental time" (Harper *et al.*, 1981). Thus when rigorous tests have been used, evidence for rhythms during the early weeks of life have not been found.

We propose that there may be a form of temporal organization that precedes developmentally the rhythmic stability seen in the older infants.

Temporal Structure of the States of Infants: Nonsequential Analysis

Eugene Yates (personal communication) has suggested that there should be systems constraints on the *amount* of time devoted to any behavior before that behavior can occur with regular rhythm. Indication of this earlier form of constraint, or organization, is seen in our finding of significant individual differences among infants in the amount of time allocated to each of the states over the early weeks of life. This is apparent in the data presented in Table 1, which shows the mean amount of time spent in each state over Weeks 2 through 5 for the first 20 subjects in the Longitudinal Project, and the reliability coefficient for each state. Reliability is calculated from the F value for individual differences (using the ANOVA) as: $r_{tt} = (1 - 1/F)$. This consistency in the amount of time spent in each state from week to week represents a major form of temporal constraint within individual infants. Such ordering is independent of any possible evidence for periodicity in the occurrence of these states.

The lack of periodicity in the states of infants during this early period is apparent in a description of their sequential distribution. Figure 1 presents such data for one infant. It shows the occurrences of episodes

Table 1. Mean Percent of the Time the Baby Was
Alone Spent in Each State on Weeks 2, 3, 4, and 5,
and Reliability Coefficient for Each State (N = 28)

	Mean	r_{tt}
Alert	6.7	.71***
Nonalert waking activity	2.8	.62***
Fuss or cry	3.4	.69***
Drowse of sleep-wake transition	6.7	.56**
Active sleep	50.3	.69***
Quiet sleep	28.1	.64***
Active-quiet transition sleep	1.9	.60**

$**p < .01.$ $***p < .001.$

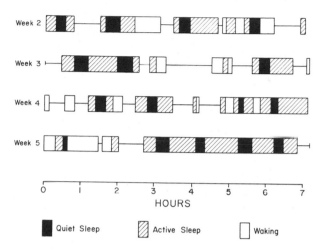

Figure 1.   For the periods of the day when the infant was alone, sequential occurrence of wakefulness, active sleep and quiet sleep periods throughout the 7-hour day, for the four successive weekly observations. Dashed portions of the figure indicate periods of the day when the infant was with the mother.

of waking, active sleep, and quiet sleep over the 7-hour day, for each of the four weekly observations (Weeks 2 to 5). Periods of the day when the baby was with the mother are indicated, but the states are distinguished only for the periods when the baby was alone. On the average, the infants in our study spent 67% of the day alone.

Several aspects of these data should be noted, namely that (a) there is variability over weeks in the patterning of periods of the day when the baby was with the mother and when the baby was alone; (b) there is recurrence of each of the states throughout the day, but their distribution is too variable to be indicative of cyclicity; and (c) there is a great deal of variability in the duration of episodes of each state within each weekly observation and across observations. Thus, despite the large amount of data recorded during each observation for this baby, these several sources of variability make it obvious that sequential analysis of the data would not yield systematic findings. And, in fact, sequential analyses did not reveal evidence for periodicity in this or any of the other subjects in the study. These findings are consistent with those of Harper et al. (1981) and Harper et al. (1983).

Figure 2 presents the sequential pattern of state periods in another infant, in order to demonstrate that the same forms of variability both within the day and over weeks are present in other infants. The general patterning of states is not too dissimilar in these two infants, and they

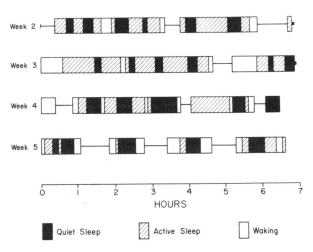

Figure 2. For the periods of the day when the infant was alone, sequential occurrence of wakefulness, active sleep and quiet sleep periods throughout the 7-hour day, for the four successive weekly observations. Dashed portions of the figure indicate periods of the day when the infant was with the mother.

do not differ dramatically from any of the other subjects. For example, active and quiet sleep recur throughout sleep, and active sleep typically precedes both quiet sleep and wakefulness. It is reasonable to conclude, then, that the findings of individual consistency over weeks in the allocation of time to each state indicate temporal controls that are not a function of either periodicity in the occurrence or regularity in the duration of any state.

Stability of State Organization over Weeks

These findings were extended by examining the consistency of overall patterning of the states in individual infants. Our objective was to examine how infants may distribute the states throughout the day, and the extent to which this overall allocation of time for the infant was consistent from week to week. For this purpose, the total amount of time spent in each state was calculated and a profile of these measures was derived for each week (for an individual infant). With four such state profiles for the infant, one for each week, it was possible to use analysis of variance to assess the similarity of these profiles—the consistency of state patterning—over the four weeks. The ANOV yields a single F-ratio, which we have called a state stability score (Thoman, Denenberg, Sievel, Zeidner, & Becker, 1981) for the infant. This statistic is used descriptively to indicate the consistency of patterning of an infant's states over weeks, not as a test for significance.

The validity of this stability statistic has been clearly demonstrated: the state stability scores for 22 full-term, normal infants ranged from 3.09 to 304.86; and the scores were found to be related to developmental outcome.

To illustrate the stability of profiles that can be found in some infants, Figure 3 shows the four weekly profiles for four subjects, based on the periods of the day when the infants were alone. The dramatic consistency in the profiles of these infants is obvious. It should be emphasized that such consistency was present despite the fact that (a) the total amount of time each baby spent alone varied from week to week, and (b) the distribution of the periods of the day when the baby was alone varied from week to week. These forms of variability are seen in the data of the infants presented in Figures 1 and 2. It is apparent that there are strong constraints on the allocation of time to the sleep and waking states from

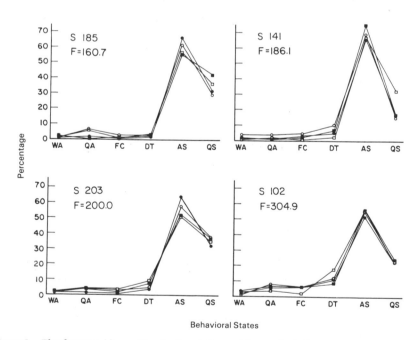

Behavioral States

Figure 3. The four weekly profiles for four infant subjects. The behavioral states are along the X-axis: WA = Nonalert waking activity; QA = quiet alert; FC = fuss or cry; DT = drowse or transition; AS = active sleep; QS = quiet sleep. The percentage of time allocated to each state while the baby was alone is shown along the Y-axis. Week 2 (●——●), Week 3 (■——■), Week 4 (○ ——○), Week 5 (□——□).

week to week, at least in these infants. They were the four most stable infants in the study.

The profiles of four other infants are shown in Figure 4. Although some consistency in state patterning is seen in these infants, these were the four most unstable subjects in our study. The consistency among state profiles of the other infants in the study ranged between the two extremes shown in Figures 3 and 4, and their state stability scores reflected this continuum of variations.

The state stability data described so far were based on the periods of the day when the infants were alone. The same state stability statistic can be derived for the total portion of the day when an infant is with the mother. During these periods the infant may be engaged in a variety of activities, and subject to a variety of forms of stimulation; and during these periods the states are distributed very differently as well. When infants are alone, they may sleep much of the time (over 80% of the day in our subjects); and when they are with the mother, for feeding, changing, social interaction, and the like, they are much more likely to be awake (over 70% of the time in our subjects). Despite the differences in states

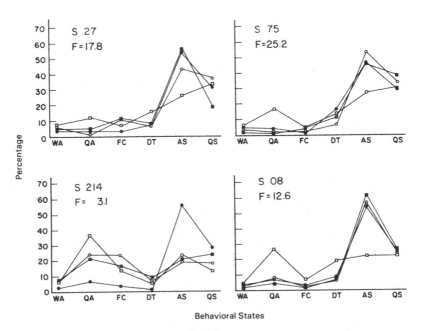

Figure 4. The four weekly profiles for four infant subjects. The behavioral states are along the X-axis, and percentage of time allocated to each state is along the Y-axis. Week 2 (●——●), Week 3 (■——■), Week 4 (O——O), Week 5 (□——□).

and marked differences in circumstances, we found a correlation of $r = 0.67$ between the state stability scores obtained when the infants were alone and the stability scores when they were with the mother. Thus, there is an overall integration of the controls for the temporal distribution of the states of sleep and wakefulness throughout the day and over the early weeks of life, an integration that is not obscured by varying environmental circumstances. Clearly there is temporal organization of the states of infants during the early weeks of life.

Implications of State Stability for the Development of Rhythmicity in States

The significance of stability in the temporal organization of behavioral states at this early age is indicated by the developmental outcome of the infants in our study (Thoman *et al.*, 1981). Each of the infants with extremely low state stability scores showed developmental disability: one died of SIDS at 3 months; one showed infantile seizures and hypsarrythmia at 7 months and subsequently was severely retarded; and one was hyperactive from a very early age, and this condition has persisted to the present time. All of the children who showed any form of developmental dysfunction at a later age had stability scores below the median for the group. Thus, a low state stability score was indicative of subtle neurological dysfunction.

There was no correlate of state stability to be found in the sequential distribution of states throughout the day. The point is highlighted by the characteristics of the infants whose data are presented in Figures 1 and 2. There is no apparent distinction between the two infants in the sequential data that would suggest greater or less organization. Yet, the infant whose sequential data are shown in Figure 1 had the highest stability score in the group of subjects (see profile for Baby BA in Figure 3); and the subject described in Figure 2 had the second lowest stability score among the 22 subjects (see Baby PL in Figure 4). The latter subject was the one who died of sudden crib death.

Clearly, it is possible to have a nonsequential kind of temporal order without apparent sequential, or rhythmic, order in the behavioral states. We would argue that this nonsequential form is the prelude developmentally to the rhythmicity in sleep and wakefulness seen in older infants and children. That is, at the earliest age, constraints on the amount of time allocated to each state is clearly expressed, but with variations both in duration of episodes of the state and in recurrence periods. With maturation of the central nervous system and entrainment from the environment, the constancy and amount of time allocated to each state changes within an hierarchical organization. Sleep periods are increasingly consolidated and are, therefore, of longer and more constant duration; whereas the categories of sleep, active and quiet sleep, are constrained in their time allocation over briefer periods of time. The developmental result is a diurnal cycle of sleep and wakefulness and, within sleep, rhythmic oscillation of the active and quiet sleep states.

This postulation for the process of emerging rhythmicity differs markedly from the notion of maturation of an "oscillator" in the brain. Whereas there may be oscillator controllers for some biological rhythms, general systems theory accommodates the notion of oscillation without the requisite of a specific oscillator. That is, in GST, oscillation is an emergent process from systems interactions that serves to maintain a dynamic equilibrium within the system. Thus, we propose that the form of temporal order measured by the state stability statistic serves as the prelude developmentally to the sequential order seen in later rhythmicity. Consistent with this proposal is the better developmental status of those infants in our study who showed greater state stability during the early weeks of life. It is, therefore, reasonable to expect that those infants with greater state stability would also develop rhythmicity in their sleep–wake states at an earlier age than the less stable infants.

DISCUSSION

A number of conceptual threads have been utilized to develop the theme that temporal order in recurring behaviors may be found in patterns other than the simple expression of sequential rhythms. That is, it has been argued, and demonstrated with data from a longitudinal project, that infants can show highly consistent patterns of temporal distribution of their states over successive observations without any evidence for periodicity in the recurrence of any of the individual states. Periodicity occurs at a later age developmentally than the kind of consistency we have found in the allocation of time among the states. We propose, therefore, that temporal stability expressed in consistency of state distributions is a developmental antecedent to temporal stability expressed in their rhythmic occurrence. This proposal derives from the general systems perspective, and assumes that the temporal order of states emerges as a function of the cooperativeness of systems components. It also assumes very widespread involvement of the central nervous sytem in the controls for state. Such a view offers an alternative to the generally accepted notion of a localized central controller for rhythmicity in sleep states and in sleep and wakefulness. Much more developmental study of the states of infants is required to link the nonsequential ordering to the more mature rhythmicity seen in the older individual. Clearly, the developing infant is the ideal subject for exploring the successive stages of biobehavioral controls in the temporal domain, in states as well as other recurring behaviors.

REFERENCES

Abraham, R. (1984). *Brain/Mind Bulletin, 9,* (2).

Bertalanffy, L. von. (1933). *Modern theories of development: An introduction to theoretical biology* (J. H. Woodger, Trans.). London: Oxford University Press.

Bertalanffy, L. von. (1968). *General system theory* (rev. ed.). New York: Brazilier.

Bunge, M. (1979). *Treatise on basic philosophy: Vol. 4. Ontology II: A world of systems.* Dordrecht: Reidel.

Denenberg, V. H. (1979). Paradigms and paradoxes in the study of behavioral development. In E. B. Thoman (Ed.), *The origins of the infant's social responsiveness* (pp. 251–290). Hillsdale, NJ: Erlbaum.

Denenberg, V. H. (1980). General systems theory, brain organization, and early experiences. *American Journal of Physiology 238 (Regulatory Integrative and Comparative Physiology)* 7, R3–R13.

Denenberg, V. H. (1981). Hemispheric laterality in animals and the effects of early experiences. *Behavioral and Brain Sciences, 4,* 1–49.

Denenberg, V. H. (1982). Comparative psychology and single-subject research. In Alan E. Kazdin & A. Hussain Tuma (Eds.), *Single-Case Research Designs* (pp. 19–21). San Francisco: Jossey-Bass.

Fowler, C. A. (1975). A systems approach to the cerebral hemispheres. *Status Report on Speech Research.* New Haven, CT: Haskins Laboratories.

Galanter, E. (1984). Timing of motor programs and temporal patterns: Discussion paper. In J. Gibbon & L. Allan (Eds.), Timing and time perception. *Annals of the New York Academy of Sciences, 423,* 193–197.

Gibbs, J. C. (1979). The meaning of ecologically oriented inquiry in contemporary psychology. *American Psychologist, 34,* 127–140.

Gunther, M. (1961). Infant behaviour at the breast. In B. M. Foss (Ed.), *Determinants of Infant Behaviour* (pp. 37–44). London: Methuen.

Harper, R. M., Leake, B., Miyahara, L., Mason, J., Hoppenbrouwers, T., Sterman, M. B., & Hodgman, J. (1981). Temporal sequencing in sleep and waking states during the first 6 months of life. *Experimental Neurology, 72,* 294–307.

Harper, R. M., Frostig, Z., & Taube, D. (1983). Infant sleep development. In Andrew Mayes (Ed.), *Sleep mechanisms and functions in humans and animals—An evolutionary perspective* (pp. 107–125). New York: Van Nostrand Reinhold.

Iberall, A. S., & McCulloch, W. S. (1969, June). The organizing principle of complex living systems. *Journal of Basic Engineering,* pp. 290–294.

Kraemer, H. C., Hole, W. T., & Anders, T. F. (1984). The detection of behavioral state cycles and classification of temporal structure in behavioral states. *Sleep, 7* (1), 3–17.

Thoman, E. B. (1975). Sleep and wake behaviors in neonates: Consistencies and consequences. *Merrill-Palmer Quarterly, 21,* 295–314.

Thoman, E. B. (1979). CNS dysfunction and nonverbal communication between mother and infant. In C. L. Ludlow & M. E. Dorqan-Wuine (Eds.), *The neurological bases of language disorders in children: Methods and directions for research* (NINCDS Monograph, pp. 43–54). Washington, DC: U.S. Government Printing Office.

Thoman, E. B. (1981). Early communication as the prelude to later adaptive behaviors. In M. J. Begab, H. C. Haywood, & H. Garber (Eds.), *Psychosocial influences in retarded performance: Vol. II. Strategies for improving competence* (pp. 219–244). Baltimore, MD: University Park Press.

Thoman, E. B. (1982) A biological perspective and a behavioral model for assessment of premature infants. In L. A. Bond & J. M. Joffe (Eds.), *Primary prevention of psychopathology: Vol. 6. Facilitating infant and early childhood development* (pp. 159–179). Hanover, NH: University Press of New England.

Thoman, E. B., Denenberg, V. H., Becker, P. T., & Freese, M. P. (1974). Analysis of mother–infant interaction sequences: A model for relating mother–infant interaction to the infant's development of behavioral states. *Maternal Infant Life Conferences.* Madison, WI: Wisconsin Perinatal Center.

Thoman, E. B., Becker, P. T., & Freese, M. P. (1978). Individual patterns of mother–infant interaction. In Gene P. Sackett (Ed.), *Observing Behavior: Vol. 1. Theory and applications in mental retardation* (pp. 95–114). Baltimore: University Park Press.

Thoman, E. B., Acebo, C., Dreyer, C. A., Becker, P. T., & Freese, M. P. (1979). Individuality in the interactive process. In E. B. Thoman (Ed.), *Origins of the Infant's Social Responsiveness* (pp. 305–338). Hillsdale, NJ: Erlbaum.

Thoman, E. B., Denenber, V. H., Sievel, J., Zeidner, L., & Becker, P. T. (1981). State organization in neonates: Developmental inconsistency indicates risk for developmental dysfunction. *Neuropediatrics, 12,* 45–54.

Walker, H. M., & Lev, J. (1953), *Statistical Inference.* New York: Henry Holt.

Weiss, P. (1969). The living system: Determinism stratified. In A. Koestler & J. R. Smythies (Eds.), *Beyond reductionism* (pp. 3–42). Boston, MA: Beacon Press.

Weiss, P. (1971). The basic concept of hierarchical systems. In P. Weiss (Ed.), *Hierarchically organized systems in theory and practice* (pp. 1–43). New York: Hafne.

Wohlwill, J. F. (1973). *The study of behavioral development.* New York: Academic Press.

CHAPTER 8

Ordinal Pattern Analysis
A Strategy for Assessing Hypotheses about Individuals

WARREN THORNGATE and BARBARA CARROLL

INTRODUCTION

Our task in writing this chapter is to outline a strategy for analyzing data that we believe to be better suited to most psychological research than the most widely used statistical techniques (e.g., t, F, and chi square tests, product moment correlation, regression, covariance, discriminant, and factor analyses). We call the strategy *Ordinal Pattern Analysis* (OPA), and derive it from a small set of first principles that deviate somewhat from those on which most classical statistical models are based. First, we assume that the goal of statistical practice is to aid in detecting and analyzing patterns in data rather than to aid in making decisions about populations given samples of data. Second, we assume that a statistic must be useful in analyzing data generated by individual subjects as well as data based on aggregations of subjects. Third, we assume that most predictions and observations in psychological research possess no more than ordinal scale properties, and that statistics employed to assess the fit between predictions and observations must be derived on the basis of this constraint.

Classical statistical tests are commonly invoked in research to answer

WARREN THORNGATE and BARBARA CARROLL • Department of Psychology, Carleton University, Ottawa, Ontario, Canada K1S 5B6.

a deceptively simple inductive question: To what extent can the data obtained from a sample of observations be generalized to a population of observations? There are at least two major problems with this question. First, in studies of individuals it usually cannot be answered. Second, in studies of individuals it usually need not be asked. When there is only one observation in the sample, or when several observations are not statistically independent, the use of these tests becomes troublesome; sampling distributions cannot be easily estimated, and the inductive rules based on these estimations cannot be followed. To deal with this problem, psychologists have usually avoided research based on single subjects or on multiple observations with unknown dependencies. Research designs have thus almost always employed multiple subjects, and analyses of results have almost always considered only aggregated data (usually averages, proportions, and variances). Studies based on observations of single subjects are usually labeled with the pejorative term *case history* and are dismissed as anecdotal and unscientific. Yet to avoid or dismiss single-subject research because of the difficulties it causes currently popular inductive engines is to allow statistical tails to wag scientific dogs. Statistical technique is supposed to be a means to an end, not a criterion for deciding which questions of phenomena are legitimate, not a Procrustean Bed onto which all psychological research should be fitted.

The inductive question that classical statistics attempt to answer is itself of little scientific merit (Edwards, 1972; Savage, 1954). Few good scientists trust the results of statistical analyses alone. Instead, some analyses will lead to attempts at empirical replication, others will lead to refinements of methods or design, and almost none will stand as the last word about some ill-defined population from which research samples came (Meehl, 1978). The reason is simple. The central question of scientific research has almost nothing to do with the extent to which one can generalize from some observations to others. Rather, it concerns the extent to which one can generalize theoretical predictions to empirical observations. Scientific method is supposed to be a means for determining how often, and under what circumstances (if any) a good fit exists between predictions and observations, between what is expected and what is obtained. Statistical methods must serve this scientific function.

Statistical methods for testing goodness of fit between predictions and observations have existed for some time, as have debates about their logic and utility (e.g., see Bush, 1963; Restle & Greeno, 1970). They are often used to test the predictions of mathematical models. But most models or theories in psychology are not mathematical and can generate only ordinal rather than point predictions of research outcomes. In addition, most measurements or indicants of psychological phenomena (e.g., stress, motivation, aggression, learning) have only ordinal relationships to the phenomena themselves. The extant goodness of fit tests require at least interval data to test point predictions, and thus cannot be used with confidence in most psychological research endeavors.

What psychology needs is a useful set of ordinal goodness of fit tests, tests that will aid in assessing the correspondence between predicted and observed (measured) ordinal patterns. Accordingly, we attempt to develop a small set of such tests below. The tests in the set are by no means exhaustive of all that may be developed and used. But they should at least serve to illustrate the style of statistical thinking that we believe is better suited to psychological research than its classical counterpart. Our plan is to define a handful of indexes that we believe are useful for assessing ordinal goodness of fit, and to demonstrate the use of each in a few research situations that are cumbersome for conventional analyses.

ORDINAL PATTERN ANALYSIS

The most important indexes of OPA are those that tell us something about how well the ordinal patterns observed in a given set of data match, or are matched by, the ordinal patterns predicted by one or more hypotheses. For example, in the simplest case, some hypothesis, X, may predict that Alice will do better than Fred on a given exam. If Alice does get a higher grade on the exam than does Fred, then X should be given credit for the correct prediction. If Fred gets a higher grade than Alice, then X should be credited with an incorrect prediction. If X can make correct predictions, say, 90% of the time, and some other hypothesis, Y, can make correct predictions only 30% of the time, then we should come to believe that X is a more valid hypothesis than Y.

Other indexes of OPA tell us something about the scope of a hypothesis in relation to a given set of data. By *scope* we mean the amount of data in the given set for which a given hypothesis makes ordinal predictions. Some hypotheses may have wide scope; they may generate predictions about all or almost all of the ordinal relationships in the data. Other hypotheses may have very narrow scope; they may address themselves only to a very small number of ordinal relationships in the data. Thus, if hypothesis X made predictions about the ordinal relations of exam scores of Fred, Alice, John, Marsha, Sam, and Lisa, whereas hypothesis Y only made predictions about the scores of Fred, Alice, and John, then X would have greater scope than Y. Alas, there is almost always a tradeoff between scope and accuracy—as one goes up the other goes down (Thorngate, 1976). Preferences for hypotheses that predict almost nothing about almost everything versus hypotheses that predict almost everything about almost nothing are matters of debate, and taste. OPA does not provide a preferential rule, but it does provide indexes relevant to establishing a preference.

A third set of indexes of OPA concerns the extent of overlap in the predictions of alternative hypotheses. Very different hypotheses often generate the same predictions about a given set of data, so one cannot establish any different measure of scope or fit. Elegant research is designed to minimize such overlap by judicious selection of experimental conditions and controls. Yet important research, especially that conducted outside the laboratory, cannot always be elegant in this way. By analyzing the ordinal scope and fit of common and unique predictions separately, OPA can aid in determining the relative merits of alternative but partially overlapping hypotheses.

Let us now consider the development of the indexes just described. To begin we define a set of observations we wish to compare with predictions:

$$Ob = \{ob_1, ob_2, \ldots, ob_n\} \tag{1}$$

To make the observations commensurable, we measure each on some common dimension so that for all ob_i and ob_j, in Ob, their measures, mob_i and mob_j, will satisfy one of the following conditions:

$$mob_i > mob_j \text{ or}$$

$$mob_i < mob_j \text{ or}$$

$$mob_i = mob_j$$

Any measure that yields numbers with weakly ordered properties will do the trick. Such measures are typically those of dependent or criterion variables.

We next consider all possible pairs of observations in Ob (ob_i, ob_j), such that

$$i > j, \text{ and} \tag{2}$$

$$mob_i > mob_j \text{ or} \tag{3a}$$

$$mob_i < mob_j \tag{3b}$$

Condition 2 requires that, in matrix form, we consider only those pairs above the main diagonal. Conditions 3a and 3b require that we consider only those pairs with measures having a definite ordinal relationship, excluding pairs with measures of equal value $(mob_i = mob_j)$. The pairs of observations satisfying Condition 2 and one of Conditions 3a or 3b form a set we call the *measured ordinal domain* or *MOD*. When there are no "ties" in our measures, the number of elements in *MOD* will be

$$\#MOD = n!/(n - 2)!2! = n(n - 1)/2$$

In contrast, when all measures are tied (i.e., all the same), then

$$\#MOD = 0$$

Thus,

$$0 \leqslant \#MOD \leqslant n(n - 1)/2$$

Now we define another set that includes the observed order relation ($>$ or $<$) of each element (pair) in *MOD*. For convenience, we put the appropriate $>$ or $<$ symbols where the commas are in each pair in *MOD*. The result is a set we call the *measured ordinal pattern 1 or MOP*. The number of its elements is *#MOP* ($= \#MOD$). One such *MOP* might appear as follows:

$$MOP = \{(ob_1 < ob_2) \ (ob_1 > ob_4), \ldots, (ob_i < ob_j), \ldots, (ob_{n-1} > ob_n)\}$$

Example

Suppose we observe a student complete five tests (T1, T2, T3, T4, and T5) in a course and note the five scores (measures) obtained: mT1 = 57%, mT2 = 68%, mT3 = 63%, mT4 = 72%, and mT5 = 68%. Then,

$$Ob = \{T1,T2,T3,T4,T5\}$$
$$MOD = \{(T1,T2)(T1,T3)(T1,T4)(T1,T5)$$
$$(T2,T3)(T2,T4)$$
$$(T3,T4)(T3,T5)$$
$$(T4,T5)\}$$

$$\#MOD = 9$$
$$MOP = \{(T1{<}T2)(T1{<}T3)(T1{<}T4)(T1{<}T5)$$
$$(T2{>}T3)(T2{<}T4)$$
$$(T3{<}T4)(T3{<}T5)$$
$$(T4{>}T5)$$

$$\#MOP = 9$$

DERIVATION OF OPA INDEXES

Let us now turn our attention to predictions of ordinal patterns generated by given hypotheses. We here define a hypothesis as a collection of one or more rules that assigns to each element of some subset of *MOD* a predicted ordinal relation: $>$ or $<$. We call the subset of *MOD* for which a hypothesis, h, makes ordinal predictions the *predicted ordinal domain* of the hypothesis: POD_h. Sometimes the *POD* of a hypothesis will contain all elements of *MOD*. In the previous example, this would occur for a hypothesis such as:

H_1: "Test performance will increase with number of tests taken"

because the hypothesis can generate ordinal predictions about the test scores of all test pairs in *MOD*, that is,

$$POD_1 = MOD$$

Often, however, the *POD* of a hypothesis will be a proper subset of *MOD*. For example, the hypothesis

H_2: "Students will do better on the third exam than on any others"

only generates the following *POD*,

$$POD_2 = \{(T1,T3)\ (T2,T3)\ (T3,T4)\ (T3,T5)\}$$

because H_2 says nothing about the ordinal relations of test scores 1, 2, 4, and 5. Similarly, the hypothesis

H_3: "Students will do better on the last test than on the first test"

generates a very restricted *POD*,

$$POD_3 = \{(T1,T5)\}$$

SCOPE

It will be useful to have an index of the relative size or extent of the predictive domains of hypotheses. One such index is easily constructed. We define the *scope*, *Sc*, of a hypothesis, h, as the number of elements in *MOD* that are included in *POD*, divided by the number of elements in *MOD*. Thus,

$$Sc_h = \#(MOD \text{ and } POD)/\#MOD \qquad (4)$$

In the example above,

$$Sc_1 = 9/9 = 1.00,$$

$$Sc_2 = 4/9 = 0.44, \text{ and}$$

$$Sc_3 = 1/9 = 0.11$$

We shall not discuss the scope of a hypothesis in any great detail here; it uses should become apparent as we proceed. Instead, we shall consider some measures of goodness of fit or, as we have called them, indexes of obtained and predictive fit.

A hypothesis, by definition, assigns an ordinal prediction to each observation pair in its predicted ordinal domain. The result is a set of ordered triples, the first and third elements of each triple represent observation pairs, and the middle element represents the predicted ordinal relation, r_{ij}, between them. We call this set of predictions the *predicted ordinal pattern, POP*:

$$POP_h = \{(ob_i, r_{ij}, ob_j)\} \text{ for all } (ob_i, ob_j) \text{ in } POD_h$$

Again, by way of example:

$$POP_1 = \{(T1 < T2)\ (T1 < T3)\ (T1 < T4)\ (T1 < T5)$$
$$(T2 < T3)\ (T2 < T4)$$
$$(T3 < T4)\ (T3 < T5)$$
$$(T4 < T5)$$
$$POP_2 = \{(T1 < T3)\ (T2 < T3)\ (T3 > T4)\ (T3 > T5)\}$$
$$POP_3 = \{(T1 < T5)\}$$

For any given hypothesis we now ask two important questions: How well is its predicted ordinal pattern, POP_h, matched by the actual ordinal pattern, *MOP?* How well is the actual ordinal pattern, *MOP*, predicted by the hypothesized pattern, POP_h? Answers to the two questions will be the same whenever the scope of the hypothesis $= 1.00$. Otherwise, some elements in *MOD* will be missing from POD_h, and the two answers will not be equal.

INDEX OF OBSERVED FIT

The first question asks how often observations match predictions generated by some hypothesis, h. In order to answer this question, we begin by counting the number of elements (i.e., the number of triples) in POP_h: $\#POP_h$. Next, we count the number of elements common to POP_h and *MOP*, that is, the number of elements in POP_h that are matched by elements in *MOP*: $\#(POP_h$ and *MOP*). We then subtract from this number the number of elements in POP_h that are not matched by elements in *MOP*: $\#(POP_h$ and not *MOP*). Now we define an *index of observed fit* of *MOP* to POP_h as:

$$IOF_h = (\#(POP_h \text{ and } MOP) - \#(POP_h \text{ and not } MOP))/\#POP_h \quad (5)$$

To continue with our examples:

$\#POP_1 = 9;$

$(POP_1 \text{ and } MOP) = \{(T1 < T2) (T1 < T3) (T1 < T4) (T1 < T5)$

$\qquad\qquad\qquad\qquad (T2 < T4) (T3 < T4) (T3 < T5)\}$

$\#(POP_1 \text{ and } MOP) = 7$

$(POP_1 \text{ and not } MOP) = \{(T2 < T3) (T4 < T5)\}$

$\#(POP_1 \text{ and not } MOP) = 2;$

$IOF_1 = (7 - 2)/9 = 5/9 = +0.56$

$\#POP_2 = 4$

$(POP_2 \text{ and } MOP) = \{(T1 < T3)\}$

$\#(POP_2 \text{ and } MOP) = 1$

$(POP_2 \text{ and not } MOP) = \{(T2 < T3) (T3 > T4) (T3 > T5)\}$

$\#(POP_2 \text{ and not } MOP) = 3$

$IOF_2 = (1 - 3)/4 = -2/4 = -0.50$

$\#POP_3 = 1$

$(POP_3 \text{ and } MOP) = \{(T1 < T5)\}$

$\#(POP_3 \text{ and } MOP) = 1$

$(POP_3 \text{ and not } MOP) = \{ \}$

$\#(POP_3 \text{ and not } MOP) = 0$

$IOF_3 = (1 - 0)/1 = 1/1 = + 1.00$

More generally, the index of observed fit will take on values between $+1.00$ and -1.00. When all the predictions of a hypothesis are matched by the measured relations, then a perfect fit is obtained and $IOF = 1.00$. When none of the predictions of a hypothesis are matched by the measured relations, then a perfect misfit is obtained and $IOF = -1.00$. When half the predictions are matched by the measurements, then $IOF = 0.00$. In this way, IOF is similar to standard correlation coefficients, particularly Kendall's *Tau* (Kendall, 1938). Indeed, when there are no ties in the measures, and when all pairs in *MOD* are found in *POD* (i.e., $Sc = 1.00$), then the *IOF* formula (5) is equivalent to the formula for Kendall's *Tau*. Of course, we are here co-relating predictions and measures rather than measures and measures. *Tau*, to our knowledge has never been used for the former purpose, although it obviously can be.

A word must be said about our exclusion of ties (equal measurements) from the *IOF* formula. The subject of ties has been a great source of debate in rank–order statistics, because the formulas for estimating the sampling distributions of these statistics are based on assumptions of continuous measurement distributions, that is, distributions that should never result in measurements being exactly equal. When ties do occur, it is common to add elaborate fudge factors to the original formulas in order to preserve their approximating characteristics (e.g., see Bradley, 1968, pp. 48–54). We have chosen to exclude them for two reasons. First, ties are unaesthetic; they add nothing to the meaning or interpretation of *IOF*, and clutter the formula. Second, we shall be estimating the sampling distributions of *IOF* by empirical means rather than the usual analytic means (cf. Diaconis & Efron, 1983). This approach allows us much greater freedom to derive indexes on the basis of aesthetics rather than formalities. It must be noted that *IOF* is a descriptive statistic, not an inferential statistic. We believe it is an easily interpreted and meaningful indicant of the goodness of fit of ordinal measurements to ordinal predictions. We have not developed it as a means of making mechanical inferences from sample to population. So the abstract, mathematical properties of *IOF* are of little concern to us at the moment.

How shall an *IOF* be interpreted? Those we have calculated from the previous example illustrate that *IOF* can have a rather straightforward meaning. Any hypothesis might be expected to predict correctly about half of the ordinal relations of the measurements by chance. When half are predicted correctly (and the other half are mismatched), then $IOF = 0$. When all the predictions of a hypothesis are matched by the measurements, we might say that the hypothesis had increased its predictive accuracy by 100% over chance; in this case, $IOF = +1.00$. Conversely, when none of the predictions of a hypothesis are matched by the measurements (all are mismatched), then we might say that the hypothesis was 100% worse than chance; in this case, $IOF = -1.00$. Therefore, IOF_1 indicates that hypothesis H_1 did 56% better than chance; IOF_2 indicates that H_2 did 50% worse than chance; IOF_3 indicates that H_3 did 100% better than chance.

Should we therefore conclude that H_3 is the best of the three hypothesis? Certainly not. Even though H_3 had the highest *IOF*, other factors must be taken into consideration in evaluating the goodness of fit of the three hypotheses, or of any hypotheses. Perhaps the most obvious factor is the scope of each hypothesis. H_1, for example, made nine times as many predictions as did H_2, and it seems reasonable to credit the former for doing so, especially as most of them were confirmed by the measurements. Another factor that seems important to consider is the concordance or overlap of the predictions of hypotheses. For example, the single and correct prediction of hypothesis H_3 appears in the set of predictions of H_1: (T1 < T5). Because H_1 is thus H_3 and more, its greater generality may be important in judging its relative merits. Yet, had this common prediction been disconfirmed by the data, we may have wished to doubt the merits of both hypotheses. Let us now consider these extra factors of hypothesis evaluation in greater detail.

INDEX OF PREDICTIVE FIT

The index of observed fit, *IOF*, tells us something of how often the ordinal pattern of measurements (our data) match the predicted ordinal pattern of a given hypothesis. It is also of interest to know how often the ordinal predictions of a given hypothesis match the ordinal pattern of our measurements. One index of the latter can be easily defined. We call it the *index of predictive fit* of a hypothesis, h, and define it thusly:

$$IPF_h = [\#(POP_h \text{ and } MOP)$$

$$- \#(POP_h \text{ and not } MOP)]/\#MOP = IOF_h \times Sc_h$$

(6)

In the three examples:

$$IPF_1 = +0.56 \times 1.00 = +0.56$$

$$IPF_2 = -0.50 \times 0.44 = -0.22$$

$$IPF_3 = +1.00 \times 0.11 = +0.11$$

The interpretation of *IPF* is also straightforward: IPF_h is the proportion of increase or decrease in predictive accuracy, relative to guessing, that hypothesis h achieves in predicting all the ordinal relations in *MOP*. In the case of hypothesis H_1, all nine elements in *MOP* are addressed, and seven of them are predicted correctly—a 56% improvement over guessing. In the case of hypothesis H_2, four of the nine elements in *MOP* are addressed, and three of these four are incorrectly predicted. We are forced to guess at the relations of the remaining five elements in *MOP* not addressed by H_2, and should guess them correctly about half the time. Thus, using H_2, we can expect to be correct in predicting the ordinal relations in MOP $(1 + 2.5)/(4 + 5) = 39\%$ of the time. This is 22% worse than we could expect to do by guessing at all nine relations. Hence, $IPF_2 = -0.22$. In the case of H_3, we can use it to predict correctly one of the nine elements of MOP; the remaining eight elements must be predicted by guessing, and we should expect to guess half ($=4$) of them correctly. Thus, using H_3, we can expect to predict correctly $(1 + 4)/(1 + 8) = 55.5\%$ of the relations in MOP, an 11% increase above chance. As a result, $IPF_3 = +0.11$.

The IPF index provides another, and perhaps better, indicant of the goodness of fit of ordinal predictions to ordinal data. Using it, we might conclude that hypothesis H_1, cited previously, is "better" than the other two—at least for predicting the ordinal relations of our single test case. Still other hypotheses, however, may generate the same IPF value as H_1, or a higher one. For example, some hypothesis, H_4, may generate predictions about five of the nine elements in MOP, all of which are correct. In this case, $IOF_4 = (5 - 0)/5 = +1.00$, $Sc_4 = 5/9 = 0.56$, and $IPF_4 = +1.00 \times 0.56 = +0.56$. How, then, should H_1 and H_4 be compared? Is it better to have greater scope and lower predictive accuracy (H_1), or smaller scope and higher predictive accuracy (H_4)? How do we assess the "goodness" of hypotheses when we suspect that additional hypotheses may be created in the future that will do at least as well as our "best" one? There are, of course, no general or widely accepted answers to these questions. Indeed, the alternative answers are matters of philosophy and taste, not matters of statistics. So far, OPA merely provides three simple indexes or scales on which hypotheses may be compared: Sc, IOF, and IPF. The way in which these indexes are used to establish hypothetical preferences is left to the discretion of their users. We can, however, partition the three indexes into some useful subindexes so that more detailed hypothetical comparisons can be made.

PARTICULAR COMPARISONS OF HYPOTHESES

Perhaps the finest examples of scientific research are those which pit two precise and detailed hypotheses, derived from two important and competing theories, against each other in a critical experiment. The hypotheses make exact opposite predictions about the outcome of the experiment. The experimental results strongly support one of the hypotheses and strongly disconfirm its competitor. The theory from which the latter was derived is then rejected, and the theory generating the former is given added credibility.

Unfortunately, such examples of scientific research are quite rare. It is often difficult to derive precise predictions from important theories. It is often difficult to design experiments that can, in principle, generate unequivocal outcomes. And even when elegant experiments can be designed to test clear and competing theories, the results of these experiments may support both theories, or neither theory, to some extent.

It is therefore desirable to develop indexes of theory–data fit that give us a detailed picture of what fits where. Consider, for example, the ordinal predictions of two hypotheses, H_i and H_j. Four subsets of their predictions may be distinguished:

1. *Convergent predictions*—the same predictions made by H_i and H_j [e.g., H_1 and H_3 both predict (T1 < T5)]
2. *Divergent predictions*—opposite predictions made by H_i and H_j [e.g., H_1 predicts (T3 < T4) and (T3 < T5), H_2 predicts (T3 > T4) and (T3 > T5)]
3. *Unique predictions of H_i*—predictions made by H_i but not made by H_j [e.g., H_2 makes predictions about (T1,T3), (T2,T3), (T3,T4), and (T3,T5), H_3 does not address itself to these four pairs of test scores]
4. *Unique predictions of H_j*—predictions made by H_j but not made by H_i [e.g., H_3 makes a prediction about (T1,T5), but H_2 does not address itself to this pair of test scores]

These four sets do not overlap; their union is simply the union of POD_i and POD_j. As a result, we may derive indexes of scope, observed fit, and predictive fit for each of the four sets, then compare the indexes across hypotheses to obtain a detailed account of the relative strengths and weaknesses of each hypothesis.

CONVERGENT PREDICTION INDEXES

How well do the common ordinal predictions of two or more hypotheses fit the ordinal patterns of data? To find out we only need to modify our existing indexes to consider these predictions. Let us therefore define the subscript pair h & i as the set of common predictions of hypotheses h and i:

$$Sc_{h\&i} = \#(POP_h \text{ and } POP_i)/\#MOP \qquad (7)$$

$$IOF_{h\&i} = [\#(POP_h \text{ and } POP_i \text{ and } MOP) - \#(POP_h$$

$$\text{and } POP_i \text{ and not } MOP)]/\#(POP_h \text{ and } POP_i) \qquad (8)$$

$$IPF_{i\&j} = IOF_{h\&i} \times Sc_{h\&i} \qquad (9)$$

Obviously, if we were considering more than two hypotheses, then we would probably wish to compare all possible pairs of them, and thus expand the number of convergent prediction indexes we calculate. We could also calculate the indexes for the intersections of three, four, or more *POP*s by logical expansion of equations 7 to 9, above. For our purposes now, however, we shall limit ourselves to hypothesis pairs and illustrate the use of equations 7 to 9 with the following examples.

$$\#(POP_1 \text{ and } POP_2) = \#\{(T1 < T3) (T2 < T3)\} = 2$$

$$\#(POP_1 \text{ and } POP_3) = \#\{(T1 < T5)\} = 1$$

$$\#(POP_2 \text{ and } POP_3) = \#\{ \} = 0$$

$$Sc_{1\&2} = 2/9 = 0.22$$

$$Sc_{1\&3} = 1/9 = 0.11$$

$$Sc_{2\&3} = 0/9 = 0$$

$$IOF_{1\&3} = \{\#[(T1 < T3)] - \#[(T2 < T3)]\}/2 = \{1 - 1\}/2 = 0.00$$

$$IOF_{1\&2} = \{\#[(T1 < T5)] - \#[ ]\}/1 = \{1 - 0\}/1 = +1.00$$

$$IOF_{2\&3} = \text{undefined}$$

$$IPF_{1\&2} = 0.22 \times 0.00 = 0.00$$

$$IPF_{1\&3} = 0.11 \times 1.00 = +0.11$$

The indexes above tell us that the three hypotheses rarely make the same predictions, and that when they do, the joint predictions fail to match the ordinal pattern of measurements much more often than chance. Our little example is, of course, being stretched a bit thin here; it is probably best to conclude that we have insufficient data in this case to reach any meaningful conclusions. In many research situations, however, different hypotheses may generate scores of predictions, dozens of which will be the same. We can then make more meaningful comparisons of these convergent predictions, and thus assess the extent to which consensual validity is matched by predictive validity. If the vast majority of convergent predictions are wrong, we may wish to doubt all hypotheses that make them, even though one of the hypotheses may be superior to others in overall predictive fit.

DIVERGENT PREDICTION INDEXES

Divergent predictions are the stuff of theoretical contests and critical experiments. Indexes relevant to evaluating them are easily defined. We label the divergent predictions of a pair of hypotheses, *h* and *i*, by the subscripts $h><i$. The set of these divergent predictions will contain all elements in POP_h [e.g., (T1 > T2)] that have their opposite in POP_i [e.g., (T1 < T2).

$$Sc_{h><i} = \#(POP_h \text{ and } POP_i')/\#MOP \tag{10}$$

$$IOF_{h><i} = \{\#(POP_h \text{ and } POP_i' \text{ and } MOP) - \#(POP_h \tag{11}$$

$$\text{and } POP_i' \text{ and not } MOP)\}/\#(POP_h \text{ and } POP_i')$$

$$IPF_{h><i} = Sc_{h><i} \times IOF_{h><i} \tag{12}$$

With a little algebra we can also derive indexes for the fit of predictions made by hypothesis i relative to the opposite predictions made by h, that is, the $i><h$ prediction set:

$$Sc_{i><h} = Sc_{h><i} \tag{13}$$

$$IOF_{i><h} = -IOF_{h><i} \tag{14}$$

$$IPF_{i><h} = Sc_{i><h} \times IOF_{i><h} \tag{15}$$

Again, we stretch our example to illustrate the calculation of these multisubscripted indexes. When comparing hypotheses H_1 and H_2:

$$\#(POP_1 \text{ and } POP_2') = \#\{(T3 < T4)\ (T3 < T5)\} = 2$$

$$Sc_{1><2} = 2/9 = 0.22$$

$$IOF_{1><2} = \{\#[(T3 < T4)\ (T3 < T5)] - \#[\ \ ]\}/2 = \{2 - 0\}/2 = 1.00$$

$$IPF_{1><2} = 0.22 \times 1.00 = +0.22$$

$$Sc_{2><1} = 0.22$$

$$IOF_{2><1} = -1.00$$

$$IPF_{2><1} = 0.22 \times -1.00 = -0.22$$

Thus, we see that hypothesis H_1 and H_2 make two opposite predictions, and that the ordinal pattern of the data match the two predictions of H_1 rather than H_2. In these circumstances, H_1 is superior to H_2. However, the scope of their competitive domain is rather small (0.22). We may wish to check the accuracy of their unique predictions before drawing any tentative conclusions about the merits of either.

UNIQUE PREDICTION INDEXES

In our example, hypotheses H_1 and H_3 do not have any divergent or competing predictions, nor do hypotheses H_2 and H_3. As a result, we cannot directly compare H_3 against the other two. This will probably occur often in real research settings, especially those of observational or archival studies; hypotheses will address different data domains. If the unique domains of several hypotheses are large and their predictions are often correct, some conceptual analyses of the assumptions under-lying the hypotheses may reveal overriding theoretical principles (as in the case of wave and corpuscular theories of light). If the unique pre-dictions of some hypothesis are consistently wrong, then time might be well spent thinking of reasons why this may be so. The indexes relevant to an assessment of the fit of unique predictions are defined below. Here we refer to predictions made by some hypothesis, h, that are not made (addressed) by another hypothesis, i, using the subscript $h\tilde{\imath}$.

$$POD_{h\tilde{\imath}} = \{POD_h \text{ and } POD_i\} \tag{16}$$

$$POP_{h\tilde{\imath}} = \{(ob_{j,\jmath jk,\jmath}ob_k)\} \text{ for all } (oh_j, oh_k) \text{ in } POD_{h\tilde{\imath}} \tag{17}$$

$$Sc_{h\tilde{\imath}} = \#(MOD \text{ and } POD_{h\tilde{\imath}})/\#MOD \tag{18}$$

$$IOF_{h\tilde{\imath}} = \{\#(POP_{h\tilde{\imath}} \text{ and } MOP) - \#(POP_{h\tilde{\imath}} \text{ and not} \tag{19}$$

$$MOP)\}/\#POP_{h\tilde{\imath}}$$

$$IPF_{h\tilde{\imath}} = Sc_{h\tilde{\imath}} \times IOF_{h\tilde{\imath}} \tag{20}$$

To illustrate the use of these indexes, consider the unique predic-tions of H_1, relative to H_2. We begin by noting the pairs of observations addressed by the former that are not addressed by the latter:

$$POD_{1\tilde{2}} = \{(T1,T2) \ (T1,T4) \ (T1,T5) \ (T2,T4) \ (T4,T5)\}$$

Using definitions of equations 17 to 20, we may calculate that

$$POP_{1\tilde{2}} = \{(T1 < T2) \ (T1 < T4) \ (T1 < T5) \ (T2 < T4) \ (T4 < T5)\}$$

$$Sc_{1\tilde{2}} = 5/9 = 0.55$$

$$IOF_{1\tilde{2}} = \{4 - 1\}/5 = +0.60$$

$$IPF_{1\tilde{2}} = 0.55 \times 0.60 = +0.33$$

These figures indicate that the scope of unique predictions of H_1 is rel-atively large, and that in its unique domain the predictions of H_1 are relatively well matched by the test scores. Similar analyses could be undertaken of the unique predictions of H_1 relative to H_3, H_2 relative to H_3, and H_3 relative to H_2 (neither H_2 nor H_3 make unique predictions relative to H_1 because H_1 addresses itself to all pairs in MOD). For the sake of brevity, we leave these calculations to the interested reader. Let us turn our attention now to the analysis of aggregated indexes.

AGGREGATED COMPARISONS OF HYPOTHESES

So far we have defined some indexes of scope, observed fit, and predictive fit that we believe are useful for evaluating the correspondence between ordinal predictions and measurements. We have applied these indexes to assess the fit of three hypotheses to one set of observations. Most research, of course, is conducted to gather several sets of observations, and it would surely be useful to extend our indexes to evaluate the aggregated fit of these sets. We may proceed in either of two ways: (a) aggregate the observations (e.g., average them), then evaluate the ordinal pattern of this aggregation using the indexes defined in the preceding sections of this chapter; or (b) evaluate the ordinal pattern of each set of observations using the preceding indexes, then aggregate the evaluations across the sets. In keeping with the spirit of OPA, we shall opt for the second alternative. We realize that the first may also be pursued, and may occasionally be preferred to the second. But here we are concerned with testing the fit of hypotheses to observations of individuals, each of whom we assume will provide us with one or more of the observation sets. So we stick to our motto: "Analysis before aggregation!" Below we present a simple means to aggregate individual analyses of scope and fit across several sets of observations.

Assume for a moment that we wish to extend our tests of the now familiar hypotheses H_1, H_2, and H_3 to observation sets derived from other students. Table 1 shows the test results for the one student so far considered (Student 1) as well as three other students (2, 3, and 4). Note that these students need not be in the same course, as nothing is mentioned in the hypotheses about such a constraint. We have purposely omitted the second test score from the results of Student 4 to illustrate how missing scores are handled in OPA.

We begin by evaluating the fit of hypothesis H_1 to these results.

Table 1. Hypothetical Student Test Scores

Student	Test 1	Test 2	Test 3	Test 4	Test 5
1	57	68	63	72	68
2	72	64	79	81	83
3	53	57	54	52	56
4	85	—	90	87	73
Average	66.8	63.0	71.5	73.0	70.0

Consider first the test scores of Student 2. There are no tied scores, so all 10 possible pairs of scores are contained in the MOD for this student, here labeled MOD2. H_1 makes predictions about all 10 of these pairs, thus for Student 2,

$$Sc_1(2) = 10/10 = 1.00$$

Recall that H_1 predicts constant improvement in test scores, so that the first element of each pair of scores should be less than the second. All pairs of test scores show this relation except (72,64), therefore,

$$IOF_1(2) = (9 - 1)/10 = +0.80, \text{ and}$$

$$IPF_1(2) = 1.00 \times 0.80 = +0.80$$

Tests of the match of H_1 to test scores for Student 3 reveals,

$$Sc_1(3) = 10/10 = 1.00$$

$$IOF_1(3) = (5 - 5)/10 = 0.00, \text{ and}$$

$$IPF_1(3) = 1.00 \times 0.00 = 0.00$$

Because of the missing score for Student 4, the number of elements in MOD(4) is reduced from 10 to 6. H_1 still makes predictions about these 6 pairs of scores, so

$$Sc_1(4) = 6/6 = 1.00$$

$$IOF_1(4) = (2 - 4)/6 = -0.33, \text{ and}$$

$$IPF_1(4) = 1.00 \times -0.33 = -0.33$$

We now wish to define aggregated indexes of scope and fit. These definiitions are given in equations 21 to 23 for some hypothesis, h, and observation sets $1, \ldots, s$.

$$Sc_h (1, \ldots, s) = [\Sigma_s \#(POD_h(s) \text{ and } MOD(s))]/ \tag{21}$$

$$[\Sigma_s \#(MOD(s)]$$

$$IOF_h (1, \ldots, s) = \{[\Sigma_s \#(POP_h(s) \text{ and } MOD(s)] - [\Sigma_s \# \tag{22}$$

$$(POP_h(s) \text{ and not } MOD(s)]\}/[\Sigma_s \#POP_h(s)]$$

$$IPF_h (1, \ldots, s) = Sc_h (1, \ldots, s) \times IOF_h (1, \ldots, s) \tag{23}$$

In the case of hypothesis H_1, above, the aggregated indexes may be calculated as

$$Sc_1 (1, \ldots, 4) = [9 + 10 + 10 + 6]/[9 + 10 + 10 + 6]$$

$$= 35/35 = 1.00$$

$$IOF_1 (1, \ldots, 4) = \{[7 + 9 + 5 + 2] - [2 + 1 + 5 + 4]\}/$$

$$[9 + 10 + 10 + 6] = \{23 - 12\}/35 = +0.31$$

$$IPF_1 (1, \ldots, 4) = 1.00 \times 0.26 = +0.26$$

Thus, on aggregate, H_1 addresses all of the pairs in the four $MODs$, and correctly predicts 31% more of them than one would expect by chance. (By way of comparison, H_1 correctly predicts the order of the five column averages in Table 1 only 5 out of 10 times for an IOF_1 of the aggregate = 0.00.) Similar aggregate indexes may be calculated for the predictions of hypotheses H_2 and H_3:

$$Sc_2 (1, \ldots, 4) = [4 + 4 + 4 + 3]/[9 + 10 + 10 + 6]$$

$$= 15/35 = 0.43$$

$$IOF_2 (1, \ldots, 4) = \{[1 + 2 + 2 + 3] - [3 + 2 + 2 + 3]\}/$$

$$[4 + 4 + 4 + 3] = \{8 - 10\}/15 = -0.13$$

$$IPF_2 (1, \ldots, 4) = 0.43 \times -0.13 = -0.06$$

$$Sc_3 (1, \ldots, 4) = [1 + 1 + 1 + 1]/[9 + 10 + 10 + 6]$$

$$= 4/35 = 0.11$$

$$IOF_3 (1, \ldots, 4) = \{[1 + 1 + 1 + 0] - [0 + 0 + 0 + 1]\}/$$

$$[1 + 1 + 1 + 1] = \{3 - 1\}/4 = +0.50$$

$$IPF_3 (1, \ldots, 4) = 0.11 \times 0.50 = +0.06$$

These aggregate indexes suggest that H_1 provides the most comprehensive and closest fit to the data in Table 1 overall. However, most of its predictive successes come from the test scores of Subjects 1 and 2. For Subject 3, H_3 generates the highest IOF (= +1.00), and the predictions of H_2 are perfectly matched by the test scores of Subject 4. This may indicate that there are consistent individual differences in theory–data fit, and that individuals may be classified according to their best fitting hypothesis rather than according to their (least squared) distance from one another.

Aggregated particular comparisons of hypotheses (not the contradiction in terms it seems) may also be undertaken. For example, to test the divergent predictions of H_1 versus H_2,

$$Sc_{1><2}\,(1,\ldots,4)\ =\ [2\ +\ 2\ +\ 2\ +\ 2]/$$

$$[9\ +\ 10\ +\ 10\ +\ 6]\ =\ 8/35\ =\ 0.23$$

$$IOF_{1><2}\,(1,\ldots,4)\ =\ \{[2\ +\ 1\ +\ 1\ +\ 0]\ -\ [0\ +\ 1\ +\ 1\ +\ 2]\}/$$

$$[2\ +\ 2\ +\ 2\ +\ 2]\ =\ \{4\ -\ 4\}/8\ =\ 0.00$$

$$IPF_{1><2}\,(1,\ldots,4)\ =\ 0.23\ \times\ 0.00\ -\ 0.00$$

These results indicate that when the divergent or competing predictions of H_1 and H_2 are pitted against the data in Table 1, the result is a draw; both hypotheses generate the same proportion of correct predictions (50%). As we have already seen, the two divergent predictions of H_1 are both matched by the test results of Student 1. In contrast, however, the two divergent predictions of H_2 are both matched by Student 4. Again, this may indicate that neither hypothesis is correct for all students, and that each is valid for some students.

TESTS OF "SIGNIFICANCE"

Whenever we have presented an outline of OPA to audiences steeped in a classical, hypothesis testing, statistical tradition, one question has invariably arisen: How does one know if a fit is "significant"? As we noted in our introduction, we do not find the question entirely proper, especially in the context of investigating individuals. But the question refuses to die, so here we attempt a partial answer in hopes of relaxing traditionalists enough to judge OPA on its own merits. Yes, one can construct significance tests for some OPA indexes. For example, if the scope of a hypothesis = 1.00, and there are no ties in the data of one individual on which the hypothesis is tested, then IOF and IPF can be treated as Kendall's Tau, and their significance can be ascertained by reference to appropriate entries in Tau significance tables. If $Sc < 1.00$, or if there are ties, then the formulas for IOF and IPF do not correspond to those for Tau, and this trick will not work. Recall, however, that significance tests of almost all rank–order statistics (such as IOF and IPF) are essentially permutation tests: one simply attempts to determine how many permutations of the numbers obtained will generate an index value at least as great as the value calculated on the obtained numbers. In the case of small sets of numbers, the permutations and calculations may be done by hand. Consider, for example, how we might proceed to determine the probability that $IOF_1(1)$ would have occurred by chance. There are $5! = 120$ possible permutations of the 5 test scores obtained by Subject #1. By our count, two of these permutations [for T1, . . . ,T5 they are: $(57,63,68_a,68_b,72)$ and $(57,63,68_b,68_a,72)$] would have given a perfect match to the predictions of H_1, so that $IOF_1 = 1.00$. Six more of these permutations [e.g., $(57,63,68_a,72,68_b)$, $(63,57,68_a,68_b,72)$, $(57,68_a,63,68_b,72)$] would have matched 8 of the 9 predictions of H_1, for an $IOF_1 = (8 - 1)/9 = 0.78$. An additional twelve permutations [e.g., $(63,68_a,57,68_b,72)$, $(68_a,57,63,68_b,72)$, and the set obtained by Student 1: $(57,68_a,63,72,68_b)$] would have matched 7 of the 9 predictions of H_1, for the obtained IOF_1 of $(7 - 2)/9 = 0.56$. In sum, $2 + 6 + 12 = 20$ of the 120 permutations of the five test scores would have yielded an IOF_1 at least as great as the one obtained. Thus, the probability of obtaining $IOF_1 = 0.56$ or greater by chance is $20/120 = 0.17$. This, of course, does not reach the traditional $p = 0.05$ level of significance, so one may be tempted to conclude that H_1 does not produce a "significant" or reliable fit of the ordinal pattern of test scores of Student 1. But we trust that we have demonstrated how such conclusions must be seen in the light of other information; no single test should be used to draw conclusions from OPA.

Other tests of significance may be generated for the other hypotheses and students. Each may be based on somewhat different permutation formulas, depending on the number of ties, missing data, changes to *POD*s resulting from different permutations, and so forth. Again, for the sake of brevity we leave such calculations to the reader. In testing larger data sets, it is usually infeasible to count permutations by hand. Computer programs can easily be written to perform this tedious chore.

DATA ANALYSIS USING OPA

Let us now consider how the OPA indexes we have developed can be used in the analysis of some actual research results. A few years ago, the senior author (WT) and Mary Mortimer began a series of studies on the relations between life experiences and personal philosophies, or philosophies of life. As part of these studies, 100 elderly respondents were asked to draw a graph showing the "ups and downs" of their lives. A grid was provided for this purpose, with an X-axis indicating age and a Y-axis indicating general happiness or unhappiness (100 = "extremely happy" to 0 = "extremely unhappy"). The plot made by each subject looked something like a polygraph tracing; some plots went straight across the page at the top of the scale (always extremely happy), or across the middle (always pretty neutral), but most did show variations in happiness over time. Table 2 shows the ratings of three males and three females as interpolated from their plots.

Table 2. Happiness Ratings across the Life Span

Subject (sex)	Year of birth	Age								
		A = 0–9	B = 10–19	C = 20–29	D = 30–39	E = 40–49	F = 50–59	G = 60–69	H = 70–79	I = 80–89
1(M)	1900	55	60	62	35	65	55	60	65	
2(M)	1910	60	55	42	68	68	70	70		
3(M)	1920	60	50	75	55	60	62			
4(F)	1890	65	68	60	55	40	45	55	60	55
5(F)	1900	48	42	55	50	40	50	55	50	
6(F)	1910	45	50	35	40	50	55	58		
Average		55.5	54.2	50.5	53.8	56.2	59.6	58.3	55.0	

What might account for each person's changes in happiness ratings? One hypothesis, H_1, is suggested by a popular quip, "The young don't know; the old have forgotten." Happiness may be related to knowledge, ability, or success. Knowledge and ability likely develop in youth and may begin to peak about 30 years of age; they may then last until perhaps 60, then decline steadily. Success resulting from the use of knowledge and ability should follow roughly the same trajectory, and so too will the happiness it accrues. Thus, happiness should rise until 30, remain more or less constant until 60, then decline. These are the ordinal predictions of H_1.

A second hypothesis is suggested by somewhat different reasoning. Until about age 20 most people are sheltered from the major worries and tragedies of life: getting and keeping a job; raising children; confronting personal limitations, death and major disappointments. During the age of worry (20 +), several changes in self-image and reassessments of life goals are often necessary. Because these changes usually follow traumatic events, they cannot sooth the pain of these events, so happiness would decline. At about age 50, enough changes have occurred to buffer the effects of most remaining life storms. A philosophy of life is consolidated by this age, and it serves to increase a sense of well-being from then on. Thus, happiness should be relatively high in youth, relatively low from about age 20 to 49, and relatively high again after 49. These are the predictions of hypothesis H_2. A third hypothesis is much less subtle than the first two. Happiness, it is assumed, belongs to the young. More specifically, happiness will increase steadly until about age 29, then decrease steadily until death. Such are the dismal predictions of H_3.

These three hypotheses are by no means the only ones we could muster to account for trends in happiness ratings; indeed, they are not even particularly insightful or well developed. However, they do allow us to generate some ordinal predictions that can be tested against the data in Table 2. To illustrate this, consider the *POP* each hypothesis generates for the first male in Table 2, Subject 1.

$$POP_1(1) = \{(A < B)\ (A < C)\ (A < D)\ (A < E)\ (B < C)$$
$$(B < D)\ (B < E)\ (B < F)\ (C < D)\ (C < E)$$
$$(C < F)\ (D > G)\ (D > H)\ (E > G)\ (F > G)$$
$$(F > H)\}$$

$$POP_2(1) = \{(A > C)\ (A > D)\ (A > E)\ (B > C)\ (B > D)$$
$$(B > E)\ (C < F)\ (C < G)\ (C < H)\ (D < F)$$
$$(D < G)\ (D < H)\ (E < F)\ (E < G)\}$$

$$POP_3(1) = \{(A < B)\ (A < C)\ (B < C)\ (C > D)\ (C > E)$$
$$(C > F)\ (C > G)\ (C > H)\ (D > E)\ (D > F)$$
$$(D > G)\ (D > H)\ (E > F)\ (E > G)\ (F > G)$$
$$(F > H)\ (G > H)\}$$

[Recall that we exclude from each *POP* set those pairs with ties in the data; the pairs predicting some relation between (A,F), (B,G), and (E,H) are thus excluded, because A = F, B = G, and E = H in the data of Subject 1.]

To calculate the scope and fit of these sets, we first note that

$$\#MOP(1) = (\#\text{pairs}) - (\#\text{ties}) = (8 \times 7) - 3 = 53$$

Thus,

$$Sc_1(1) = 16/53 = 0.30$$
$$Sc_2(1) = 14/53 = 0.26$$
$$Sc_3(1) = 17/53 = 0.32$$

Counting the hits and misses of the three hypotheses, we obtain,

$$IOF_1(1) = (7 - 9)/16 = -0.12 \qquad IPF_1(1) = 0.30 \times -0.12 = -0.04$$

$$IOF_2(1) = (6 - 8)/14 = -0.14 \qquad IPF_2(1) = 0.26 \times -0.14 = -0.04$$

$$IOF_3(1) = (8 - 9)/14 = -0.07 \qquad IPF_3(1) = 0.32 \times -0.07 = -0.02$$

Table 3 presents the results of corresponding tests for all six subjects on the sample.

Clearly, as can be seen in Table 3, none of the three hypotheses does an outstanding job of predicting the ordinal patterns of the happiness ratings. Of the three, hypothesis H_2 seems to do better than the other two; its aggregated-fit indexes are the only ones with positive values. Even so, it can predict the ordinal relations of all $MOPs$ only 6% better than chance, and does worse than the other two hypotheses in predicting the rating patterns of Subjects 1 and 3. If the subjects had mid-life slumps of happiness, as suggested by H_2, they did not reveal it clearly in the ratings they made.

Do the data suggest other hypotheses that might produce much better fits? The three hypotheses considered so far all predict changes in happiness ratings based more on aging itself than on critical life events. Interviews that accompanied the rating task suggested that these events are at least as important as age in determining happiness. Because history provides these events more often than logic or design, we decided to consider two additional hypotheses derived from historical considerations. The interviews produced many stories about tragedies resulting from the Great Depression; we thus hypothesized that happiness ratings would be lowest during the Depression years (roughly 1930–1939). This is hypothesis H_4. In addition, many men confessed that despite the tragedies of their comrades, their involvement in World War II gave them the most exciting times of their lives. In contrast, many women considered the war years to be almost as upsetting as the years of the Depression. Thus, we might expect that happiness ratings would be highest for men (H_5), and second lowest for women (H_6), during the war decade from 1940 to 1949.

Table 3. OPA Indexes for Hypotheses about Happiness as a Function of Age

Subject	Sc_1	IOF_1	IPF_1	Sc_2	IOF_2	IPF_2	Sc_3	IOF_3	IPF_3
1	0.30	−0.12	−0.04	0.26	−0.14	−0.04	0.32	−0.07	−0.02
2	0.35	+0.29	+0.10	0.30	−0.33	+0.10	0.28	−1.00	−0.28
3	0.38	+0.09	+0.03	0.28	−0.25	−0.07	0.31	+0.11	+0.03
4	0.31	−0.81	−0.25	0.22	+0.47	+0.10	0.29	+0.10	+0.03
5	0.33	−0.18	−0.06	0.23	+0.17	+0.04	0.27	+0.14	+0.04
6	0.34	+0.00	+0.00	0.27	+0.82	+0.22	0.32	−0.85	−0.27
Aggregate	0.33	−0.18	−0.06	0.25	+0.25	+0.06	0.30	−0.26	−0.08

In order to test these three hypotheses more easily, the data from Table 2 were moved about to align columns according to the appropriate decades of the century. The result is shown in Table 4.

Indexes of scope and fit were then calculated for H_4, H_5, and H_6 in manners analogous to calculations of indexes for the first three hypotheses. To illustrate, consider the calculation of relevant indexes for Subject 1:

$$POP_4(1) = \{(B > E)\ (C > E)\ (D > E)\ (E < F)\ (E < G)\ (E < H)$$

$$(E < I)\}$$

$$POP_5(1) = \{(B < F)\ (C < F)\ (D < F)\ (E < F)\ (F > G)\ (F > H)\}$$

$$Sc_4(1) = 7/53 = 0.13$$

$$Sc_5(1) = 6/53 = 0.11$$

$$IOF_4(1) = (7 - 0)/7 = +1.00 \qquad IPF_4(1) = 0.13 \times 1.00 = +0.13$$
$$IOF_5(1) = (6 - 0)/6 = +1.00 \qquad IPF_5(1) = 0.11 \times 1.00 = +0.11$$

As one can see from Table 5, all three hypotheses do quite well in their own predictive domains. Hypothesis H_4 makes predictions that are very well matched by the data of 5 out of 6 of the subjects; only Subject 5 does not fit its predictions. Similarly, the predictions of H_5 and H_6 are well matched by the ordinal patterns of the data. Though the scopes of these hypotheses do not match those of the previous three, their improved IOFs more than compensate for their relative specificity.

Can H_4 and H_5/H_6 be compared? Is one better, or preferable, to the others? An examination of their particular subsets reveals that very little overlap exists in their predictions. In the case of H_4 versus H_5, for example,

Table 4. Happiness Ratings as a Function of Historical Era

Subject (sex)	Year of birth	A = 1890s	B = 1900s	C = 1910s	D = 1920s	E = 1930s	F = 1940s	G = 1950s	H = 1960s	I = 1970s
1(M)	1900		55	60	62	35	65	55	60	65
2(M)	1910			60	55	42	68	68	70	70
3(M)	1920				60	50	75	55	60	62
4(F)	1890	65	68	60	55	40	45	55	60	55
5(F)	1900		48	42	55	50	40	50	55	50
6(F)	1910			45	50	35	40	50	55	58
Average		65.0	57.0	53.4	56.2	42.0	55.5	55.5	60.0	60.0

Table 5. OPA Indexes for Hypotheses about Happiness as a Function of Historical Era

Subject	Sc_4	IOF_4	IPF_4	Sc_5	IOF_5	IPF_5	Sc_6	IOF_6	IPF_6
1	0.13	+1.00	+0.13	0.11	+1.00	+0.11			
2	0.15	+1.00	+0.15	0.12	+0.20	+0.02			
3	0.17	+1.00	+0.17	0.17	+1.00	+0.17			
4	0.12	+1.00	+0.12				0.12	+1.00	+0.12
5	0.10	−0.20	−0.02				0.13	+0.71	+0.09
6	0.15	+1.00	+0.15				0.15	+1.00	+0.15
Aggregate	0.13	+0.91	+0.12	0.13	+0.75	+0.10	0.13	+0.90	+0.12

no divergent predictions are made, so the two hypotheses cannot be directly pitted against each other. The same two hypotheses make only one convergent prediction (E > F), and it was confirmed by the ratings 3 out of 3 times. Similarly, H_4 and H_6 make no divergent predictions, and make one one convergent prediction (E < F), which is correct 2 out of 3 times. Comparisons between H_5 and H_6 cannot be made as they address themselves to different data sets. In effect, the three hypotheses are complementary rather than competing, and in view of the good match the relevant data made to the predictions of each, there seems to be no reason to prefer one over the others.

Obviously, the relatively good fits of the ratings to the predictions of H_4 through H_6 do not guarantee that these hypotheses are correct. Still other hypotheses may make substantially the same predictions, or make other predictions with equal accuracy. However, in view of the success of these hypotheses, one would probably want to investigate them further with new data. It is clear that H_4 through H_6 are not conceptually independent, even though their predictions almost never overlap. All three are derived from an assumption of historical influence, and sister hypotheses more relevant to other situations may make better tests of the assumption in other data sets. We stress again that the purpose of OPA is not to generate inductions from data, but to guide further research.

ADDITIONAL TOPICS

ORDERS OF DIFFERENCES

Hypotheses often generate predictions about orders of differences of observations, and when measurements of the observations satisfy the properties of an interval scale, OPA may be applied to test these predictions as well. For example, if the happiness scale in the example above were an interval scale (which it is not), then we could test predictions such as, "Fluctuations in happiness will grow smaller as an individual grows older." One method of testing this prediction would be to find the absolute distance between each successive pair of ratings. In the case of Subject 1, these would be:

$$d1 = |55 - 60| = 5, \quad d2 = |60 - 62| = 2, \quad d3 = |62 - 35| = 27,$$
$$d4 = |35 - 65| = 30, \quad d5 = |65 - 55| = 10, \quad d6 = |55 - 60| = 5,$$
$$d7 = |60 - 65| = 5$$

The seven distances would then be compared to the predictions of the hypothesis:

$$POP = \{(d1 > d2)\ (d1 > d3)\ (d1 > d4)\ (d1 > d5)\ (d2 > d3)\ (d2 > d4)\ (d2 > d5)$$
$$(d2 > d6)\ (d2 > d7)\ (d3 > d4)\ (d3 > d5)\ (d3 > d6)\ (d3 > d7)\ (d4 > d5)$$
$$(d4 > d6)\ (d4 > d7)\ (d5 > d6)\ (d5 > d7)\}$$

A special case of predictions about differences occurs under what is traditionally called the null hypothesis. In weakened form, a prediction of no difference between the results of two conditions is equivalent to a predictions that this difference will be smaller than the differences between all other conditions. Usually, this places quite a constraint on the predicted ordinal pattern. Thus, if we predict that no difference will be found in happiness ratings during the first two decades of life, then in the case of Subject 1 we are predicting that at very least

$$POP = \{(d1 < d2)\ (d1 < d3)\ (d1 < d4)\ (d1 < d5)\}$$

Predictions about differences in variance may also be entertained. For example, the previous hypothesis that fluctuations in happiness will decrease with age can lead to the following prediction: Happiness variability will be greater in the first half of life than in the second half. As variance increases, so too does the expected distance between any two scores. Thus, we would predict that if we took any two ratings of happiness in the first half of a subject's life, then took any two ratings in the last half, the absolute difference of the first two would be greater than the absolute difference of the second two. In the case of Subject 1, there are four ratings in the first half of his life, and four ratings in the last half. In each of these sets, $4!/(2! \times 2!) = 6$ differences may be calculated; we would predict that the first six would be greater than the last six—a total of 36 predictions or POP elements.

HYPOTHESIS DEVELOPMENT

Although OPA is designed to examine the fit of data to ordinal hypotheses, it can also be useful in the development of hypotheses. Research is often undertaken to answer questions about the degree of relationship between sets of observations without previously stated predictions. "Fishing expeditions" are then conducted to determine what might vary with what. Though these expeditions are often rightly discouraged on the grounds that they capitalize on chance, they are equally often of some use in directing subsequent research. Ordinal expeditions are no exception. For example, if we examine adjacent columns of Table 2, we find none that show consistent increases or decreases in happiness for all six subjects. We do find, however, that the two most recent decade ratings of 5 out of 6 of the subjects are higher than the ratings of their first two decades, suggesting that present happiness may in general be higher than recalled happiness of childhood. This certainly runs contrary to the popular notion that childhood memories tend to be the sweetest, and may well be worth pursuing in subsequent research. Table 4, however, reveals that this increase is confounded with era, and that we may also wish to test the hypothesis that life has become easier or better for most people as this century has progressed. Note that 6 out of 6 ratings of happiness in the 1960s are greater than ratings of happiness on the 1950s, but that further rating increases occurred in the 1970s for only 3 out of 6 of the subjects. Economic indicators seem to have shown a similar trajectory over these decades; in retrospect, the 1960s were years of growth and plenty, the 1970s were years of oil crises and economic woes. This correspondence, in conjunction with the good fit of H_4, suggests that we delve more deeply into the relation between economic and psychological well-being.

In sum, a search down the columns of Tables such as 2 and 5 may reveal consistencies not addressed by prior hypotheses. They may be uninterpretable, or they may lead us to superstitious hypothesizing behaviors. But they may also be of great use in telling us what we should pursue next.

ORDINAL PATTERNS IN FREQUENCY TABLES AND TRADITIONAL RESEARCH DESIGNS

The use of OPA is not limited to data generated by single individuals. It is often possible to employ the now familiar OPA indexes of scope and fit in examining the correspondence between predictions and observations of frequencies (as a supplement to chi square or similar tests), or between predictions and observations of group averages or data sets. Consider, for example, the prediction that Canadian women (W) tend to vote Liberal (L) in federal elections, and Canadian men (M) tend to vote Conservative (C). Ignoring for a moment those Canadians who vote for the New Democratic Party, a sample of 100 voters in the last federal election might reveal that #WL = 32, #WC = 21, #ML = 8, and #MC = 39. We may wish to perform a 2 × 2 chi square to judge the statistical significance of these contingencies, but it would tell us only if the LC proportions were significantly different between males and females, and would not address the issue of the order or direction of these proportions. From the perspective of OPA, the prediction would generate a two element $POP = \{(WL > WC) \ (ML < MC)\}$; four additional pairs would comprise the MOP [(WL > ML) (WL < MC) (WC > ML) (WC < MC)], so

$$Sc = 2/6 = 0.33$$

$$IOF = (2 - 0)/2 = 1.00, \text{ and}$$

$$IPF = 0.33 \times 1.00 = 0.33$$

This result is not terribly exciting, but it does illustrate the use of OPA in contingency tables. Its use becomes far more sensible in the analysis of large contingency tables, for example, those having five dimensions of four levels each. Traditional analyses of such tables (e.g., see Bishop, Fineberg, & Holland, 1975) is cumbersome at best; OPA provides a rather elegant alternative.

The analyses of continuous variables in factorial or related designs has long been the traditional domain of the analysis of variance, but even here OPA may find a use in assessing theory–data fit. To illustrate, consider another simple experiment in which 10 males and 10 females are asked to rate their liking of three political candidates, including one from the NDP (N), in a split-plot factorial design. One hypothesis predicts that for females the average rating of the Liberal will be higher than the average for the Conservative, and that both these averages will be higher than that of the NDP; for males, the average Conservative rating will be higher than the average Liberal rating, and both will be higher than the average rating of the NDP. In traditional analysis of variance terms, the hypothesis predicts a main effect for candidates, and a sex-by-candidate interaction. It also predicts certain directions of mean differences, and these are usually tested with the eyeball and complex formulas for simple effects. In using OPA, we simply note that the design generates six averages (three for males and three for females), so that of the 15 possible pairs of these averages, the hypothesis makes the following predictions about 6 of them:

$$POP = \{(FL > FC) \quad (FL > FN) \quad (FC > FN) \quad (MC > ML) \quad (MC > MN)$$
$$(ML > MN)\}$$

A quick look at the ordinal pattern of the averages should tell us how many of these predictions came true. If we do not wish to deal in averages, we can examine the 10 pairs of ratings relevant to each of the six predictions, and form partial hit indexes for each. For example, if the ratings of 8 out of 10 of the females revealed that $(FL > FC)$, then we could assign a hit score of 0.80 to this prediction; similar scores could be calculated for the remaining five predictions. In either case, the result would be an index of ordinal fit, and we think it would be far more meaningful than an F ratio, p level, or estimate of amount of variance accounted.

Much could also be said about the use of OPA for stimulus contingent predictions, for predictions of multiple dependent or criterion variables, or for time series analyses. In future articles we hope to cover these topics in detail.

CONCLUSIONS

Now that we have outlined the major features of OPA and demonstrated some of their uses, we believe it is worthwhile to reconsider our original arguments for the development of OPA, and the reasons why it should be employed in the analysis of psychological research. These considerations would perhaps not be necessary if OPA were the only analytic tool around. But psychologists are faced with a glut of statistical procedures, indeed, so many procedures that it would take all of a researcher's time just to learn them, and no research would get done. Some convincing reasons must therefore be given for learning and using OPA in hopes that they will persuade psychologists to learn and use it.

It is not uncommon for a researcher to learn a handful of statistical procedures in graduate school (often under duress), then employ only these in subsequent research. This strategy would be defensible if the procedures learned were appropriate for most, if not all, domains of research enquiry. But we believe that those commonly learned (variants of the general linear model, chi square, and a few rank order statistics) are not appropriate for most domains of research in psychology. The general linear model in particular seems ill suited for a large proportion of psychological research, especially that involving single subjects or repeated measure designs. It decomposes variation into main effects surrounding the grand mean and interactions surrounding the main effects and other interactions in much the same way as Ptolemy's model of the solar system decomposes the orbits of planets into cycles and epicycles.

This conception is awkward at best for testing predictions about complex ordinal relations of averages. The averages themselves do not necessarily tell us anything about the general patterns of our results (nomothetic patterns), or the particular patterns generated by individuals (idiographic patterns). And the omnibus tests of significance derived from the general linear model inform us only of the reliability of mean differences, not the goodness of fit between the direction of these differences and the direction predicted by some theory or hypothesis.

In the end, it is the pattern of data, not the statistical significance of differences, that determines research importance. Recognition of this fact has prompted several statisticians to reconsider the centrality of descriptive statistics in the analysis of research results (e.g., Hildebrand, Laing, & Rosenthal, 1977; Tukey, 1977), and to relegate inferential statistics to their proper secondary role. In developing OPA, we follow the lead of these statisticians. Its indexes are descriptors of predictive generality and fit, rather than of accountable variance. It has been intentionally developed to describe the concordance of patterns, rather than their statistical significance as traditionally defined.

Like all descriptive statistics, the indexes of OPA make no assumptions about the processes underlying the production of data to be analyzed. Interpretations of indexes are thus predicated on a researchers' knowledge of where the data come from and how they are obtained. This can be both a curse and a blessing. It is a blessing because a researcher need not be concerned with establishing random samples, independent observations, normal distributions, counterbalanced designs, or complete data sets in order to make good use of OPA. It is a curse because it requires the researcher to give considerable thought to alternative interpretations of results that were obtained without these usual requirements. Of course, it is unfair to label thinking a curse. The thinking required to make good use of OPA in observational or correlational research is nothing more than the thinking required for good detective work: using one's noodle to cull spoiled hypotheses by imaginative analyses of existing or subsequent data. Such thinking is only a curse to those who have not learned how to do it.

Even though OPA indexes can be tested for significance, their real value lies in the interpretations that can be made of them. But how shall they be interpreted? How wide is a wide scope? What is a good fit? These questions have no objective answers. Like the interpretive traditions that have arisen in the use of classical statistics, interpretations of OPA indexes will develop in conjunction with their use in particular research areas. For some areas, a scope of 0.10 or a fit of + 0.20 will probably be seen as quite acceptable; for others, a scope of 0.80 or a fit of + 0.60 will regularly occur. This is really no different than the $p = 0.05$ or $p = 0.01$ levels of significance that have become the norm in many research areas, or the joy or sorrow with which researchers in different areas greet a correlation of, say, $+0.50$. The objectivity sometimes claimed by the use of traditional statistics rarely, if ever, exists. What is claimed as objective is almost always nothing more than consensual. In this regard, interpretations of OPA indexes are no more subjective than interpretations of more established statistical indexes.

Perhaps the most troublesome aspect of OPA has nothing to do with the procedure itself, but with the development or derivation of hypotheses that may be subjected to it. Hypotheses are relatively easy to construct; what is difficult is to show how they may be deduced from theoretical premises or assumptions. Whereas the methodological skills of psychologists have grown enormously in the past few decades, skills in theory construction and hypothesis derivation have, if anything, shown precipitous decline. The age of grand theories in psychology is long past. What now passes for scientific research in the discipline is largely scientism, the amassing of numbers to be poured into giant computer-driven statistical engines in hopes of separating truth from falsehood. If the discipline is to once again flourish, it must undergo a renaissance of theoretical interest and skill. By employing OPA as an analytical alternative to the omnivariate statistical procedures that now dominate our discipline, one will quickly confront the theoretical issues that are so well hidden by averages, mean squares, alpha levels, beta weights, eigenvalues, and the like. We do not pretend that OPA will resolve these issues, but we do claim that it will help challenge psychologists to try.

REFERENCES

Bishop, Y., Feinberg, S., & Holland, P. (1975). *Discrete multivariate analysis.* Cambridge, MA: MIT Press.

Bradley, J. (1968). *Distribution free statistical tests.* Englewood Cliffs, NJ: Prentice-Hall.

Bush, R. R. (1963). Estimation and evaluation. In R. Luce, R. R. Bush, & E. Galanter (Eds.), *Handbook of mathematical psychology* (Vol. 1, pp. 429–469). New York: Wiley.

Diaconis, P., & Efron, B. (1983, May). Computer intensive methods in statistics. *Scientific American,* pp. 116–130.

Edwards, A. W. F. (1972). *Likelihood: An account of the statistical concept of likelihood and its application to scientific inference.* Cambridge: Cambridge University Press.

Hildebrand, D., Laing, J., & Rosenthal, H. (1977). *Prediction analysis of cross classifications.* New York: Wiley.

Kendall, M. (1938). A new measure of rank correlation. *Biometrika, 30,* 81–93.

Meehl, P. (1978). Theoretical risks and tabular asterisks: Sir Karl, Sir Ronald, and the slow progress of soft psychology. *Journal of Consulting and Clinical Psychology, 46,* 806–834.

Restle, F. (1970). *Introduction to mathematical psychology.* Reading, MA: Addison-Wesley.

Restle, F., & Greeno, J. (1970). *Introduction to mathematical psychology.* Reading, MA: Addison-Wesley.

Savage, L. (1954). *Foundations of statistics.* New York: Wiley.

Thorngate, W. (1976). Possible limits on a science of social behavior. In L. Strickland, F. Aboud, & K. Gergen (Eds.), *Social psychology in transition* (pp. 121–139). New York: Plenum Press.

Tukey, J. (1977). *Exploratory data analysis.* Reading, MA: Addison-Wesley.

PART III

Toward the Study of Individual Subjects: Contributions from Different Fields in Psychology

CHAPTER 9

Academic Diagnosis
Contributions from Developmental Psychology

HERBERT P. GINSBURG

The aim of this chapter is to show how the findings, theories, and methods of cognitive developmental psychology can provide important applications to the problem of *academic diagnosis*—the understanding and
fostering of individual children's school learning. The thesis is that two
branches of cognitive developmental psychology, namely Piaget's clinical
interview method and recent research and theory on the development
of mathematical thinking, provide a distinctive view of academic diagnosis, offer a framework for the critique of existing diagnostic techniques,
and in some areas provide the substance and methods for new diagnostic
approaches. These should be useful not only for the diagnosis of children's arithmetic, but for the assessment of academic knowledge and
cognition in general. Academic diagnosis is one example of how scientific
psychology can contribute to the understanding of and to the aid of the
individual. At the same time, the new approaches to academic diagnosis
suffer from major shortcomings because developmental psychology has
failed to deal successfully with key issues of performance, learning, and
personality. The chapter concludes with a discussion of how recent experience with cognitive diagnosis sheds light on the strengths and weak-

HERBERT P. GINSBURG • Department of Developmental and Educational Psychology,
Teachers College, Columbia University, New York, New York 10027. Preparation of this
chapter was supported in part by a Public Health Service grant from the National Institute
of Child Health and Human Development (1 R01 HD16757–01A1).

nesses of scientific psychology and suggests directions for its future development.

THE NATURE AND GOALS OF DIAGNOSIS

In the academic setting, the process of diagnosis is usually initiated in response to a practical problem, as when a classroom teacher identifies a child as having difficulties in learning ordinary school arithmetic. The evidence for failure is typically both abundant and clear: the child performs poorly in classroom activities, getting wrong answers or giving no answers at all when called on in class; he or she receives low grades in arithmetic and in most other school subjects; and usually the child knows quite well—and so do peers—that he or she is failing. Yet outside of the classroom, in the playground, the child's everyday behavior often reveals at least average intelligence. The child does not appear to be retarded or emotionally disturbed. Wanting to know why the child is having problems in arithmetic, the teacher requests answers to some key questions: How badly is the child doing? Why is the child having trouble? Is the child's memory faulty, can he or she deal with abstractions, think logically? The teacher wishes a diagnosis—an understanding that can provide insight into the child's performance, identify those cognitive and other psychological factors responsible for the failure to learn, and suggest practical remedies. In the typical educational setting, almost any form of thinking or technique that is even remotely plausible may be used to obtain a diagnosis—practitioner's knowledge, intuition, hunches, general knowledge of psychology, informal observation, stereotypes, and standard tests. The educator must cast a wide net in order to develop a practical understanding of a student experiencing difficulty.

What is diagnosis from the standpoint of scientific psychology? The essence of diagnosis—or assessment or testing—seems to be the systematic attempt to obtain reliable information leading to the development of a sound psychological theory of the child's functioning in academic situations. Diagnosis employs psychological concepts and methods to conduct what might be called research in miniature—research focusing on the individual and aimed at obtaining data that can provide the basis for a theory of individual functioning. Diagnosis is not primarily concerned with producing general scientific knowledge, presumably of wide scope, but aims instead at obtaining a deep and practical understanding of a particular individual at a given historical moment. At the same time, such research in miniature may result in new general knowledge; the facts of the individual case may be incongruent with existing theory and therefore require new explanatory concepts. Similarly, because existing methods may not be sufficiently sensitive to investigate the individual, new ones, ultimately of value to general psychology, may have to be invented.

Given that diagnosis is the attempt to develop a sound and practically useful theory of the individual, the next question is, What kind of theory should the diagnostician attempt to create? What should be its goals? There seem to be five distinct types of diagnosis that can be briefly characterized as follows.

Traditionally, diagnosis has adopted the goal of *normative comparison.* This involves a comparison of the child with peers on some psychological attribute or dimension. The main question to be answered is, How does the individual compare with peers on key indexes of achievement and ability? The main technique employed in answering this question is standardized testing. The child is given various tests, like the Key Math (Connelly, Nachtman, & Pritchett, 1976) and the teacher is informed that the child is in the 25th percentile of "computational skill" or the 34th percentile of "conceptual ability." Normative diagnosis of this type is the prevailing practice.

A second type of diagnosis adopts the goal of *testing for performance.* This involves determining whether the child can perform a certain activity or solve a particular task, like solving addition problems involving "carrying." The main question to be answered is, Does the individual succeed or fail on a particular task? The main technique employed in answering this question is the standarized presentation of carefully selected tasks of educational interest. This is sometimes called criterion referenced testing.

A third goal for diagnosis may be termed *cognitive analysis:* it involves the discovery and measurement of specific cognitive processes involved in academic performance. The main question to be answered concerns what cognitive activities—processes, operations, structures, and so forth—generate the observed behavior, namely difficulty in school arithmetic. The aim is not to compare children or to identify successful or unsuccessful performance, but to characterize, with as much precision as possible, the specific cognitive processes of the individual child. Although some success in answering such questions may be achieved by standarized testing, a more powerful method is the clinical interview.

A fourth goal for diagnosis is the *assessment of learning potential.* This involves discovering whether the individual can learn certain kinds of material and identifying the conditions that facilitate such learning.

The aim is not to characterize the child's current status—in terms of normative standing, absolute performance, or cognitive process—but to discover what he might be able to do under favorable conditions.

A fifth goal for diagnosis is obtaining an *integrative portrait* of the individual. This involves a description of the configuration of psychological processes that characterize the individual's functioning. The main questions to be answered concern the nature of the unique constellation of cognitive and learning processes characterizing the individual. How are these processes integrated into the systems of personality and motivation? To answer questions like these, a variety of methodologies need to be employed, ranging from standardized testing to naturalistic observation.

These then are five types of diagnosis—normative comparison, performance testing, cognitive analysis, assessment of learning potential, and the integrative portrait. They may be illustrated by a series of simple questions concerning early arithmetic: Does a child add with less facility than do peers? Can the child add 2 digit numbers? How does the child do 2 digit addition? Can the child learn to do 3 digit addition? And how are these activities integrated in the child's cognition and personality as a whole?

CONTRIBUTIONS OF DEVELOPMENTAL PSYCHOLOGY

We next consider how cognitive developmental psychology, new and old, contributes to achieving the normative and cognitive goals of diagnosis. The contributions involve a theory of academic knowledge and a sensitive measurement technique—Piaget's clinical interview method.

NORMATIVE DIAGNOSIS

In the case of a child obviously failing in school work, the diagnostician begins with the normative aim of determining how the individual compares with peers on key indexes of ability. The chief approach to achieving the normative aim has involved the standardized measurement of ability. Several diagnostic tests (e.g., the Key Math, Connelly, Nachtman, & Pritchett, 1976) have attempted to provide measures of general mathematical ability or more specific abilities like computational or conceptual skill. From such tests, we might learn that a child ranks at the 15th percentile in computational skill and at the 30th percentile in conceptual ability.

Such comparisons are valuable to the educator to the extent that they provide information illuminating important psychological characteristics of the child and leading to effective educational activities. Judged by this standard, most normative tests are of limited utility. One reason for this is that the theories underlying them to fail to involve clear analyses of specific cognitive processes. Typically these theories postulate notions like quantitative thinking, general mathematical skill, conceptual understanding of mathematics, or computational skill. For the teacher, comparisons involving notions like these are useful mainly in designating broad areas of strength or weakness that deserve closer attention. Thus, the normative test may show that a child is particularly adept in quantitative skill but not in concepts. This information may be valuable in suggesting to the teacher that there is a problem with respect to concepts and that the child exhibits proficiency in some areas of mathematics. Yet the normative test explains little if anything about mathematical performance. It fails to reveal in what ways the child is skilled at quantity and in what respects he does not understand concepts. The lack of such information usually prevents typical normative tests from offering practical benefits. What good does it do the teacher to know that the child is weak on general mathematical ability or on computational skill? Can this do more than remind the teacher of what is usually well known, namely that the child needs help in mathematics in general and computation in particular?

Fortunately, recent cognitive research leads to attempts to improve on normative tests (Glaser, 1981). Cognitive theorists do not find it useful to conceptualize mathematical thinking in terms of general mathematical ability or similar traits, such as computational skill. Instead, they describe in detail the cognitive processes underlying mathematical performance. Thus, recent theories (as represented, for example, in Ginsburg, 1983) postulate information processing (e.g., Resnick, 1983), mathematical (Groen & Parkman, 1972), or computer-based models (VanLehn, 1983). By providing new and specific conceptions of mathematical thinking, this research permits the attempt to construct normative tests comparing children on theoretically important aspects of mathematical knowledge. Consider first the nature of this research and then the kind of normative tests that derive from it.

A large body of recent research evidence, only part of which can be alluded to here (much is summarized in Ginsburg, 1982, 1983), suggests that early mathematical thinking is extremely complex and may be characterized by several different cognitive systems operating simultaneously. Among these are the informal concepts and procedures that are developed before children enter school or outside the context of formal education, and the invented procedures that children create themselves in the context of the classroom. Each of these notions has important implications for the development of diagnostic methods.

Recent research reveals what may be a startling finding, namely that children possess an informal mathematics, quite apart from what they may learn in school. An example is the perception of "more," studied originally by Binet (1969) in 1890. Given two randomly arranged collections of blocks, a child of 3 or 4 years of age can easily determine which of the two collections has more than the other, at least when relatively small numbers of elements are involved. We have shown that young children exhibit the perception of more at a young age in both the United States (Ginsburg & Russell, 1981) and Africa (Posner, 1982). Another example of informal knowledge can be found in the procedures that children employ to solve simple addition and subtraction problems. Gelman and Gallistel (1978) have shown that children as young as 2 and 3 years understand some basic principles underlying addition and subtraction; many writers agree that young children often employ counting to solve addition and subtraction problems (Fuson & Hall, 1983). Diagnostic tests would do well to provide measures of important informal notions such as these.

Recent research also shows that once children enter school, they do not always do arithmetic in the ways they are taught. The child's representation of these procedures is not necessarily identical with the procedures themselves. The child assimilates what is taught into an already existing cognitive framework. Instead of using the official standard algorithms or number facts, they often employ invented procedures (Ginsburg, 1982; Groen & Resnick, 1977), various tricks that are often based on informal knowledge. Thus, instead of remembering the sum 6 + 7 as one is supposed to, the child may solve the problem by counting on the fingers, a method which the child (and perhaps the teacher) sees as cheating. In the case of written calculation, the invented procedures usually employ some forms of simplification and counting. For example, in solving a column addition problem like 43 + 76, the child might simplify the terms into 40 + 70 and then 3 + 6, and might solve part of the problem by a counting procedure (e.g., 3 + 6 is 6, 7, 8, 9). In view of the prevalance of invented strategies, diagnostic methods need to be sensitive to the child's construction of mathematics, not just to its conventional definition.

These are but a few representative features of the new findings concerning the development of mathematical thinking. Their use in test development may be illustrated by consideration of the Test of Early Mathematics Ability (TEMA), which we have recently developed (Ginsburg

& Baroody, 1983). Based on recent cognitive research, the TEMA attempts to measure aspects of informal mathematical knowledge and written procedures learned in school. For example, the test contains items dealing with the addition of concrete objects, the enumeration of dots, mental addition and subtraction, and the mental number line. The TEMA, which is given in a standardized fashion and has been normed on a large sample, yields measures of specific formal and informal concepts and procedures, not general mental traits. The test compares children not only on what they have learned in school (like alignment procedures for column addition) but on their unsuspected strengths, the informal procedures developed outside the context of schooling. The resulting profile describes aspects of informal and formal knowledge. Such a diagnosis is intended to provide a rounded view of the child experiencing difficulties in school arithmetic, a view that sheds light on both what the child can and cannot do. Whereas others will have to evaluate the TEMA, it is an example of the kind of test construction that can result from recent cognitive developmental research and that may prove to be of some practical utility.

Although tests like the TEMA seem to represent progress, they suffer from several limitations. Because their aim is primarily normative—the comparison of children with peers—they cannot accomplish much in the way of precise measurement of cognitive process. The TEMA measures several areas of mathematical thinking—areas which seem to be more interesting that those covered by traditional normative tests—but does not have the capability to examine any in depth. For example, the TEMA contains mental addition problems because recent research has shown that mental arithmetic is a key aspect of children's mathematical thinking. Presumably the TEMA's inclusion of such items represents an improvement over tests based on simple trait theories. Yet, the TEMA only manages to compare children's performance on such items; it does not go far in uncovering the cognitive processes actually employed in solving them. It seems impractical to accomplish both tasks, the normative and the cognitive, at the same time. As we shall see in the following, the cognitive aim may be accomplished by specially constructed standardized tests or by the clinical interview method.

Another limitation of tests like the TEMA is that the logic underlying standardized testing prevents them from consistently obtaining an accurate assessement of competence. Originally, standardized testing was developed in a praiseworthy attempt to avoid mislabeling children: "the interests of the child demand a ... careful method. To be a member of a special class can never be a mark of distinction, and such as do not merit it, must be spared the record" (Binet & Simon, 1916, p. 10). In particular, the aim was to eliminate the inconsistency produced by testers who were subjective in approach, whose methods of examination varied in apparently idiosyncratic ways and therefore resulted in unreliable diagnoses. Eliminating this inconsistency would insure that all children receive fair treatment. Consequently, the essence of standardized testing is presenting the same items to all children in the same manner. No flexibility or variation in questioning is permitted. The rationale is that if the presentation of test items is thus held constant, then subsequent differences in performance can be attributed not to variations in the questions but to real differences in the subjects' psychological characteristics. Such standardization of procedure, it is argued, provides for fairness; it permits the objective and reliable ranking of children with respect to peers on a given psychological trait.

Standard tests do indeed provide the fairness intended. The lack of variation in questioning eliminates certain kinds of bias, for example, that which might be shown by an examiner who judges a child to be retarded (or not retarded) on the basis of physical appearance or race alone. But there is a price to pay for the rigidity producing this kind of color blindness. To the extent that tests are standardized, they lack sensitivity to the individual and may therefore misjudge competence in many cases. For example, the standard tester cannot make the adjustments in questioning necessary to eliminate minor misunderstanding of instructions, and this may result in the child's failing to exhibit an existing skill or concept. Beyond superficial attempts at rapport ("we are now going to play a game"), the standard tester cannot present the problems in ways uniquely designed to motivate the individual child. Promoting understanding and motivation, at least for some children—and perhaps in the most relevant children, namely those doing badly in school—requires a flexibility of approach that by definition cannot be provided by standardized testing. In its attempt to be fair by maximizing objectivity, standardized testing regulates and narrowly focuses the examiner's activities, and in the process diminishes sensitivity to the child.

In brief, normative diagnostic tests have typically been based on general notions that limit the explanatory and practical value of the comparisons obtained. Recently, attempts have been made to use advances in research on children's mathematical thinking to improve normative diagnostic tests. But such efforts can achieve only limited success. The normative aim of these tests, even when informed by cognitive theory, prevents them from providing precise accounts of the cognitive process. Also, the standardized nature of these tests robs them of the flexibility necessary to measure competence in many cases.

COGNITIVE ANALYSIS

Suppose the diagnostician wishes to inquire into specific cognitive processes underlying the child's work. For exmple, why does the child get wrong answers in written calculation? One way of approaching a question like this is by means of a standardized test focused on specific cognitive processes. Consider first the cognitive processes underlying errors of written calculation and then how standard tests can be devised to measure them.

Recent research makes an important contribution in providing detailed analyses of the cognitive processes responsible for children's difficulties in arithmetic. For example, research shows that children's errors of written computation are often caused by underlying bugs (Brown & Burton, 1978) or systematic error strategies (Ginsburg, 1982). These are procedures that are fundamontally defective and in certain situations lead to regular, predictable errors. The child applies the systematic error strategy quite conscientiously and accurately, but it is so fundamentally flawed that it must lead to error in certain problems. Some error strategies are frequent and well known. For example, in subtraction, one common bug is when the child operates according to his own rule, "Always subtract the smaller from the larger." Applied to 23 − 16, this bug results in the answer 13, because the child subtracts the 3 from the 6 in the units column. Sometimes, this bug may lead to a correct response, as in the case of 25 − 11, as both numbers in the subtrahend are smaller than those in the minuend. The notion of the systematic error strategy or bug asserts that wrong answers may not be random or meaningless. Instead, mistakes may be generated by use of misguided but systematic procedures, many of which have been identified in this country and others (Brown & Burton, 1978; Ginsburg, Posner, & Russell, 1981). Moreover, these bugs usually have sensible origins (Ginsburg, 1982) in the child's learning history. For example, the rule, "Always subtract the smaller from the larger" may have originated in early subtraction lessons; the child's mistake was to generalize it beyond its proper domain. In brief, many calculational errors are not random; they derive from systematic error strategies that often have sensible origins.

The bugs identified by the recent research can be measured by standardized procedures. Indeed, Brown and Burton (1978) have managed to describe bugs so clearly as to be able to develop a computer program that measures them in a standardized fashion. The basic logic involves presenting the child with written calculational problems carefully selected so as to reveal the operation of key bugs. For example, suppose one wishes to identify the error strategy "Always subtract the smaller from the larger." To measure this bug, it is necessary to present the child with problems of the form $82 - 28$, where the subtrahend (28) has a number in the units position (8) that is larger than the units number (2) in the minuend (82). On problems like these, use of the bug results in a clearly predictable and distinctive answer—an answer that is most likely to have been generated by the bug in question. Thus, on the problem cited, the bug generates the answer 66, whereas the standard algorithm produces the answer 54. If carefully selected problems of this type are employed, then the child's overt response—the calculated answer to the written problem—provides useful evidence concerning the underlying strategy. Thus, in the above example, the answer 66 leads one to infer that the operative bug was "smaller from larger." To make the inference convincing, it is also necessary to demonstrate that the child consistently uses the identified bug on problems designed to reveal it (e.g., like $43 - 25$ to get 22).

Although useful in measuring clearly defined cognitive processes, a standard test of this type suffers from several limitations. One difficulty involves validity of the measure: it may not be clear that the intended bug is really being measured because observed responses can often be generated by a number of bugs. For example, if a child does $26 + 20 = 40$, one reasonable explanation is that the child used the "zero makes zero" bug. That is, the child added $6 + 0 = 0$ because of the belief that zero added to something makes zero, and then added $2 + 2 = 4$, to get the answer 40. But there are other possibilities as well. For example, the child could have been using a multiplication bug, multiplying by units. Operating with this bug, the child begins by multiplying units by units $(6 \times 0 = 0)$ and then tens by tens $(2 \times 2 = 4)$ and also gets the answer 40. Because both bugs generate the same answer, the interpretation of the child's response is ambiguous: the answer itself is obviously not an infallible guide to the underlying bug.

Another difficulty arises when the child employs a bug that the test is not designed to measure. Brown and Burton (1978) have shown that there are literally hundreds of distinguishable bugs. It is obviously impossible for a standardized diagnostic test, which must be adminstered in about 30 minutes, to measure many, let alone all of them. Hence, on a test of this type, those answers generated by bugs that were not intended to be measured must remain uninterpretable.

One should be aware of limitations such as these in evaluating any standard test focusing on specific cognitive strategies, concepts, and the like. It is possible that a strategy different from the one postulated can produce the observed behavior. And if an individual employs an unanticipated strategy or concept, the test may not be able to identify it, beyond categorizing it as "other."

Several solutions to these kinds of difficulties have been proposed. One solution—use of the talking aloud procedure—attempts to supplement the data normally obtained by standardized testing. Another solution—Piaget's clinical interview method—relaxes, or even abandons, the rules of standardization in order to introduce a degree of flexibility to test administration. Consider each solution in turn.

In the talking aloud procedure, the child is asked to "think aloud" or "say everything that comes into your head" while solving the problem. The rationale is simply that verbalizations concerning the solution process often remove ambiguities of the types previously described and provide positive evidence concerning the operation of bugs. If a given written answer could have been generated by at least two different bugs (as in the example $26 + 20 = 40$ given earlier), then the child's verbalizations ("I added 6 and 0 and got 0") may decide the issue. If a given written answer could not have been generated by a bug that was intended to be measured, then the child's verbalizations may reveal the operation of an obscure bug, or at least one previously unknown to the tester. The talking aloud procedure is not new; it has a long history (Ginsburg, Kossan, Schwartz, & Swanson, 1983), and has been used with considerable success in problem-solving research (Newell & Simon, 1972).

Clearly, it would be useful to supplement standardized procedures of the type developed by Brown and Burton (1978) with a talking aloud method. This in fact was done many years ago. The pioneers in this area were Buswell and John (1926), who used carefully selected problems and children's verbalizations to measure the "habits" of arithmetic. These habits included what we would now call bugs, as well as invented strategies. Recently, we have developed the Diagnostic Test of Arithmetic Strategies (Ginsburg & Mathews, 1984), which also uses specially designed problems and talking aloud to measure the most frequent bugs in addition, subtraction, multiplication, and division, and also to identify informal strategies that lead to sound solutions. (This test also uses elements of Piaget's clinical interview method to be described later.) Tests like these, which attempt to provide accurate measures of key cognitive processes employed in children's arithmetic, are intended to be of practical value for education.

Despite their virtues, standard tests of bugs or similar cognitive processes, even when supplemented by talking aloud procedures, suffer from major limitations. One is that tests like these, because of their standardized nature, do not permit the tester to explore possibly interesting responses. If the tester adheres to standardized testing procedures, he may learn something from an unusual written response and an intriguing verbalization, but is not free to explore matters productively, to examine mental activities in adequate detail.

Fortunately, Piaget (1929) long ago developed a research procedure, the clinical interview method, designed to overcome limitations like these. For many years, American researchers considered the clinical interview method to be inadequate for purposes for serious research. It was thought to be sloppy and imprecise, suitable only for exploratory pilot work. This view, it is argued, is mistaken. The clinical interview method is based on a sound theoretical rationale that makes it useful for both basic research and diagnosis. Indeed, the method has recently begun to assume an important role in research (e.g., Gelman, 1980), and has long been used by sensitive clinicians in diagnosis.

Implicit in the clinical interview method are three distinguishable purposes, each requiring different techniques and each justified by different arguments. The clinical interview aims at discovering cognitive activities; specifying them with precision; and evaluating levels of competence. In any given instance, the distinctions among the aims may be blurred and more than one aim may be involved. But the aims and the techniques of the clinical interview method are vital for the cognitive purpose of diagnosis.

Piaget's first aim, discovery, is the least controversial of the three. In Piaget's view, one basic task of cognitive developmental psychology is to discover the cognitive processes actually used by children in a variety of contexts. Piaget felt that it is unproductive to begin research with an *a priori* definition of cognition; our first interest should be in discovering what cognitive processes are operating. To accomplish this, Piaget designed the clinical interview method, which he described as an unstructured and open-ended procedure intended to give the child a chance to display his "natural inclination," and to let the examiner engage in a kind of naturalistic observation of unanticipated results.

When discovery is the aim, the examiner employs an open-ended task, asks questions in a manner contingent on the child's previous responses, and requests a good deal of reflection on the part of the child. The interviewer attempts to avoid putting words in the child's mouth or suggesting answers—these are both clear dangers of the clinical interview method. The clinical method is intended to draw from the child, in an unbiased manner, rich material allowing the discovery of unsuspected cognitive processes.

Piaget's aim of discovery, and his clinical interview method, appear to have important implications for diagnosis and for research. If the child in a diagnostic session displays an unusual or unexpected answer, what could be more useful for the examiner than to follow up on it in a manner that is deliberately unstandardized— that is, contingent on the child's response? If the child's answer suggests an unusual bug, then the appropriate tactic is to engage in open-ended questioning designed to identify it. For example, a third grader, Butch, said that he was working with fractions in school. The interviewer asked, "Fractions? Can you show me what you are doing with fractions?" Butch then wrote $8\sqrt{16}$—apparently a division problem. Asked "What do you do with it?" Butch replied, "You add it up and put the number up there." He wrote:

$$\begin{array}{r} 23 \\ 8\sqrt{16} \end{array}$$

Obviously, Butch was doing something unusual. In this brief segment, he confused division with fractions, and worked the division problem as addition, making a minor error that resulted in the answer 23 instead of 24. The clinical interview method was effective in discovering Butch's unusual procedures; it is hard to see how standardized testing could accomplish as much because Butch's procedures were so rare and unexpected. For purposes of discovery, the clinical interview method should play a major role in the diagnosis of mathematical thinking.

The aim of specification may also be served by this method. In Piaget's view, once interesting intellectual phenomena have been discovered, the cognitive processes underlying them need to be specified—that is, measured and described in detail. If, for example, we learn from exploratory work that a child adds by counting, we next need to describe the strategy in detail. This is often difficult to accomplish because underlying cognitive activities are complex. An addition strategy, for example, may involve a sequence of several different activities, like counting on the fingers, using remembered number facts, and reasoning from general principles (e.g., commutativity). To capture such complexity we require the freedom to ask questions contingent on previous responses, to clarify misunderstood questions, and to probe ambiguous responses. Furthermore, identifying the operative strategy can be a difficult inferential process because a given response can be produced by any one of several underlying processes (as, for example, when a given answer could have been generated by one of three different bugs). To cope with this situation too we need the flexibility of the clinical interview. It allows us—on the spot—to modify the questioning so as to create and examine hypotheses concerning alternative processes that could generate the current response.

When specification is the aim, the interview involves the techniques already discussed, namely asking the child to engage in extensive reflection, and employing contingency of questioning. The clinical interview also employs a technique that is usually not recognized as characteristic of it, namely the experimental method, in the sense of holding some variables constant while deliberately varying others. If, for example, the interviewer feels that the child experiences difficulty with the materials employed in the task, the experimenter may systematically vary them while holding language constant. If the experimenter feels that language is the problem, then the materials may be held constant and the language systematically varied. This is essentially a factorial arrangement, and if undertaken with a large enough number of subjects and an appropriate counterbalancing procedure it would be an example of traditional experimental design, necessary for deciding among alternative hypotheses. Such experimentation is extremely common in the clinical interview. In fact, one may describe clinical interviewing for the purpose of specification as the flexible and extremely demanding application of experimental methodology in a manner contingent on individual reactions. Furthermore, this notion of the clinical interview method as experimentation suggests a related interpretation of validity: the clinical interview method is valid to the extent that the experiments embedded in it employ adequate controls, rule out alternative hypotheses, lead to logical conclusions, and the like. In other words, the validity or accuracy of the clinical interview method may be judged—at least in part—by the same criteria as those employed to evaluate ordinary experiments. The clinical interview method is valid to the extent that it provides reasonable tests of the hypotheses in question.

When the aim is specification, the clinical interview method should play an important role in diagnosis and research. Because mathematical cognition is complex, sophisticated and flexible methods are required to measure it. Mathematical cognition does not consist of a collection of simple responses or memorized facts. It is a knowledge system of great complexity (Davis, 1983), involving ensembles of operations and extensive semantic networks (Resnick & Ford, 1981). Moreover, the mathematical knowledge of individuals is often manifested in idiosyncratic ways that are themselves complex. Because mathematical cognition is characterized by great complexity, sophisticated and flexible methods are required to measure it. The clinical interview seems ideally suited for coping with the complexities of mathematical knowledge in general and for deciphering its idiosyncratic manifestations.

A third aim of the clinical interview is the assessment of competence. Quite early on, Piaget (1929) arrived at the apparently simple insight that children's behavior does not always reflect the true extent of their knowledge. It is important to focus not on merely typical behavior, but on competence—the child's best performance at his current stage of development. Doing this requires a concern with issues like motivation. A prerequisite for the expression of competence often seems to be a positively motivated state. Especially when the problem is challenging, the child must be properly motivated to employ complex cognitive processes and hence exhibit his highest possible level of performance. Motivating a child requires dealing with the child on an individual basis, determining whether he or she is interested in a particular task and understands the instructions, and tailoring the task to his individual needs. Flexible methods are required to establish motivation in many children; standardized procedures, especially when group tests are used, are apt to be ineffective.

One method for establishing motivation is the flexible construction and presentation of tasks. For whatever reasons, sometimes obscure, a child may find one form of a task boring or overly difficult, but not another. One set of instructions, but not another, may captivate a child. The measurement of competence therefore requires the flexible and innovative construction and presentation of tasks. Moreover, this must be done, often on the spot, in response to the needs of individual children.

The situation is no different in diagnosis. The tester in the classroom or the clinic needs to establish the student's competence. It is usually all too well known that the child in question is performing poorly. The relevant questions are, What does he really know? What is he capable of? Answers to these questions depend in part on insuring that the child is properly motivated to work on the mathematical problems that it is necessary to administer. The clinical interview method can adjust the method of presentation to promote motivation in the individual; the very logic of standardized testing prevents it from employing such flexibility.

These then are some ways in which the clinical interview method can be used to accomplish the three aims of diagnosis. To this point, the virtues of clinical interviewing have been stressed, particularly at the expense of standardized testing. This was done because these virtues too frequently go unrecognized and because the value of standardized testing is often overrated. At the same time, the weaknesses of the clinical interview method should not be ignored. Any method that relies so heavily on the skill and sensitivity of the individual examiner is subject to many pitfalls. For example, it is possible for the examiner to ask leading questions, to bias the child's response, or to make the child feel incompetent. Moreover, little is known concerning the psychometric properties of the clinical interview method. In principle the method can be evaluated by the standard psychometric criteria of reliability and validity (Ginsburg et al., 1983), but in practice this has not been done to any significant extent. Although the clinical interview seems to be a powerful tool for research and diagnosis, the method needs to be studied intensively so that its reliability and validity in a variety of situations can be determined. Also, because the clinical inteview method is a rather artificial situation in that it involves an unusually high degree of adult intervention, it may need to be supplemented by other techniques, including the observation of activity in the natural setting, to be discussed later. In general, a continuum of assessment situations may be necessary to obtain a balanced view of the child's abilities.

In summary, recent research in cognitive developmental psychology offers a rich portrait of the ways in which children solve or fail to solve mathematical problems. This analysis suggests cognitive processes that diagnostic procedures ought to consider. One way of measuring these processes is by means of standard tests. These are often useful but may suffer from the defects of being unable to offer unambiguous interpretations of cognitive processes or to explore them in sufficient detail. Piaget's clinical interview method attempts to overcome these difficulties by employing flexible techniques to discover and specify cognitive processes and to establish competence. The cognitive analysis of academic knowledge seems to have a promising future because it can provide educators with practical information about individual children. And if an attempt is made to "teach to the test," the effects may be salutary because then teaching will focus on cognitive processes.

NEW DIRECTIONS

We next consider some possible new directions in academic diagnosis. Here are some speculations on how developmental psychology might contribute to the diagnosis of performance and learning, and to integrative diagnosis. We conclude by turning the tables to consider implications of work on academic diagnosis for scientific psychology.

THE DIAGNOSIS OF PERFORMANCE

In the course of academic diagnosis, the examiner is usually interested in whether the child can perform successfully on key tasks, often specified by the school curriculum. Can the child produce on demand the simple number facts? Can the child do multiplication with two digits? The examiner is not concerned with how the child does or fails to do these things (that would be the aim of a cognitive diagnosis) but is interested in establishing whether the child can do them. This may be called the diagnosis of *performance*, because its chief focus is on overt performance or behavior with respect to an important task. Sometimes such assessment is called "criterion referenced testing" because its aim is not to compare the child with peers but to determine whether the child meets a clear behavioral criterion, regardless of what peers may or may not do. Thus, according to Wallace and Larsen (1978):

> In *criterion referenced testing,* as opposed to normative referenced testing, an individual's performance is evaluated in terms of an absolute or specific criterion. . . . Because they are so specific, criterion referenced tests provide the teacher with *exact* information, which is directly applicable to programs of instruction. (p. 33)

In general, it does not seem too difficult to develop criterion referenced tests of this type. Usually, the examiner determines the main topics to be learned in school (e.g., 2 digit addition), develops specific tasks illustrating those topics (e.g., 23 + 46), and then observes whether the child solves the taks. An example is the Basic Educational Skills Inventory—Math (BESI) (Adamson, Shrago, & Van Etten, 1972, quoted in Wallace & Larsen, 1978), which measures such performances as "naming printed numbers, counting pictured objects, counting orally, writing numbers, number sequencing—before and after" (Wallace & Larsen, 1978, p. 454). Performance tests like these serve a useful function. For example, knowledge that the child succeeds at naming printed numbers—that is, can provide conventional names ("four") to ordinary numerals (4) appropriate to the age level in question—may be more valuable for instructional purposes than the information that the child is in the 15th percentile of computational skill.

At the present time, the diagnosis of performance typically uses tasks that are considered basic to the school curriculum—2 digit addition, number facts, and the like. No doubt mastery of many of these tasks is indeed crucial—it is important to know whether the child can add 2 + 2. But curriculum derived tasks are not sufficient: they are not the only ones useful for children to master. (And conversely, some curriculum tasks may not be important for students to master. Is there any sound reason, other than avoiding school failure, for learning the tedious mechanics of long division?) Recent research on the development of mathematical thinking has shown that informal mathematical tasks are important too. On their own, children acquire key informal mathematical skills and concepts not covered in the curriculum. For example, without the benefit of school instruction, children often develop rather complex mental addition strategies that are used in a more powerful manner than the poorly mastered algorithms taught in the standard curriculum (Ginsburg, 1982). Furthermore, these informal mental addition strategies may serve as a sound foundation on which to base instruction in the algorithm (Baroody, 1986).

The clear implication of this recent research is that criterion referenced tests should include informal tasks, because these tap unschooled skills and concepts that are important in themselves for dealing with quantity in the ordinary enviroment and can also be used to improve instruction. If this suggestion were implemented, then criterion referenced tests would assume a new character: they would include tasks dealing with such topics as estimation, mental calculation, finger counting, and the like.

In brief, developmental psychology can contribute to the diagnosis of performance, as to normative diagnosis, by providing theory-based information specifying the nature of important tasks. Recent research suggests that criterion referenced tests should include not only curriculum-based tasks, but informal tasks as well. One reason is that the informal skills and concepts tapped by these tasks are important in themselves. A second is that they can form the basis for effective school learning. And yet another reason is that knowledge of informal strengths may lead the educator to take a more optimistic view of the child's potential for learning than would be likely otherwise.

IDENTIFYING LEARNING POTENTIAL

Suppose that a normative diagnosis shows that the child is in the 20th percentile of quantitative skill; that a performance diagnosis reveals that he cannot do column addition; and that a cognitive diagnosis suggests that a particular bug is responsible for the failure in column addition. The next question that arises is whether the child can learn column addition. Indeed, for purposes of education, learning potential is the crux of the matter, and although valuable, all other information—normative, performance, and cognitive—must be of secondary interest. After all, the educator's ultimate aim in seeking diagnostic information is to promote the child's learning.

Papert (1980) even suggests that a focus on the current unsatisfactory situation—in our terms, normative, performance, or cognitive diagnosis—may be counterproductive:

> The invention of the automobile and airplane did not come from a detailed study of how their predecessors, such as horse-drawn carriages, worked or did not work. Yet, this is the model for contemporary educational research.... There are many studies concerning the poor notions of math or science students acquire from today's schooling. There is even a very prevalent "humanistic" argument that "good" pedagogy should take these poor ways of

> thinking as its starting point. . . . Nevertheless, I think that the strategy implies a commitment to preserving the traditional system. It is analogous to improving the axle of the horse-drawn cart. But the real question, one might say, is whether we can invent the "educational automobile." (p. 44)

From Papert's perspective, one might argue that it is less important to know what informal knowledge children posses at age 4 or why third graders in the current schools make addition errors than it is to discover what children can do under more nearly ideal circumstances. Current cognitive processes, as they are shaped by the typical school environment, may be almost irrelevant to the issue of learning in more stimulating circumstances. Whether the child can or cannot count at age 4 or employs some error strategy may not be of great relevance for what he or she can accomplish in the atypical classroom.

Much of the literature on radical educational reform may be taken to support this point. Many years ago, educators like Kohl (1967), in his *36 Children*, showed that unusual classrooms could produce atypically fine learning in poor children who as a group show dramatic failure in ordinary classrooms. More recently, Papert (1980) and his colleagues have shown that the LOGO computer environment can produce dramatic learning in physically handicapped children, some of whom are even judged to be retarded in ordinary classrooms.

Despite this evidence, it seems as if Papert underestimates the utility of knowledge concerning the child's current status: as the accuracy of this information increases, so does the ability of the driver to steer the "educational automobile" safely and to determine how far and in what direction it can be driven. Of course, research is required to settle the issue. Empirical investigations need to establish the degree to which diagnostic knowledge concerning the child's current abilities assists in discovering and guiding what the child can learn and in establishing the limits on this learning. Nevertheless, Papert's emphasis on what can be learned in favorable circumstances is useful in leading to a focus on the diagnosis of learning potential.

What do we know about this kind of diagnosis? We may begin by noting the depressing fact that the field of educational assessment has devoted almost no attention to learning potential: formal procedures for evaluating learning potential in the key academic subject matter areas—for example, reading, writing, and arithmetic—do not seem to exist. No doubt sensitive clinicians attempt to examine learning potential, but they must employ informal means to do so.

What does developmental psychology have to contribute to the diagnosis of learning potential? Rather little. For the most part, cognitive psychologists have focused attention on a narrow aspect of mental life—

the current cognitive structures and processes of the individual—or more properly, groups of individuals. Traditional cognitive research has examined such issues as the nature of counting strategies in different groups of young children at ages X versus Y. Even the few studies employing as subjects the same children at different age levels typically examine existing cognitive structures in a static fashion and do not focus directly on learning or developmental processes. To be sure, there are exceptions. For example, Brown and VanLehn (1982) have proposed an intriguing theory of how bugs are generated in the child's attempt to make sense of instruction. But the fact remains that most cognitive developmental psychology is not directly concerned with the processes of learning and development, but with the characterization of differences in current structures or processes at various age levels.

Fortunately, there are now signs of renewed interest in learning potential. In recent years, developmental psychologists have been influenced by Vygotsky's (1978) notion of the "zone of proximal development" (ZPD) (see also Rogoff & Wertsch, 1984) and by Feuerstein's (1979) "dynamic assessment" of learning potential. Both Vygotsky and Feuerstein have stressed the social aspects of learning potential. For Vygotsky, the ZPD refers to

> the distance between the actual developmental level as determined by independent problem solving and the level of potential development as determined by problem solving under adult guidance or in collaboration with more capable peers. (p. 86)

For Feuerstein, the "mediated learning experience" is crucial, where "mediated" refers to the influence of a parent, peer, or other social agent. In both cases, then, the focus is on learning that might occur when facilitated by social influence.

Clearly, the Vygotsky and Feuerstein perspectives are fruitful and should lead to improvement in diagnosis. Indeed, several investigators (e.g., Campione, Brown, Ferrara, & Bryant, 1984) have drawn on this general perspective to develop innovative procedures for the diagnosis of learning potential. At the same time, the Vygotsky and Feuerstein perspectives raise several issues requiring further examination. On the independent variable side, both Feuerstein and Vygotsky stress the salutary influence of social factors in promoting learning. Although this is a useful emphasis, little is known concerning how these factors operate. We need to discover, for example, what kind of help and how much help is given by adults or peers, how the type of help given is influenced by the nature and requirements of the task, as well as by the child's cognitive status and personality. Also, on the dependent variable side, we need to consider the different types of learning that may result from different types of social interaction. Do certain types of help produce more transfer, domain-specific learning, memorization, invention, discovery, speedy learning, or spontaneous use of what has been learned than do other types of help? Research on issues of this type has only just begun (see, for example, Greenfield, 1984; Rogoff & Gardner, 1984).

In addition to attempting to clarify the role of social influences, one should not lose sight of the fact that other factors may influence learning potential as well. Thus, the provision of a stimulating environment may by itself promote learning (as when the introduction of a rich musical environment stimulates the learning of an individual perhaps predisposed to creativity in that area). Another factor may involve personality reorganization: through therapy or other means an individual may remove internal barriers and thereby liberate a potential for learning (for example, an individual overcomes a feeling of anxiety connected with mathematics and now displays a potential to learn it). No doubt, nonsocial factors influence learning potential and we need to learn more about how they operate.

In conclusion, the notion of learning potential is crucial for diagnosis and for general theory in developmental psychology. Advances in our understanding of learning potential and the various influences on it should eventually lead to important and eminently practical improvements in diagnostic practice.

INTEGRATIVE DIAGNOSIS

Another aim of diagnosis is to obtain an integrated portrait of the individual. The educator and clinician are always faced with the necessity of understanding "the whole child"—for example, a child who not only performs poorly in school arithmetic, but also feels anxiety about it, believes that he is a "dummy," dislikes school, is not doing very well in reading and other academic subjects, withdraws in the face of criticism, finds it hard to concentrate when difficult problems are presented, is dependent on the teacher for problem-solving suggestions, is afraid to show his grades to his parents, lives in a family where the parents have a strong fear and dislike of mathematics in particular and schooling in general, and so on. A comprehensive understanding of the child, one who is failing or excelling in schoolwork, must include consideration of general performance in school, attitudes and motivation, cognitive style, relations with parents and other adults, personality structure, and social environment. Such noncognitive factors obviously play a major role in the child's academic performance. Thus, the understanding of motivation may be at least as useful for educational practice—for remediation—as knowledge of cognitive structure or process. Diagnosticians must and do perform integrative analyses all the time: in the real world, children's academic cognition is embedded in a larger psychological and social context.

What kind of contribution can developmental psychology make to integrative diagnosis? At the present time, the required theory and methods are lacking. Although psychologists generally agree in principle that an integrated portrait is desirable, they also argue that in practice one can study thinking effectively only by focusing on limited sets of variables or isolated processes—that is, only through a process of analysis. Hence, contemporary research has focused mainly on what might be called the technical aspects of mathematical cognition—for example, invented strategies, bugs, semantic networks. In general, these processes have been considered in isolation. Only a few writers have attempted to suggest alternative approaches. For example, Papert (1980) stresses the affective component in learning mathematics. He argues that falling love with mathematics—one's relationship with the subject matter—is as crucial as knowledge of the subject matter itself. Allardice and Ginsburg (1983) present case study material illustrating the roles of motivation and personality in mathematics learning difficulties. For example, one important feature of difficulty in mathematics learning is a feeling of helplessness. Cole and Scribner (1974) stress the cultural context of mathematical and other forms of thinking. Shoenfeld (1983) demonstrates how belief systems and metacognitions are essential components of competent mathematical peformance.

Yet, despite recent attempts like these to broaden the conceptual framework in which academic cognition is considered, a good deal remains to be done. We need to develop new theoretical models and appropriate methods for understanding the whole child in the environment. For example, we need to understand the links between achievement motivation and cognitive process, between locus of control and knowledge, between social interaction and learning. Such attempts at integration are as important for basic theory as they are for diagnosis: an adequate account of the child's psychological functioning must explain the living synthesis between the cognitive and noncognitive.

In attempting to produce a theory of this type, we should remember that the "new" emphasis is a return to perennial themes in psychology. For example, Freud's cognitive psychology involved an examination of the interplay between personality components—repression, id impulses, and the like—and the workings of the mind as expressed in such phenomena as dreams and verbal errors (Ginsburg & Gannon, 1983). Using a case study aproach, Luria (1968) analyzed memory in the context of the psychological functioning of the whole person. Henry Murray (1938) attempted to develop diagnostic procedures for capturing both personality and cognitive characteristics of the individual. Perhaps contemporary psychology can make progress by tackling once again some of these enduring issues.

CONCLUSIONS

We have seen how developmental psychology, new and old, makes a contribution to the applied problem of diagnosis—research aimed at understanding the individual. We have seen that to understand the individual case, we need to draw on different types of evidence—normative, performance, cognitive, learning—and ultimately must attempt to produce an integrated portrait of the individual's functioning. Next we can ask, What are the general implications of this work on academic diagnosis for scientific psychology?

An answer to this question depends on one's view concerning the relation between diagnosis and research. According to the view proposed in this chapter, diagnosis is research in miniature—research designed to achieve an understadning of the individual case. As such, diagnosis is not fundamentally different in method from what is ordinarily considered to be research. The main distinction between the two is in the degree of generalization intended. Research is meaningless unless it can be generalized beyond the initial source of data; by contrast, the aim of diagnosis is to comprehend the original source of data, not to generalize beyond it. Because it aims at producing general knowledge, research must be concerned with representativeness of subjects (or with the representativeness of phenomena, as Thorngate and Carroll argue in this volume). Because it aims at understanding the individual, diagnosis has no concern with representativeness. Otherwise, research and diagnosis are quite similar. Both use, or aspire to use, reliable techniques to arrive at valid psychological theory, whether of the individual as such, or of people or phenomena in general.

If diagnosis and research are similar in the way described, then the lessons derived from diagnostic experience may be informative with respect to research. Thus, it is not only the "basic science" that makes a contribution to the applied; the reverse is true too. The lessons are several.

First, standardized methods of testing, which form the backbone of most research studies, are overrated because they are basically insensitive with respect to competence. Diagnosticians know this because their often lengthy informal contact with individual children repeatedly demonstrates that standard tests can be misleading. Researchers too need to learn this lesson and its correlate, namely that to obtain an accurate portrait of the individual, whether for the purposes of diagnosis or scientific psychology, we need to employ flexible methods that are useful for assessing competence. One such procedure is Piaget's clinical interview method, which has an ironic history. Whereas its roots were in psychiatric diagnosis, Piaget elaborated on it for purposes of research. Nevertheless, most researchers have not accepted it as a legitimate research method. Now the method is highly valued in diagnostic work and is only beginning to be appreciated as a research technique.

Second, just as group data can help us to understand the individual, so individual data can help us to understand the group. Another way of saying this is that diagnosis has led to interesting hypotheses, often confirmed by large-scale research, concerning the general psychology of mathematical thinking. As previously maintained, diagnosis is research in miniature that does not aim at generalization beyond the individual subject. But when such generalization is made, it often turns out to be quite accurate. The history of psychology confirms this proposition: for example, Piaget's observations of his own three infants have been widely replicated, even across wide cultural differences. Other examples of researchers making replicable hypotheses on the basis of few subjects include Freud, Wertheimer, Skinner, Chomsky, and Roger Brown. So the intensive study of the individual, whether for purposes of research or diagnosis, may lead more reliably than one might expect to general propositions concerning psychological functioning.

Third, diagnostic work demonstrates the need for more basic knowledge concerning learning potential and the ways in which cognitive processes are integrated into the personality as a whole. In the real world of children failing in school, diagnosticians must be concerned with these fundamental issues. Researchers should take them seriously too.

Finally, the diagnostic experience shows how valuable are the theories and techniques developed in developmental psychology generally. Key aspects of developmental psychology—for example, recent research on mathematical thinking and Piaget's clinical interview method—have been shown to be relevant to the very practical concerns of diagnosis. This is perhaps a kind of validation. If, as Lewin claimed, "there is nothing so practical as a good theory," then the soundness of the application proves the goodness of the theory.

ACKNOWLEDGMENTS

The writer wishes to thank Jane Knitzer, Kathy Gannon, and Karen Letourneau for their helpful comments on an earlier draft of this paper and Jaan Valsiner for his editorial advice.

REFERENCES

Adamson, G., Shrago, M., & Van Etten, G. (1972). *Basic educational skills inventory: Math (Level A and Level B).* Olathe, KS: SelectEd.

Allardice, B. S., & Ginsburg, H. P. (1983). Children's psychological difficulties in mathematics. In H. P. Ginsburg (Ed.), *The development of mathematical thinking* (pp. 319–350). New York: Academic Press.

Baroody, A. J. (1986). The value of informal approaches to mathematics instruction and remediation. *Arithmetic Teacher, 33,* 14–18.

Binet, A. (1969). The perception of lengths and numbers. In R. H. Pollack & M. W. Brenner (Eds.), *The experimental psychology of Alfred Binet* (pp. 85–98). New York: Springer.

Binet, A., & Simon, T. (1916). *The development of intelligence in children.* Vineland, NJ: Training School.

Brown, J. S., & Burton, R. B. (1978). Diagnostic models for procedural bugs in basic mathematical skills. *Cognitive Science, 2,* 155–192.

Brown, J. S. & VanLehn, K. (1982). Towards a generative theory of "bugs." In T. P. Carpenter, J. M. Moser, & T. A. Romberg, (Eds.), *Addition and subtraction: A cognitive perspective* (pp. 117–135). Hillsdale, NJ: Erlbaum.

Buswell, G. T., & John, L. (1926). *Diagnostic studies in arithmetic* (Supplementary Educational Monograph, No. 30). Chicago, IL: University of Chicago.

Campione, J. C., Brown, A. L., Ferrara, R. A., & Bryant, N. R. (1984). The zone of proximal development: Implications for individual differences and learning. In B. Rogoff & J. V. Wertsch (Eds.), *Children's learning in the "zone of proximal development"* (New Directions for Child Development, No. 23, pp. 77–91). San Francisco: Jossey-Bass.

Cole, M., & Scribner, S. (1974). *Culture and thought.* New York: Wiley.

Connelly, A. J., Nachtman, W., & Pritchett, E. M. (1976). *Key math diagnostic arithmetic test.* Circle Pines, MN: American Guidance Service.

Davis, R. B. (1983). Complex mathematical cognition. In H. P. Ginsburg (Ed.), *The development of mathematical thinking* (pp. 253–290). New York: Academic Press.

Feuerstein, R. (1978). *The dynamic assessment of retarded performers.* Baltimore, MD: University Park Press.

Fuson, K. C., & Hall, J. W. (1983). The acquisition of early number word meanings. In H. P. Ginsburg (Ed.), *The development of mathematical thinking* (pp. 49–107). New York: Academic Press.

Gelman, R. (1980). What young children know about mathematics. *Educational Psychologist, 15,* 54–86.

Gelman, R., & Gallistel, C. R. (1978). *The child's understanding of number.* Cambridge, MA: Harvard University Press.

Ginsburg, H. P. (1982). Children's arithmetic. Austin, TX: Pro Ed.

Ginsburg, H. P. (Ed.). (1983). *The development of mathematical thinking.* New York: Academic Press.

Ginsburg, H. P., & Baroody, A. J. (1983). *The test of early mathematics ability (TEMA).* Austin, TX: Pro Ed.

Ginsburg, H. P., & Gannon, K. E. (1983, June). *Sigmund Freud: Cognitive psychologist.* Paper presented to the Piaget Society, Philadelphia.

Ginsburg, H. P., & Mathews, S. C. (1984). *Diagnostic test of arithmetic strategies.* Austin, TX: Pro Ed.

Ginsburg, H. P., & Russell, R. L. (1981). Social class and racial influences on early mathematical thinking. *Monographs of the Society for Research in Child Development, 46* (6, Serial No. 193).

260

HERBERT P. GINSBERG

Ginsburg, H. P., Posner, J. K. & Russell, R. L. (1981). Mathematics learning difficulties in African Children: A clinical interview study. *The Quarterly Newsletter of the Laboratory of Comparative Human Development, 3,* 8–11.

Ginsburg, H. P., Kossan, N. E., Schwartz, R., & Swanson, D. (1983). Protocol methods in research on mathematical thinking. In H. P. Ginsburg (Ed.), *The development of mathematical thinking* (pp. 7–47). New York: Academic Press.

Glaser, R. (1981). The future of testing: A research agenda for cognitive psychology and psychometrics. *American Psychologist, 36,* 923–936.

Greenfield, P. M. (1984). A theory of the teacher in the learning activities of everyday life. In B. Rogoff & J. Lave (Eds.), *Everyday cognition: Its development in social context* (pp. 117–138). Campbridge, MA: Harvard University Press.

Groen, G. J., & Parkman, J. M. (1972). A chronometric analysis of simple addition. *Psychological Review, 79,* 329–343.

Groen, G. J., & Resnick, L. B. (1977). Can preschool children invent addition algorithms? *Journal of Educational Psychology, 69,* 645–652.

Kohl, H. (1967). *36 children.* New York: New American Library.

Luria, A. R. (1968). *The mind of a mnemonist.* New York: Basic Books.

Murray, H. A. (1938). *Explorations in personality.* New York: Oxford University Press.

Newell, A., & Simon, H. (1972). *Human problem solving.* Englewood Cliffs, NJ: Prentice-Hall.

Papert, S. (1980). *Mindstorms: Children, computers, and powerful ideas.* New York: Basic Books.

Piaget, J. (1929). *The child's conception of the world.* New York: Harcourt.

Posner, J. K. (1982). The development of mathematical knowledge in two West African societies. *Child Development, 53,* 200–208.

Resnick, L. B. (1983). A developmental theory of number understanding. In H. P. Ginsburg (Ed.), *The development of mathematical thinking* (pp. 109–151). New York: Academic Press.

Resnick, L. B., & Ford, W. W. (1981). *The psychology of mathematics for instruction.* Hillsdale, NJ: Erlbaum.

Rogoff, B., & Gardner, W. (1984). Adult guidance of cognitive development. In B. Rogoff & J. Lave (Eds.), *Everyday cognition: Its development in social context* (pp. 95–116). Cambridge, MA: Harvard University Press.

Rogoff, B., & Wertsch, J. V. (Eds.). (1984). *Children's learning in the "zone of proximal development"* (New Directions for Child Development, No. 23). San Francisco: Jossey-Bass.

Schoenfeld, A. (1983). Beyond the purely cognitive: Belief systems, social cognitions, and metacognitions as driving forces in intellectual performance. *Cognitive Science. 7,* 329–363.

VanLehn, K. (1983). On the representation of procedures in repair theory. In H. P. Ginsburg (Ed.), *The development of mathmetical thinking* (pp. 197–252). New York: Academic Press.

Vygotsky, L. S. (1978). *Mind in society: the development of higher psychological processes.* Cambridge, MA: Harvard University Press.

Wallace, G., & Larsen, S. C. (1978). *Educational assessment of learning problems.* Boston, MA: Allyn & Bacon.

CHAPTER 10

A Method for the Analysis of Patterns, Illustrated with Data on Mother–Child Instructional Interaction

BARBARA ROGOFF and MARY GAUVAIN

Psychologists have long debated philosophies of science and appropriate approaches to the analysis of behavior (Altman & Rogoff, in press; Mishler, 1981). Some approaches stress the importance of examining noninferential aspects of individual characteristics or behavior, with the hope that the meaning of the data will become clear as the field amasses sufficient data. Other approaches stress that one must attend to the meaning of human actions and to do so requires attending to how individuals function as part of larger contexts. From this perspective, human actions are intricately interwoven with the context of performance; psychological events derive from human actions in context.

The contrasting philosophies of science underlying these views may be characterized on the one hand in terms of focus on individual traits or the interaction of separate individuals as the basic units to be studied in order to understand human behavior, and on the other hand, in terms of focus on the interrelations between individuals inseparably embedded in the larger context (Altman & Rogoff, in press; Rogoff, 1982). Though

BARBARA ROGOFF • Department of Psychology, University of Utah, Salt Lake City, Utah 84112. MARY GAUVAIN • Department of Human Development, University of Pennsylvania, Philadelphia, Pennsylvania 19104.

some psychologists, notably Kimble (1984), feel that prospects for achieving epistemological harmony between such conflicting perspectives are not encouraging, Rogoff (1982) and Altman and Rogoff (in press) suggest that different approaches for understanding human behavior serve complementary purposes. They suggest that psychologists could benefit from expanding their traditional research and theoretical approaches focusing on isolated individual characteristics, useful in some analyses, to include approaches focusing on how individual functioning is embedded in social context. The latter approaches have until recently been more common in other disciplines, such as anthropology and sociology.

Our aim in this chapter is not to compare the utility of methods focusing on separate individuals or their rich relationships in social context—each has its uses—but to show how the approaches may be synthesized and extended to provide analyses that are systematic in the investigation of meaningful communication. The purpose of this chapter is to demonstrate how the analysis of individual actions in context can extend our knowledge of human behavior, and to show that such analyses can be done systematically and thereby satisfy important methodological concerns regarding clarity, generality, and reliability.

We illustrate this approach in the field of social interaction, with observations of mother–child instruction. The field of social interaction has struggled with issues of the interaction of separately defined individuals and contexts versus the maintenance of the temporal and interactional context in the analysis of individuals' actions (see Lamb, Suomi, & Stephenson, 1979, for several approaches to this issue).

Our attempts to examine interactional processes in instructional communication were enriched by the work of Vygotsky (1978), Bruner and colleagues (Wood, Bruner, & Ross, 1976), and Wertsch (1979) in this area. Especially useful was Vygotsky's notion of the *zone of proximal development*, in which children's skills and knowledge are seen as developing through participation in activities with adults, as children internalize the processes practiced jointly to advance their individual skill.

In this chapter, we describe how our methods for studying mother–child instructional interaction evolved in our investigation of the role of instruction in children's learning. First, we used frequency counts of superficial aspects of the mothers' and children's individual behavior, summed over the session or examined in short sequences. Next we shifted perspective to focus on the dyadic context rather than on separate individuals as the unit of analysis, through in-depth ethnographic analysis of the joint activity of a few mother–child dyads. Finally, to systematize and generalize our single-case analyses, we developed the approach that is the main focus of this chapter: the systematic analysis of patterns of joint activity appearing in the interactional context of multiple cases. In our analysis of patterns of joint activity, we used variables tied functionally to the purpose of the instructional interaction, that (a) allow consideration of the actions of the participants as interrelated rather than separate, and that (b) incorporate changes over the course of instruction rather than impose a static conception of the participants' actions.

The same data base is used throughout the chapter to illustrate the different analytical approaches used to understand the process of instructional interaction: 32 videotapes of mothers teaching their 6- to 7- or 8- to 9-year-old children. The mothers were asked to prepare their children to be able to organize independently either groceries on kitchen shelves as if they were involved in a home chore (home task), or photographs of household items in compartments as if they were doing homework (school task). Both tasks involved the mother and child in sorting 18 items into 6 shelves or compartments. The mother was shown the correct placement before the child was present, and she had a cue sheet indicating the correct placements of items for reference if necessary during the instruction. The mother was not informed of the rationale for categorizing the objects, or even told that there was one. She was simply told to prepare the child for an upcoming memory test involving arranging some of the same items and some new ones. The mother was asked to use whatever means she would ordinarily use in instructing the child. The child was also informed of the test, so both participants knew that the child needed to learn the organization of objects.

CODING INDEPENDENT BEHAVIORS OF MOTHER AND CHILD

The analysis of the videotapes began with a coding scheme to determine the overall frequency of specific independent behaviors of the mother and of the child and to compare instructional interaction at two ages and in home versus school classification tasks (Rogoff, Ellis, & Gardner, 1984). The aim of the study was to determine whether mother–child instruction was tailored to the interaction of the child's age and the task (such that the younger children in the school task would be involved in more intensive instruction than children in the other 3 groups) or whether the instruction was independently adjusted for the child's age and task (in which case the literature would lead to predictions that the older children and those in the school task would receive more open-ended questions whereas the younger children and those in the home task would receive more directives and nonverbal instruction).

The coding focused on occurrences of maternal directives and open-ended questions, and mothers' and children's verbal and nonverbal references to the grouping of items, the task requirements, and the upcoming memory test. The following stretch of transcript, involving a mother and her 6-year-old son sorting photographs in the school task, indicates how the mother's (M) and the child's (C) behaviors were coded.

MOTHER/CHILD	VARIABLE CODED
The mother picks up the picture of a bucket and holds it in front of the child, asking "What's that?"	M: open-ended question
The child replies, "It's a bucket and it helps you carry things and ..."	C: item label
The mother fills in, "Yeah, and it helps you clean," and looks at the child.	M: category label
The child nods.	
The mother asks, "OK, what else, do you see something else that helps you clean?" as she adjusts the broom in the box for cleaning items.	M: Open-ended question, category label, informative gesture
The child watches his mother's hand on the bucket card, then points to the bucket and then to the broom.	C: informative gesture, identification of associated item
The mother agrees, "The broom. So it should be put in here," as she holds the bucket in the cleaning box.	M: directive, identification of associated item, informative gesture
The child takes the bucket from his mother's hand and places it in the correct box.	C: item placement

The frequencies obtained from such coding gave information regarding the methods of instruction used by mothers, and the child's participation.

Using the frequency data, we examined differences in the mother's and child's behaviors depending on the age of the child and the task (home or school version). We found that many of the behaviors varied according to the interaction of task and child's age. More instruction occurred (i.e., most of the variables were coded more frequently) with the younger children in the school task, which seemed to be regarded as more difficult for the younger children.

These results were of interest in clarifying the relation of these instructional variables to differences in children's age and task. However, further questions regarding the process of mother–child instruction seemed to require viewing the data from a different analytical perspective. The behavior frequency count had separated the mother's and child's actions rather arbitrarily from each other and from the sequence of interaction. This yielded what we thought was an incomplete picture of the data, and we sought a way to maintain the context in the coding system. (See also Valsiner, 1982, and Zetlin & Sabsay, 1980, for discussions of similar issues.)

Of central concern was the loss of meaning resulting from identical coding of events that are on the surface identical but serve different purposes (and therefore mean different things) in the context of the interaction. For example, the mother's question "What do you think goes next?" may mean one thing if a child has just solved a similar problem in the experimental task, but quite another if it follows a succession of errors. Neither meaning is captured by our coding category of open-ended question.

With a related data set, Holly and Rogoff (unpublished data) attempted to examine the interactional context by considering the temporal sequence of behavior, examining the contingencies between successive independent behaviors of mothers and children similar to those used in the analysis of overall frequencies. For example, the transcript presented above would be segmented into sequences according to mother's turn, child's turn, mother's turn, and so on, with four turns by each participant in that particular stretch of transcript. Each segment would then be coded in terms of the instructional content (i.e., the variables appearing in the right-hand column) expressed by the mother or the child, as well as errors by the child and feedback by the mother. The contingencies were used to examine the relationship between adjacent mother and child behaviors, such as whether errors by the child were more likely following directions or questions by the mother.

However, it became clear that analyzing contingencies between independently defined maternal and child behaviors could not, in principle, handle the question of interest. Although the contingencies allowed the examination of events in the immediate context of what preceded and followed them, our coding still separated the mother's and child's behaviors, treating them independently. It is important to point out that this is not a necessary aspect of sequential or contingency analyses— they can be applied to behaviors of the dyad instead of to behaviors of individuals (Bakeman, 1983), but at that point we considered only individuals as the unit of analysis.

In addition to the problem of our separating the behaviors of mother and child, our contingency analysis was unable to handle the fact that at any point in an interaction a given act may mean something different than at another point in the interaction. Each coded event was assigned the same meaning wherever it occurred, but an inherent feature of communication is that the meaning of actions changes as circumstances change. The analysis of sequences could not reinstate the meaning and the context that had been removed from the coded data. Hence the coding of the mother's and child's behavior independent of each other and of preceding events still removed the interactional context from the coded categories, missing the participants' adaptations to the changing circumstances of the problem as they developed its solution.

Furthermore, the separation of the behaviors from the interactional context required each event to be coded in terms of surface characteristics rather than in terms of the purpose it served for the participants. Going from coded behavior to interpretation of data requires a large step, often involving inference of the subjects' purposes underlying the pattern of data. This step is problematic if the behaviors coded do not refer to their purpose in the interactional event, because any particular action may serve several alternative or simultaneous purposes. For example, a mother's glance at her child may function to maintain her child's attention, evaluate the pace of instruction, maintain her social status relative to the child, and work on completion of a specific component of the task (plus any number of goals less relevant to instruction).

However, in interpreting the results we wished to understand the purpose, not just the surface characteristics of the subjects' behavior, and we depended on earlier direct but unsystematic observation of the interactions—during piloting, running the study, and viewing the tapes—before this information was reduced to the coded behaviors. The behavior coding scheme had carefully removed the behavior from the context required for adequate interpretation.

The same problem is identified by Adamson and Bakeman (1982), who state that if researchers

> attempt to discern the organization of infants' interactions with their environment...[by coding] occurrences of discrete behavior patterns, interobserver agreement may be high but analyses of such frequency count data may leave them with a sense that little has been retained about the dynamics of the interactive process. (p. 1)

A researcher's interpretation of the theoretical meaning of statistical results generally draws on a deeper understanding of the data base than is captured in the summary of results. If this understanding is achieved haphazardly rather than through systematic attention, there is a danger that the interpretation of the data is not valid. Though measures of reliability ensure that several observers independently agreed on what a specific action meant, the basis of the researcher's interpretation of the pattern of data is less commonly systematically scrutinized or checked against the realities of specific cases. In other words, although researchers take care in clarifying and delimiting the coding of observed behaviors, less attention is paid to the process by which the researcher interprets these data. Such checking might involve examining the independent interpretation of two researchers given the same detailed results before the analyses are sifted to make sense of the data. Or the researcher's interpretation could be checked against the records of representative or prototypical individual cases to determine how well the researcher's portrayal of the event generalized across the sample fits with the event as it occurred in specific cases.

Despite the fact that examining sequences of behavior did not resolve the question of integrating context in the analysis of interaction, the attempt did help clarify the question. We began to examine alternate approaches, including ethnography, ethnomethodology, and hermeneutics to see how other researchers attempted to handle the problem of incorporating meaning directly in observations of human activity. The next sections describe some attempts to apply ethnographic analysis to this data.

ETHNOGRAPHIC APPROACHES TO STUDYING INDIVIDUALS IN INTERACTION

The studies described in the previous section (and many other psychological studies) attempt to separate different influences in order to relate discrete behaviors to discrete causes (Hay, 1980). These approaches may be useful for determining the frequency and ordering of particular behaviors by an individual or individuals, and to examine their differences as a function of subject characteristics or experimental situations. But our intent was to understand how individuals achieve meaningful communication together, within the context of an evolving interaction. In interaction, participants jointly construct a framework of meaning through which they communicate (McDermott, Gospodinoff, & Aron, 1978). This meaning may be obscured by dividing the cooperative actions of mother and child into behaviors for which only one is credited, and by forcing the same, single interpretation onto superficially similar acts regardless of the purpose they serve in the interaction.

Our attempts to examine the interaction through an ethnographic perspective were influenced especially by the work of Bremme and Erickson (1977), Cazden, Cox, Dickinson, Steinberg, and Stone (1979), Cicourel (1972), McDermott et al. (1978) and Mehan (1979). They argue that participants in social interaction provide explicit evidence regarding the

meaning of their actions to each other, informing each other of their
intentions through jointly created discourse and action, including clar-
ifications in cases of ambiguity. This evidence is essential to the achievement of understanding between participants, but also provides researchers with evidence regarding the meaning of actions. In other words,
researchers can obtain information regarding the meaning of actions by
attending to the evidence that participants provide to each other as a
necessary aspect of joint activity. McDermott *et al.* (1978) argue that

> we can use the ways members have of making clear to each other and to
> themselves what is going on to locate to our own [the researchers'] satisfaction
> an account of what it is that they [the participants] are doing with each other.
> In fact, the ways they have of making clear to each other what they are doing
> are identical to the criteria which we use to locate ethnographically what they
> are doing. (p. 247)

Similarly, Shotter (1978) contends that

> whatever methods we may propose that mothers and children use in their
> attempts to make sense to one another, and develop their capabilities in their
> exchanges with one another, are also methods that we may use in our attempts
> to make sense of them. (pp. 45–46)

This perspective seemed promising as a way to examine the interaction of the dyad as a unit, in the evolving context of the instructional
session. Gardner and Rogoff (1982) used it to analyze the transcript of a
single dyad, looking in depth at how the mother and child together
prepared the child for the memory test on the school task. They argued
that the mother and her 8-year-old child jointly accomplished the child's
cognitive performance. The mother's work to establish and maintain a
common framework for the processing of information (through involving
the child in her understanding of the task, and linking the experimental
task to more familiar classification problems) seemed essential for the
dyad's joint accomplishment of the task.

Using a dozen or so transcripts from the same data set, Rogoff and
Gardner (1984) examined the interactions of the mothers and children
for evidence of how they jointly arranged for the transfer of information.
They proposed that mother–child instruction has the following features:
The adult assists the child in transferring relevant information from more
familiar contexts to the new problem, and assists the child by structuring
the problem so that the child can address manageable subgoals. Both
the adult and child work to transfer the responsibility for problem solving
to the child over the course of the session. The child, as well as the
mother, is active in managing the pace and nature of instruction.

Using the segment of transcript presented above, Rogoff and Gardner

(1984) noted that both the mother and child worked to transfer the responsibility for managing the task over the course of the interaction. The mother subtly tested her child's readiness for more involvement by reducing the amount of direction and organization she provided for components of the task. If the child made an error or indicated a lack of readiness through hesitation or other cues, the mother quickly reerected the "scaffolding" (to borrow Wood, Bruner, & Ross's term, 1976) that she had momentarily removed, reestablishing her support for the child's performance. These negotiations of responsibility were often transacted through glances and other nonverbal cues.

MOTHER/CHILD

The mother picks up the picture of a bucket and holds it in front of the child, asking "What's that?"

The child replies, "It's a bucket and it helps you carry things and..."

The mother fills in, "Yeah, and it helps you clean," and looks at the child.

The child nods.

The mother asks, "OK, what else, do you see something else that helps you clean?" as she adjusts the broom in the box for cleaning items.

The child watches his mother's hand on the bucket card, then points to the bucket and then to the broom.

The mother agrees, "The broom. So it should be put in here," as she holds the bucket in the cleaning box.

The child takes the bucket from his mother's hand and places it in the correct box.

It appears that the mother was searching for more than the name of the item, because she was not satisfied with the child's initial reply. He provided the name of the item, but not the appropriate category label. So she provided the category label, and returned the responsibility to him, waiting for him to place the item in the appropriate box. When action was not forthcoming from the child, the mother increased her support for his performance by making his job more concrete: to find another item that would fit in the same category. She also subtly suggested the answer to her question, offering him a candidate by handling an item that would be an appropriate response to her question. The child took this cue and gave the appropriate response. He indicated the item she had been adjusting, and they finally got the item placed in its appropriate category.

Rogoff and Gardner supported their speculations about the nature of joint cognitive activity and instruction by presenting segments of transcript (and their interpretation) illustrating each point. The quoted transcripts were selected for their vividness in communicating points that appeared characteristic of the sample in general. However, such selection of data for presentation often makes readers uncomfortable about potential biases of interpretation of the data as a whole. Hence, it is worth discussing how rigorous ethnographers provide checks for their interpretation of the data.

With analyses involving a single case (or a few cases), researchers may balance the small sample of individuals with a more intensive analysis of the observations in order to examine all relevant data provided in the event. This may be contrasted with psychological research involving statistical methods with large samples, in which most of the variation observed is regarded as random and relegated to the error term. In such approaches, the goal is to fit the same statistical model to each individual. In their analyses of a small number of cases, ethnographers often attempt to account for a great deal of the observed behavior, rather than being satisfied with much unexplained variation in the observed events. For example, Mehan (1979) emphasizes the aim of constructing a model that accounts for the organization of each and every relevant instance. Rather than regarding anomalies as random error, Mehan attempts to account for such instances. The goal here is to tailor the explanation to each case.

The problem of assuring reliability of findings is also handled differently in ethnographic research than in conventional psychological research. In one common method, the researcher trains several observers to a criterion of agreement regarding how to label what is seen, reports a number indicating their level of agreement on labels, and bases the interpretation of the data on the labels assigned by those observers. However, the fact that several observers can achieve consensus on what to call a behavior does not make their label true. It simply means that if another person were similarly trained (enculturated) to see events according to the same perspective, they would likely call that behavior by the same label. Hence objectivity may be no more than shared subjectivity, in conventional approaches as in all others.

Because ethnographic researchers often do not abstract the reported data as far from the observed events, they focus on making the evidence for their interpretations of the event more explicit and subject to the reader's scrutiny. The reader is frequently given excerpts of the event to check the investigator's interpretation of instances, and in some cases, the entire corpus of raw data (in the form of transcripts) is made available to the reader to examine (e.g., Green & Wallat, 1979; Mehan et al., 1976). Similarly, researchers using protocol analysis for modeling cognitive processes sometimes report an entire protocol or protocols for the reader's examination (e.g., Anzai & Simon, 1979; Shrager & Klahr, 1983). This gives the reader the opportunity to consider alternative interpretations of the same material, rather than requiring the reader to rely on the interpretations of the researchers.

Although some researchers using traditional approaches may not recognize the extent to which all research involves selectivity and interpretation, Cicourel (1974) points out that researchers cannot avoid interpretation in any kind of research, because they must rely on knowledge of the context and of norms for behavior in order to recognize the relevance of the observed behavior for the theory being tested. In traditional psychological research, the investigator's interpretation of the data may be casual and not open to examination. Interpretation may occur in piloting, as the researcher attempts to choreograph relevant behavior, or in formulating the discussion section to make sense of the pattern of behavior recorded in the data. Cicourel urges that the procedures by which researchers interpret their observations be made explicit, a process required in careful ethnographic analysis. By presenting transcripts of the data alongside a step-by-step tracing of the interpretation, the ethnographer provides the reader with an opportunity to perform an independent examination of the reliability of the interpretation.

A number of ethnographers extend their analyses beyond single cases, and attempt to characterize the generalities or patterns appearing in a variety of similar cases (i.e., individuals, dyads, classrooms, events). Mehan (1979) is a prime example, with a careful constitutive ethnography of the organization of 9 classroom lessons, composed of a total of 590 interactive sequences.

In the next section we report our attempt to maintain the meaning of individual actions in temporal and interactional context, while extending and organizing the data base used in our ethnographic analyses. We used some of the information gained from the contextually rich ethnographic analyses to systematically examine the generality and variations in patterns of joint problem solving appearing in the 32 mother–child dyads.

THE ANALYSIS OF FUNCTIONAL PATTERNS OF INSTRUCTIONAL INTERACTION

To assess the generality of the findings regarding instructional communication obtained from the ethnographic analyses conducted by Rogoff and Gardner (1984), and to examine the variety of approaches used by the entire sample, we devised a method for systematically analyzing patterns using multiple cases. This method builds on the contextually rich analysis offered by ethnographic approaches, but it is not an ethnographic method focusing on single cases and reporting observational transcripts in detail. Rather, it involves considering the transcripts of a substantial number of cases in detail, and may involve the provision of illustrative transcript material to describe how the coding categories are derived. Just as the approach is not ethnography, neither does it use conventional methods, such as statistics, in the usual manner. The statistics employed in the following are a descriptive tool, helping to characterize the data in conjunction with graphic analysis of a number of individual cases. Thus our analysis attempts to synthesize and extend some aspects of ethnography and conventional psychological approaches.

We maintained stress on the joint contribution of the mother and the child to solving the problem over the course of the session, but sought a language of data reduction that would allow the use of the same terms to characterize each dyad. Thus each dyad's approach was examined in its own context, but the abstracted account of their approach was cast in terms appropriate for all the dyads. This meant that the reduced data did not maintain the rich idiosyncratic context of each dyad, but attempted to maintain the focal variations in approaches to the problem through inclusion of the primary alternatives in the abstracted language used for all dyads.

Determining the variations in approach that would be relevant to the selected level of analysis required both intimacy with the data and a functional task analysis of how the dyads could handle the problem of preparing children for a learning test involving a classification system. The instructional session was thus viewed as a problem-solving task, with specified pieces of information that the child needed in order to be prepared for the memory test.

Specifically, preparation for the memory test required the child to understand the basis for grouping items together. Minimal preparation would involve exposing the child to the correct placement of items. More effective, however, would be providing the child with a rationale for the groups of items that would be useful for dealing not only with the current items but with the new items to be presented in the test. Although this provision of category structure could be provided entirely by the mother, we expected that children who participated in the classification decisions would be better prepared for the test. (In fact, we expected that the best form of participation would be gradually increasing responsibility on the part of the child, but as will be seen in the following, we found that large and constant participation was also closely related to good performance on the test of learning.) Finally, reviewing the information or quizzing the child in specific preparation for the test would help ensure that the child understood the classification system. We devised a coding scheme focusing on such variables to characterize the functional efforts of the participants, and then examined commonalities and differences in the approaches used by the various mothers and children to accomplish similar goals.

Such a functional approach requires that a particular action be understood in the context of a goal. For example, a child's action of putting an item on a shelf in the grocery task (coded as "item placement" in our earlier frequency analysis) could serve the goals of learning how the items are organized, obeying the mother's orders, or simply working to complete the minimal task in order to leave the room. In our analysis, we focused on the goal set by the experimenter, teaching and learning how the items are organized. This does not preclude the operation of other goals in the individual or dyad, but was used as a means of focusing the analysis and thus excluding actions (such as merely placing an item on a shelf) that are not involved in teaching or learning the organization of items. Similarly, a goal can be accomplished by a variety of substitutable actions, which may be treated as equivalent. For example, learning how the items are organized could involve watching the mother attentively, asking questions, or trying to place items on shelves. If the goal of learning the organization of items, set by the experimenter, seemed not to be adopted by the participants, it would of course not be an appropriate simplification of the interaction. With our data, it appeared to be a primary goal of all or almost all dyads, though they evidenced other goals as well that are not analyzed here.

CODING AND RELIABILITY

The variables we coded were determined on the basis of our earlier ethnographic analyses along with the task analysis of what is necessary to prepare children to classify items independently in the test: provision of information regarding the nature of classification, the basis for classifying each group and the correct grouping of items, the involvement of the children in decision making, and preparation for the test.

The data were coded in the format illustrated in Figure 1. On each variable, we assessed reliability of judgment by independently coding 25% of the transcripts. The judgments are reliable, with Spearman *rho* correlations and Pearson correlations falling between .94 and 1.00 for each variable except one with a *rho* of .84.

In order to simplify presentation, the variables involving discussion of what similarity is, the need to sort, or the existence and presence of new items in the test will not be discussed, as they did not relate to test

MOTHER-CHILD INSTRUCTIONAL INTERACTION
(M = Mother; C = Child; M/C = both)

Dyad number _____ Age _____ Condition _____ Coded by _____ Date _____ Page _____

Orientation: (M or C or M/C)

_____ What similarity is, or the need to sort

_____ Existence of test, or presence of new items on test

Item Placement: (M or C or M/C)

_____ What similarity is

_____ Existence of test, new items in test, or actual preparation for test

Introductory category labels: _____ _____ _____

By whom: _____ _____ _____

Item Placement decision episodes:

Page of transcript

Item number
and location

Who determined
category label

Who determined group
and location for item

Test Preparation: (M or C or M/C)

_____ Need to study, need to remember groups, presence of new items in test

_____ Review of groups or locations, quiz, mnemonic for location

Comments on the general approch used by dyad:

Figure 1. Simplified version of coding sheet.

performance or to other variables in an interesting manner in any of our analyses.

The variables thus include the extent to which category labels were provided and by whom; the extent of the child's participation in item placement decisions and the eventual correctness of placement; and the existence of preparation for the test and who managed it. The coding of the variables involving category labels and item placement involved the sequential matrix shown in Figure 1.

The matrix was organized according to the sequence of decision episodes involving the provision of a category label or determining the grouping of items, in the order in which each dyad handled the items. Each episode was identified in terms of the item involved (e.g., 3a was the first item of Group 3) or the category rationale provided (e.g., 3 for a label for the items belonging in Group 3). For each episode involving an item, the occurrence of decisions regarding its group or location was indicated, as well as whether these decisions were correct or not. An example of a correct decision would be indicated by writing "3a in 3," to indicate that the first item of Group 3 was placed in the correct location. An example of an incorrect decision would be written "3a in 5," if the item were put where Group 5 items belong. For all this information, the person or persons responsible for the decision (not the physical handling of the item or repetition of category label) was indicated.

For example, the coding for the episode represented in the example transcript presented previously would indicate that the bucket is the eighth item decision made by this dyad. It is item 4b, the second item considered from Group 4. The mother labels the category, "it helps you clean," and correctly determines the group and location in which the item belongs.

Note that a great deal of interaction is overlooked in the summary of the segment provided in the preceding paragraph. The previous ethnographic analysis of the same segment considered the mother's fine-tuned support of the child's participation, which is not covered in the coding above. The richness of the negotiations of responsibility between mother and child no longer appear in the data. However, if the fine-grained negotiations of responsibility involved in *each* decision were the focus of analysis, the coding scheme could be designed to represent this information in an analogous fashion. Indeed, this would be an interesting analysis, complementing the present focus.

In the present focus, the data in the sequential matrix provided information for judgments of the child's participation in the decisions regarding group membership and location for each item: none, total responsibility without guidance from mother, small and sporadic responsibility, noticeable responsibility occurring only at the beginning or in the middle of the process, noticeable responsibility increasing as the session proceeded, or great responsibility spread throughout the session accompanied by guidance from mother.

It should be noted that whereas our analysis identifies the child or the mother (or both) as being responsible for managing specific decisions, this separation of child's and mother's roles has a quite different nature than that employed in the original behavior frequency coding scheme. There, each person's behavior was coded without regard for the other person's concurrent behavior and other features of the context. In the present analysis, on the other hand, coding for the responsibility for decisions requires consideration of information available in the context, including the other person's actions. Thus, in the present attribution of responsibility, the coding itself emphasizes rather than excludes the way in which each person's actions fits with those of the other person and the unfolding context of their interaction.

GRAPHIC AND STATISTICAL ANALYSIS PROCEDURE

To examine the relationship between the way each dyad prepared the child for independent classification, we first graphed the relationship between each variable obtained during the instruction session and test performance. These univariate histograms were followed by correlational tests examining the strength of relationships observed in the graphs. The correlations between variables were calculated separately for each of the four groups (two ages in two tasks), and the average correlation was found using Fisher's r to z_r approach. The correlations portray the same information as the graphs for relationships that are linear, but the graphs allow closer examination of patterns that are not linear.

We then graphed multivariate displays in which the identity of individual dyads was maintained to track patterns across variables considered simultaneously. The multivariate displays were organized in terms of the children's performance on the test, so that each variable from the instructional session and the relations between them could be examined for predictiveness of child's independent performance. Simultaneous consideration of the instructional variables as predictors of test performance allowed consideration of meaningful patterns of links between instructional variables, such as whether one variable substitutes for or accompanies another in producing good performance with a subgroup of the dyads. For example, is equally good test performance shown by dyads in which the child participates in item placement decisions or in which test preparation occurs, or do the children who perform best have both of these experiences? Correlations between the instructional variables themselves were also considered in the multivariate analysis.

This approach to analyzing data is similar to that advocated by Tukey (1977) for exploratory data analysis. Tukey stresses that exploratory and confirmatory data analysis can and should proceed side by side. He notes that graphic displays strengthen comparisons of the data by permitting the researcher to see how the data behave generally, as well as how points deviate from the general pattern. Tukey claims that exploratory data analysis is like detective work, with graphs forcing the researcher to notice the unexpected.

Working from the ethnographic tradition, two recent books also propose related methods of qualitative data analysis (Goetz & LeCompte, 1984; Miles & Huberman, 1984). Miles and Huberman argue that it is possible to systematize ethnographic data and compare across several cases. They feel that displaying data in a systematic fashion is essential to sorting and making general statements about ethnographic or qualitative data, as it is to analyzing data in quantitative form. They advocate abstracting data from the idiosyncratic details of a case, and argue for the use of numbers when appropriate to characterize the facts—but with the numbers never separated from the textual data they represent. They argue that it is always necessary to check back from the generalized statements to the raw data in order to be sure that generalized statements are grounded in the individual cases. Miles and Huberman demonstrate how data displays may be used to condense the information to the conceptual core or for arranging cases and observations in some conceptual order to derive comparisons that may relate to other variables.

UNIVARIATE RELATIONS BETWEEN INSTRUCTION AND TEST PERFORMANCE

In this section we present the results found using our graphic and statistical analyses with the variables reflecting the dyads' attempts to prepare the child for the test, discussing how each of the instructional variables relates to the child's performance on the test of learning. In the following section we consider the instructional variables together, to examine the pattern of instruction of the group of children who performed most poorly, compared with the group with intermediate performance and the group with excellent performance. Figures 2 and 3 provide the univariate data displays and the arrangement of instructional variables used for the multivariate analysis.

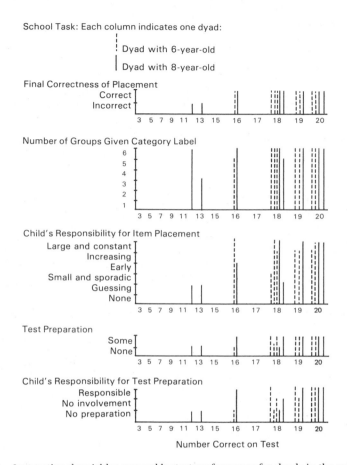

Figure 2. Instructional variables arrayed by test performance for dyads in the school task.

Correctness of Final Item Placement

In the dyads in which item placement was concluded with an incorrect arrangement of items, the children performed more poorly on the test of learning than in the dyads in which items were eventually correctly organized [$r(30) = -.86$, $p < .01$]. Only 3 dyads (9%) concluded the session with the items misplaced. The children in these three dyads received the two lowest scores in the school test, and the lowest score in the home test (3, 12, and 13 items correct).

Interestingly, the number of errors during item placement prior to achieving a final arrangement appeared unrelated to children's test performance. This suggests that the presence of errors is not related to poor

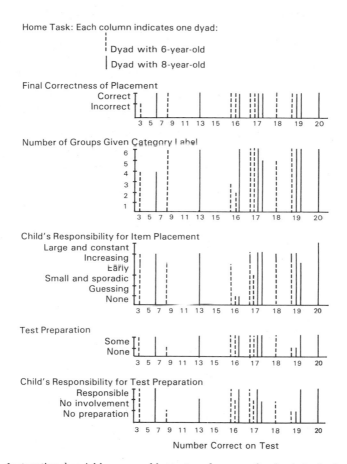

Figure 3. Instructional variables arrayed by test performance for dyads in the home task.

learning, as one might presume. For some dyads, the errors might indeed indicate difficulties in learning, but for others, the errors may serve a purpose, such as allowing the learner greater involvement in the decision making, or informing the teacher as to when the learner needs assistance. What appears to matter is whether the dyad eventually achieved the correct organization.

Number of Groups Given Category Labels

The extent to which category labels were provided relates to children's test performance, $r(30) = .52$, $p < .01$. Most of the dyads (70%)

labeled all 6 groups with a comprehensive category label that encom-
passed all members of the group and could be used to make decisions
regarding the placement of new items, including phrases indicating function (e.g., these are things you put on hot dogs) as well as single-word
labels (e.g., condiments). The other 30% of the dyads provided category
labels for 2 to 5 of the groups, and either did not attempt to label the
other groups or occasionally listed associations with individual items
(e.g., that goes with the mustard and the relish). The mean number of
items correct on the test for those who labeled all 6 categories was 17.4
(range 8–20), compared with 14.2 (range 3–18) for those who labeled fewer
categories.

Children's Participation in Labeling

There was not a significant relationship between children's participation in determining category labels and test performance, $r(30) = -.28$,
n.s. Children were responsible for designating at least one of the 6 labels
in 33% of the dyads, and the performance of these dyads averaged 15.8
correct (range 6–20), whereas the average performance of the children
whose mothers provided labels was 17.5 correct (range 3–20).

When children did participate in labeling the categories, fewer category labels were provided overall $[r(30) = -.62, p < .01]$. Of the dyads
in which the mother contributed the labels, only 24% failed to label all
6 groups. In contrast, 45% of the dyads in which the child participated
in labeling did not label all 6 groups. It is likely that the mothers' efforts
to label categories would be more effective and complete, in that they
had a cue sheet and prior exposure to the correct arrangement of objects,
as well as possibly greater categorization skills.

Children's Participation in Decisions Regarding Group Membership

Of particular interest was whether the mother or child or both were
responsible for deciding which items went with which group, and how
the relative responsibility for this cognitive work shifted over the course
of placing the 18 objects. We expected that a pattern of increasing participation by the child would be most highly related to good test performance, on the basis of our earlier ethnographic examinations of
mother–child instruction and the concept of the zone of proximal development.

Children's participation in item sorting appeared to be different in
the home and school tasks. Children were more involved in the school
task—7 of 16 children were involved to a large and constant extent in

the school task, compared with only 1 of 16 in the home task. An additional 3 children in the school task were involved to such an extent that decisions regarding item placement were virtually their responsibility alone (they were required to guess), compared with no children required to guess in the home task. All children in the school task were involved to some extent, but 2 of the children in the home task were not involved at all in decision making regarding item placement.

Children in the school task who participated to a large and constant extent or who showed increasing participation over the session produced higher test scores ($n = 7$, mean score $= 18.7$, and $n = 4$, mean score $= 19.0$, respectively) than those whose participation was sporadic ($n = 1$, score $= 18$), decreasing ($n = 1$, score $= 16$), or involved guessing without guidance ($n = 3$, scores 12, 13, and 18, mean $= 14.3$). In the home task, the pattern was less systematic: the one child who participated to a large and constant extent scored 20 correct on the test, the nine children who participated to an increasing extent ranged from 3 to 19 correct (mean $= 14.3$), and the six children involved in the other types of participation produced scores ranging from 8 to 19 (mean $= 15.3$). The correlation between playing a large and constant role or an increasing role (versus all other roles) and test performance was $r(14) = .62, p < .01$ for the school task, and $r(14) = .37, p < .10$ for the home task. The less systematic results for the home task may have derived from expectations that this familiar situation was easy to learn or not important to learn well, or that the appropriate division of responsibility included a more dominant role of the mother.

The relation between level of children's participation in item placement decisions and test performance at least in the school task is consistent with our expectations stressing children's participation, which arose from our earlier ethnographic analyses and Vygotsky's notion of the zone of proximal development. We expected the performance of the children who participated to an increasing extent to be good, but we had not considered the performance of the children who participated to a large and constant extent. Although it is possible that the excellent performance of the children who played a large and constant role resulted from being more skilled in classification before the instructional session began, the fact that the analysis brought the extent of participation to our attention indicates one way that this analysis extends our previous understanding.

A suggestion that guidance from the mothers needed to accompany participation by the children is available in comparing the performance of the children who were required to guess with that of another group of children who also were given responsibility for determining item placement, but after having been provided with rationales for each category. The children who were required to guess were given total responsibility for the decisions, without maternal guidance regarding the rationale for sorting. An example of such an approach is provided by the mother's statements in one dyad, after 37 (mostly erroneous) decisions. For the placement of one item, the mother gave the following minimally helpful hints, interspersed between each of the child's unsuccessful attempts:

This is in the wrong color [box] ...

No, that's still the wrong color ...

Try another color.

Just try one arbitrarily, just see if it's right or wrong ...

No, that isn't it either ...

There. OK, yup.

All 3 of the guessing cases involved older children in the school task. The mothers may have presumed that these older children already possessed task-relevant skills, and adopted the role of teachers who were to test the children. However, without guidance regarding the rationales for grouping items, the children's performance was quite poor (averaging 14.3).

Much better scores were produced by three children (older children in the school condition, like the guessers) who were involved in a related but more effective approach (coded as large and constant participation). They received a preliminary lesson by the mother in which the rationale for sorting the items was provided, and then figured out which items fit which groups. These children averaged an impressive 19.7 items correct on the posttest (scores were 19, 20, and 20), suggesting that maternal guidance regarding category labels may have been important in preparing children for the test.

Test Preparation and Responsibility

After the completion of item placement, most of the dyads (77%) devoted attention to memory activities in preparation for the test. They reviewed which items fit which groups and which groups fit which locations, developed a mnemonic to help remember the location for each group, or quizzed the child to determine whether the child had learned the information. A significant correlation was found between the extent to which preparations were made for the test and subsequent test performance, $r(30) = .36$, $p = .05$.

In those dyads that prepared for the test, the children were likely to participate: in 71% of the 24 dyads preparing for the test the children were involved in managing the memory activities. There was a significant correlation between the extent to which children were involved in test preparation and their test performance, $r(30) = .38, p < .05$.

MULTIVARIATE GRAPHIC ANALYSIS:
OVERALL PATTERNS IN THE TEST PERFORMANCE

The patterns appearing when variables are considered together provide clarification of the overall pattern, suggesting how the instructional variables fit together and relate to test performance. Figures 2 and 3 present each dyad's scores on the instructional variables, one variable below the next, with the dyads arranged according to the child's test performance. We graphed the dyads participating in the school and home tasks separately, because the univariate analyses showed some differences for the two tasks.

To discuss the pattern we will compare three groups: the dyads receiving the lowest scores (6 children scored 3 to 13 items correct), the intermediate scores (15 children scored 16 to 18 items correct, none scored 14 or 15), and the highest scores (11 children scored 19 or 20 items correct out of 20). Though the intermediate-scoring children performed quite well, there appear to be instructive differences between them and those who achieved a perfect or near-perfect score (20 or 19). If the children had not as a group done so well on the test, we might have been able to make even finer distinctions. Average performance was 16.41 correct out of 20, with a standard deviation of 4.16 (range 3–20). With this ceiling effect, it is impressive that relationships were found and proved to be statistically significant in the correlations.

Lowest Test Scores. Three of the six children receiving the lowest test scores were the only ones in the sample who were never exposed to the correct organization of items during instruction, as mentioned earlier. Two of these three children did not have all six category labels provided. (The correlation between final correctness of placement and provision of all six category labels was $r(30) = .40, p < .05$.) Two of the three children did not receive guidance in their decisions regarding the grouping of items (they were required to guess), and two of them had no test preparation.

The other three children who received very low test scores were all in the home task. They were provided with an item organization that was correct, and two of the three had all six category labels available. However, none were involved in item placement decisions to a large and constant extent, one received no test preparation, and another did not participate actively in the test preparation that was provided.

Intermediate Test Scores. The dyads in which children achieved test scores from 16 to 18 correct were intermediate on the instructional variables, between the dyads of the lowest scoring and highest scoring children. Whereas all 15 were exposed to the correct arrangement of items by the end of item placement, six of them (40%) did not have all six category labels available. Only three (20%) were involved in item placement decisions to a large and constant extent. Of the remainder, five (33%) were either involved sporadically, not involved, or carried the whole responsibility without guidance (guessing). Most of the dyads (73%) did test preparation, but only half of the children took an active role in the test preparation.

Highest Test Scores. The 11 children who scored all or almost all correct on the posttest received the most guidance in learning how to handle the task, and participated to the greatest extent. They all were exposed to the correct arrangement of items by the end of item placement, and they all had the full set of six category labels available. Almost half of them (46%) participated in decisions involving item placement to a large and constant extent; the remainder were involved to a moderate extent (five with increasing involvement over the session, one with involvement only during the middle of the session). Most of the dyads (82%) carried out test preparation, and in 89% of those, the children participated to a substantial extent.

The correlations indicated a significant relation between the extent of children's involvement in decisions regarding item placement and children's participation in test preparation, $r(22) = .71, p < .01$, consistent with the suggestion that the extent of children's participation in problem solving under maternal guidance related to their learning of the information.

SUMMARY: METHODS AND RESULTS OF THE FUNCTIONAL PATTERN ANALYSIS

The general characteristics of the method we used in this functional pattern analysis are as follows:

1. The analysis focuses on the sequence of development of the purposive acts within an event, by examining shifts in strategies for handling the task.

2. The variables are functionally defined as they relate to the goal of the event as a whole. They do not involve superficial behaviors independently defined and separated from the context in which they occur. Rather they focus on acts that are defined in terms of their utility in

reaching the goal, and that consider the surrounding context. Depending on which goal is identified for analysis, different acts by the participants become relevant. The occurrence of a given goal-related act is judged by considering the behavior of both participants and the unfolding nature of the event. Such acts can be reliably coded by independent observers.

3. Patterns are analyzed with statistical methods and with careful examination of graphic arrays of the relations between variables. The graphic arrays can involve the examination of univariate relationships between variables as well as the tracking of patterns across multiple variables to account for anomalous or similar cases. Anomalies are pursued as informative for the results as a whole, and the raw data are referred to in the attempt to account for these cases, perhaps calling attention to unexpected variables, and to substantiate general patterns by looking within cases.

We stress that statistical approaches, examination of graphic arrays of the data, and ethnographic analyses of individual cases can supplement each other's strengths. They provide different and complementary vantage points for interpreting the data. We feel that our understanding of these instructional interactions would be less complete if any of these pieces of information were omitted.

The specific results of this analysis were presented as a means of illustrating the method rather than to examine questions of content. However, as a way of validating the method, it is worth noting that our exercise of using this method with these data corroborate and add to our knowledge of the relationships in the data. Specifically, the analysis supports the idea explored in Rogoff and Gardner's (1984) earlier ethnographic analysis that an important aspect of instructional interaction may be the guided participation of the child in an activity with a more skilled person. The importance of the mother's guidance was suggested in the relation to test performance of the eventual correctness of the item arrangement, the availability of all six category labels, and the provision of category rationales for children who then took over responsibility for item placement. The importance of the child's participation was suggested by the relation to test performance of the extent of and increase in the child's participation in item placement decisions, and responsibility for explicit preparation for the test. This analysis brought to our attention the importance of considering the extent to which children participate in instructional interaction, and supported our interest in the role of children's increasing responsibility over the course of learning in social interaction.

This analysis does not of course indicate causal direction. Although we speculate that guided participation plays a role in children's learning, the possibility must be kept open that some other variables make the connection between the sort of instructional interaction we observed and children's performance. For example, the children who are consistently involved in making item placement decisions may be more skilled in classification, and this may account for their better test performance. The results of this study must be taken as an attempt to draw connections between variables. We are encouraged by the fact that relationships between instructional variables and children's performance were found. This contrasts with a failure to find relationships between test performance and the instructional variables used in our earlier frequency analysis of independent behaviors of mothers and children (Rogoff, Gardner, & Ellis, unpublished data).

In sum, the ethnographic investigations directed our attention to considering the role of guided participation in instructional interaction. The present analysis forced us to specify the ways in which we could observe guided participation in these data, and allowed us to examine systematically the relationship between guided participation and children's learning.

CONCLUSION

Having completed these analyses, we are using this approach in present work with different data sets by determining coding categories through identifying the goals and possible problem-solving strategies of the participants, by closely examining the interaction of a few dyads, and by examining the data closely through graphic arrays and statistical procedures focusing on patterns of individual performance. It is certainly possible to carry out such analyses in a less self-conscious fashion, or to use parts of the approach as relevant to a particular study (e.g., see Ellis & Rogoff, 1986, for an examination of adult versus peer instruction on the same home and school tasks).

A number of other researchers have developed related ways of systematizing ethnographic or qualitative data. Among them are investigators who have also added to knowledge through the ethnographic examination of single cases (e.g., Cazden et al., 1979; Green & Wallat, 1979; Mehan, 1979). Mehan and Riel (in press) analyzed instructional chains from their ethnographic work in which a classroom teacher taught a child how to do a task, and the child then taught other children. The data were abstracted in a table to allow examination of patterns appearing across 12 cases, in teachers' versus students' instructional moves and the extent of the learner's participation. Similarly, Moore (1981) used ethnographic field notes and interview transcripts regarding the work experiences of high school students to construct systematic descriptions of the students' task-related learning. He focused on tasks as problems for the student to solve, and examined the relative involvement of students and supervisors in functionally related actions.

Other systematic analyses of functional patterns in interactional data derive from alternative scholarly traditions. The work of the Grossmans (Grossman, 1981; Grossman & Henke, 1983; Grossman, Schwan, & Grossman, in press) stems from the ethological tradition, and dentifies meaningful patterns in infants behavior when reunited after separation from their mothers and in toddlers' reactions to a strange toy while in their mothers' company. Als, Tronick, and Brazelton (1979) discuss the analysis of face-to-face interaction in infant–adult dyads using narratives and graphic displays of behavior. And from the hermeneutic tradition, Kreppner, Paulsen, and Schuetze (1982) developed an analysis of patterns appearing over time in individual cases (families) and across cases as families integrate a now baby into their interactional system in phases corresponding to the baby's growing role.

An example of pattern analysis applied to individuals rather than groups is provided by Shrager and Klahr (1983) in their investigation of how individuals attempt to understand complex systems without the benefit of instruction. Protocols of adult subjects' attempts to work an electronic toy and their talk-aloud explanations of their thinking were systematically analyzed by summarizing transcripts in general terms that describe the structure of the segment of the protocol, focusing on goal generation and satisfaction, hypothesis formation, and experimental observations leading to acceptance or revision of the hypothesis. In this way, Shrager and Klahr worked from contextually rich individual cases to more abstracted functional accounts whose characteristics were more easily examined.

To summarize, this chapter has presented a chronology of research methods we employed in investigating teaching and learning interactions. Our aim was to describe a systematic analysis of multiple cases based on contextually rich data. The analysis that we describe focuses on abstracting regularities and significant variations in how specific cases (dyads, in our data) deal with problems or tasks. The analysis stresses using the sequence of acts within the event and the contextual relationships between the acts of different participants to develop a functional analysis of the workings of the event. We advocate using graphic displays of the abstracted acts, reference to ethnographic raw data to follow typical or anomalous cases, and statistical tests to support the patterns found in the visual displays of abstracted data. Such analysis may provide one way to analyze the complex web of human actions with a focus on the embedded and meaningful relations between an individual's actions and the context of those actions, including the mutual actions of other individuals.

ACKNOWLEDGMENTS

We wish to acknowledge Shari Ellis's and William Gardner's roles in the early discussion and development of these ideas, their involvement in developing the tasks and preparing data, and the impetus provided by their master's theses. In addition, Irwin Altman, William Gardner, Jamie Germond, Artin Goncu, Donald P. Hartmann, Jayanthi Mistry, Barbara Radziszewska, and Jaan Valsiner provided very helpful comments on earlier drafts that assisted us in presenting these ideas.

REFERENCES

Adamson, L. B., & Bakeman, R. (1982, March). *Encoding videotaped interactions: From counts to context.* Paper presented at the International Conference on Infant Studies, Austin, TX.

Als, H., Tronick, E., & Brazelton, T. B. (1979). Analysis of face-to-face interaction in infant–adult dyads. In M. E. Lamb, S. J. Suomi, & G. R. Stephenson (Eds.), *Social interaction analysis: Methodological issues* (pp. 33–76). Madison, WI: University of Wisconsin Press.

Altman, I., & Rogoff, B. (in press). World views in psychology and environmental psychology: Trait, interactional, organismic, and transactional perspectives. In D. Stokols & I. Altman (Eds.), *Handbook of environmental psychology.* New York: Wiley.

Anzai, Y., & Simon, H. A. (1979). The theory of learning by doing. *Psychological Review, 86,* 124–140.

Bakeman, R. (1983). Computing lag sequential statistics: The ELAG program. *Behavior Research Methods and Instrumentation, 15,* 530–535.

Bremme, D., & Erickson, F. (1977). Relationships among verbal and non-verbal classroom behaviors. *Theory into Practice, 5,* 153–161.

Cazden, C. B., Cox, M., Dickinson, D., Steinberg, Z., & Stone, C. (1979). "You all gonna hafta listen": Peer teaching in a primary classroom. In W. A. Collins (Ed.), *Children's language and communication: 12th Annual Minnesota Symposium on Child Psychology* (pp. 183–231). Hillsdale, NJ: Erlbaum.

Cicourel, A. (1974). *Cognitive Sociology.* New York: Free Press.

Ellis, S., & Rogoff, B. (1986). Problem solving in children's management of instruction. In E. Mueller & C. Cooper (Eds.), *Process and outcome in peer relationships* (pp. 301–325). New York: Academic Press.

Gardner, W. P., & Rogoff, B. (1982). The role of instruction in memory development: Some methodological choices. *Quarterly Newsletter of the Laboratory for Comparative Human Cognition, 4,* 6–12.

Goetz, J. P., & LeCompte, M. D. (1984). *Ethnography and qualitative design in educational research.* Orlando, FL: Academic Press.

Green, J., & Wallat, C. (1979). What is an instructional context? An exploratory analysis of conversational shifts across time. In O. K. Garnica & M. L. King (Eds.), *Language, children, and society* (pp. 159–188). Oxford: Pergamon Press.

Grossman, K., & Henke, B. (1983, August). *Behavior pattern analysis of child-mother-jack-in-the-box interactions of 47 two year olds.* Paper presented at the meetings of the International Society for the Study of Behavioral Development, Munich.

Grossman, K. E. (1981, April). *Infant and social environment interaction: Epistemological considerations behind the ethological approach.* Paper presented at the meetings of the Society for Research in Child Development, Boston, MA.

Grossman, K. E., Schwan, A., & Grossman, K. (in press). Infants' communications after brief separation: A reanalysis of Ainsworth's Strange Situation. In P. B. Read & C. E. Izard (Eds.), *Measuring emotions in infants and children* (Vol. 2).

Hay, D. F. (1980). Multiple functions of proximity-seeking in infancy. *Child Development, 51,* 636–645.

Kimble, G. A. (1984). Psychology's two cultures. *American Psychologist, 39,* 833–839.

Kreppner, K., Paulsen, S., & Schuetze, Y. (1982). Infant and family development: From triads to tetrads. *Human Development, 25,* 373–391.

Lamb, M. E., Suomi, S. J., & Stephenson, G. R. (Eds.). (1979). *Social interaction analysis: Methodological issues.* Madison, WI: University of Wisconsin Press.

McDermott, R. P., Gospodinoff, K., & Aron, J. (1978). Criteria for an ethnographically adequate description of concerted activities and their contexts. *Semiotica, 24,* 245–275.

Mehan, H. (1979). *Learning lessons.* Cambridge, MA: Harvard University Press.

Mehan, H., & Riel, M. M. (in press). Teachers' and students' instructional strategies. In L. L. Adler (Ed.), *Issues in cross-cultural research.* New York: Academic.

Mehan, H., et al. (1976). *Texts of classroom discourse* (Report No. 67a). La Jolla, CA: Center for Human Information Processing, University of California at San Diego.

Miles, M. B., & Huberman, A. M. (1984). *Qualitative data analysis.* Beverly Hills, CA: Sage Publications.

Mishler, E. (1979). Meaning in context: Is there any other kind? *Harvard Educational Review, 49,* 1–19.

Moore, D. T. (1981). Discovering the pedagogy of experience. *Harvard Educational Review, 2,* 286–300.

Rogoff, B. (1982). Integrating context and cognitive development. In M. E. Lamb & A. L. Brown (Eds.), *Advances in Developmental Psychology* (Vol. 2, pp. 125–170). Hillsdale, NJ: Erlbaum.

Rogoff, B., & Gardner, W. P. (1984). Adult guidance of cognitive development. In B. Rogoff & J. Lave (Eds.), *Everyday cognition: Its development in social context* (pp. 95–116). Cambridge, MA: Harvard University Press.

Rogoff, B., Ellis, S., & Gardner, W. (1984). The adjustment of adult–child instruction according to child's age and task. *Developmental Psychology, 20,* 193–199.

Shotter, J. (1978). The cultural context of communication studies: Theoretical and methodological issues. In A. Lock (Ed.), *Action, gesture and symbol: The emergence of language* (pp. 43–78). New York: Academic.

Shrager, J., & Klahr, D. (1983). *Instructionless learning: Hypothesis generation and experimental performance.* Manuscript submitted for publication.

Tukey, J. W. (1977). *Exploratory data analysis.* Reading, MA: Addison-Wesley.

Valsiner, J. (1982, March). *Strategies of dyadic problem-solving with infants in a simulated laboratory situation.* Paper presented at the International Conference on Infant Studies, Austin, TX.

Vygotsky, L. S. (1978). *Mind in society.* Cambridge, MA: M.I.T. Press.

Wertsch, J. V. (1979). From social interaction to higher psychological processes. *Human Development, 22,* 1–22.

Wood, D., Bruner, J. S., & Ross, G. (1976). The role of tutoring in problem solving. *Journal of Child Psychology and Psychiatry, 17,* 89–100.

Zetlin, A. G., & Sabsay, S. (1980). Characterizing verbal interaction among moderately mentally retarded peers: Some methodological issues. *Applied Research in Mental Retardation, 1,* 209–225.

CHAPTER 11

The Role of the Case Study in Neuropsychological Research

JANE DYWAN and SIDNEY J. SEGALOWITZ

Although a relatively new branch of psychology, neuropsychology has inherited and built upon the clinical-theoretical traditions of the European neurologists and the operational-statistical leanings of American psychology. Applied interest in brain–behavior relationships gained momentum during the wars as psychologists were called on to screen soldiers for duty and to assess the brain-injured and behaviorly disturbed veterans to facilitate appropriate case management. As is often the case in a new field, however, clinician and researcher were usually one and the same individual.

An interest in brain–behavior relationships in the normal brain was aroused among experimental psychologists around 1960. This enlivened interest in developing a general model of neuropsychological functioning was stirred by two events. One was the work of Sperry and his colleagues (e.g., Gazzaniga, Bogen, & Sperry, 1962, 1963, 1965) who were able to demonstrate remarkable functional asymmetries in the capabilities of the cerebral hemispheres following section of the corpus callosum and anterior commissure for the treatment of intractable epilepsy. The other

---

JANE DYWAN • Department of Psychiatry, McMaster University, Hamilton, Ontario, Canada L8N 3Z5. SIDNEY J. SEGALOWITZ • Department of Psychology, Brock University, St. Catharines, Ontario, Canada L2S 3A1. This work was supported in part by a Research Fellowship from the Medical Research Council of Canada awarded to Jane Dywan, and Grant A0695 from the Natural Sciences and Engineering Research Council of Canada awarded to Sidney J. Segalowitz.

291

event was the discovery that hemispheric asymmetry could be detected and unobtrusively measured in the intact normal individual. Kimura (1961a, b, 1967) pioneered in this area with the development of the dichotic listening task that allowed for differential responsiveness to verbal or melodic stimuli depending on the ear to which the subject attended. Although there was some overlap between these two events in terms of the clinical-experimental distinction, Sperry's work was mainly representative of the clinical-theoretical approach whereas Kimura's was more typically experimental, that is, she assessed the reliability of subtle differences in subjects' responses on the basis of statistical probability. Both methods have and continue to enrich our understanding of brain–behavior relationships.

The experimental neuropsychologist interested in brain–behavior relations has an unenviable choice to make: to perform an intensive study of individuals with certain methodological, statistical, and interpretive problems, or to use group designs that are often experimentally more elegant and statistically safer, but that ignore variations between people in brain organization. Although the case study still features more prominently in the field of neuropsychology than it does in most other areas of psychological research, it is nonetheless the case that most researchers opt for group designs.

The reluctance to rely on the case study is based on a number of concerns that follow from some generally held premises. The first is based on the notion that single-organism research is basically unsound because such studies necessarily lack appropriate control conditions. The second is that the variability found in both brain structure and in behavior preclude the possibility of making valid inferences from the single case. The third premise is that case studies based on the performance of brain-damaged individuals cannot be used as a basis to build a model of normal brain function. The fourth is that the single case does not lend itself easily to statistical analysis, further limiting the generalizability of observations. We will examine these difficulties as they relate to the methods of inference in neuropsychology and the merits of various approaches.

PREMISE ONE

The first premise—that the single case lacks proper scientific control—suggests that the single-case investigation provides no constraints in the range of explanation one can engender based on the data collected. Although once expounding this view himself, Donald T. Campbell, the noted statistician, has come to accept the case study as a legitimate method for scientific enquiry (Campbell, 1975). He points out that although hypotheses cannot be confirmed on the basis of a single observation, they can be tested through the use of theory-based predictions. In studying the scientific process as it occurs in the intensive study of the single case, he found that the researcher rarely makes inferences from a single observation but engages in a complex process of pattern matching. It is rarely a case of one interesting datum being explained by culling a convenient theory from the entire repertoire available within the particular discipline. Unless the scientist is so theory bound that he ignores disconfirming data, he will keep his entire data set in mind. Through a qualitative approach to the analysis of a single case, the researcher can find that his prior beliefs and theories were wrong. Campbell concludes that the ability to reject theories speaks to the value of the single case.

He points out, however, that the same cautions apply to single-case research as apply to other forms of investigation. Briefly these are as follows:

1. The researcher must engage in active hypothesis testing, that is, it is important to record observations that do not fit one's theory as well as those that do. The cumulative literature aids in this process through the publication of disconfirming case-study evidence.

2. Attempts to confirm nonintuitive, theory-based predictions allows one to reach stronger conclusions than can be reached through observation alone.

3. The cautions involving experiment-wise error rate* hold for the case study as well as the mutlifactored group design. One must limit the number of hypotheses that are to be tested on any one set of data.

4. A problem-centered rather than a method-centered approach enables a researcher to triangulate on the target by using multiple methods to focus on the same construct.

An example of a current, problem-centered case study that exemplifies these aspects of good research design is that of Nolan and Caramazza (1983). The study consists of a detailed analysis of reading, writing, and repetition in a case of deep dyslexia. Deep dyslexia is a condition that is characterized by striking impairments in oral reading. For example, such patients will substitute semantically related words and read *girl*

---

* The term *experiment-wise error* refers to the situation where each hypothesis being tested within a study increases the probability that an effect will reach statistical significance solely by chance. To control for this, one must require more stringent tests of significance for each hypothesis so that the overall criterion or experiment-wise error rate remains within reasonable limits.

when presented with *daughter*. Also, they can read nouns better than verbs and usually cannot read functors such as *the* at all.

Careful evaluation of the patient's history, and the extent and location of physiological damage were reported for this case as were the data from the various assessment techniques that were used. They did not rely solely on clinical judgment to decide whether the patient's performance on one task was significantly worse than on another, but used a d' analysis or chi square whenever appropriate. Moreover, they laid out their conclusions in a clearly testable form. Their careful documentation of data and their explicit articulation of their theoretical claim allow for replication, reinterpretation, or disconfirmation. Neuropsychologists, whether in the clinic or laboratory, can compare this patient with patients with whom they have had experience or with other data that speak to the cognitive mechanisms of reading and decide whether Nolan and Caramazza's conclusions are warranted when assessed in this larger framework. Thus, this single case study becomes part of a progressive research program attempting to develop an understanding of the mechanisms involved in reading by studying the ways in which this usually integrated function can break down.

It is clear that a scientific approach can be taken with the single case. One could argue, therefore, that the single-case approach in neuropsychology cannot be faulted on the basis of its validity as a form of scientific enquiry. The case study can, however, be criticized on the same basis that one can criticize any scientific methodology, that is, for not following the rules of good research design. Are there, however, issues specific to neuropsychological research that bring inferences from the single case into question? This issue is the basis of the following discussion.

PREMISE TWO

The second premise on which the single case is questioned as an appropriate base on which to build a theory of brain–behavior relationships is that the variability inherent in both brain structure and in behavior preclude the ability to make valid inferences from the single case. Kolb and Whishaw (1980), in their textbook on neuropsychology, use this argument to caution against an overreliance on the single case. As an illustrative example of variability in the normal brain, they point out that the massa intermedia is missing in the brains of nearly one quarter of the population. The massa intermedia is a mass of gray matter that connects the left and right thalami across the midline and there is no known relationship between the presence or absence of this structure and behavior. Conversely, they illustrate how behavioral variability adds further complexity to the problem of making valid inferences. For example, the brains of two individuals can appear grossly similar in structure whereas the results of a neuropsychological evaluation could find one individual to be normal and the other to be in the impaired range. Early group studies on aphasia illustrated this behavioral variability quite dramatically.

In an attempt to discover whether aphasia was in fact the consequence of a general disruption in patients' ability to use symbolic logic, Weisenburg and McBride (1935) ushered in the modern age of neuropsychological research with a group study that served as a model for subsequent research. Incorporating the notions of psychometric evaluation and experimental psychology, Weisenburg and McBride gave an extensive battery of tests, both verbal and nonverbal, to 60 aphasic patients and compared their performance to that of 38 nonaphasics with unilateral brain damage and 85 non-brain-damaged subjects. They found, as expected, that aphasics performed poorly on verbal tasks and that the majority of aphasics performed poorly on the nonverbal tasks as well. They could not conclude, however, that aphasia necessarily affects general cognitive processes because some of the most severely impaired aphasics performed adequately on the nonverbal tests. Similarly, Orgass, Hartje, Kerchensteiner, and Poeck (1972), when using the Wechsler Scales (Wechsler, 1958), found that the mean performance of aphasics on visuospatial tasks was well below the norms for the population but that individual scores on these nonverbal portions of the test ranged from 50, which is in the mentally defective range, to 113, which is considered to be bright to normal. This is an incredible range when one considers that it extends from the 1st to the 80th percentile of the population. The issue, then, for those who would wish to make general inferences on the basis of an intensive investigation of any single case is that the case in question may be exceedingly atypical. It is because of these pecularities in subjects' neuroanatomy or behavior, or an interaction of the two, that Kolb and Whishaw recommend the use of statistics to test the significance of experimental findings and why they emphasize data based on group studies in their text. So, although it would seem that the problem of these enormous individual differences would appear to make the use of single-case research antithetical to theory building, we should nonetheless examine the way this variability is dealt with in the group design before we discard the case study as a legitimate approach to theory building.

Statistical analysis of data obtained from a group is, in an important sense, a multiple case-study analysis. When a psychologist runs 30 research subjects, be they rats or humans, on the same task, the implicit assumption is that it is not 30 individuals that are being tested, but rather 30 exemplars of the same system. Individual differences are dealt with by relegating them to the error term. This is considered methodologically safe. Any consistent effects that result from the measures or manipulation employed by the researcher are probably stronger than they appear because the error term against which the effects are tested has been inflated by all the individual difference factors. However, neuropsychology brings its own difficulties of interpretation to the group design. This has been illustrated recently in examining the nature of lateralization in groups that are supposed to be less lateralized for language. For example, left-handers (Segalowitz & Bryden, 1983) or women (Inglis & Lawson, 1981) are deemed to be less lateralized because as a group they show less asymmetry on standard tests, such as dichotic listening. However, the question remains as to whether they are less lateralized as individuals or whether the direction of their lateralization is less consistent across individuals in the group. Presumably we can view the relevant asymmetry scores and tell which is so, but stricly speaking we cannot interpret the individual lateralization scores because we have no measure of stability (standard deviation) for them. This is a rather elementary example and the problem can be circumvented in a number of ways.

However, the same problem arises with other issues, such as the question of complementarity. For example, is it the case that left lateralization for language implies a right lateralization for spatial skills? Group trends suggest this is so, because in right-handers, as a group, there is a functional separation of these skills. It is possible, though, that functional complementarity is a group effect only, that individuals need not show it (Bryden, Hecaen, & DeAgostini, 1983). There are many factors to bear in mind on this particular issue, but the general problem persists for the experimental neuropsychologist working with normals: the group trend may not reflect the characteristics of individuals in the group.

This problem is very similar to that experienced a number of years ago by learning theorists. The gradual learning curve, so familiar and so intuitively appropriate, in fact applies to groups well but not to individuals (e.g., Sidman, 1952). Only intensive case studies of individuals can show this. In neuropsychology, we are not interested in a group brain. If the results do not apply to individuals, they are not of interest. This problematic variability does not go away with the use of the group design, but rather it is relegated to the error term and may overshadow some very real effects. This confounding of error of measurement with individual variability is serious because this variability is critical in understanding brain–behavior relationships.

Taking a different tack, Marshall and Newcombe (1984), reflecting the theoretical perspective of Dukes (1965) and Campbell (1975), point out that the "theoretical significance (in the nonstatistical sense) of a clinical datum is not logically linked to the number of patients in whom that datum can be observed." They argue that a single case that does not conform to a hypothesized model of functional organization is sufficient to "falsify any strong form of the original hypothesis." They also refute the suggestion that one cannot build a theory of brain–behavior relationships from the study of the single case. As evidence for this view they point to the progress made from the late 19th century until the 1920s based on the case studies of clinical neurologists who studied the selective deficits of language, perception, action, and memory. It was on the basis of these case studies that some quite explicit and sophisticated theories of higher cognitive function evolved. A brief digression will illustrate this point.

CASE STUDY RESEARCH IN EARLY NEUROPSYCHOLOGY

The 19th century studies undertaken by behavioral neurologists were based on the observation of individuals who had sustained focal brain lesions or who were demonstrating some focal dysfunction or unusual dissociation of function (Benton, 1981; Lecours, Lhermitte, & Bryans, 1983; Young, 1970). For example, Bouillaud (1796–1881) concluded from observation of brain-damaged patients that the brain must consist of separate centers otherwise it would be impossible to understand how a lesion of one part of the brain could cause paralysis of some muscles of the body without affecting others. Further observations, made during his study of encephalitis, led him to think that even more complex mental functions, such as speech, could be accommodated within the localizationist model. Based on Bouillaud's work, Auburtin (1825–?) proposed that a lesion of the anterior lobes would be found in a patient who was, at that time, suffering from aphasia at the hospital of Bicetre. This prediction was confirmed by the surgeon and physical anthropologist Paul Broca (1824–1880). Autopsy revealed a lesion in the posterior third of the inferior frontal convolution of the left hemisphere. A second patient suffering from a disturbance of articulated speech was found to have an even more discrete lesion in the same region of the left hemisphere. Broca concluded that the seat of language was in the frontal lobes although he did not as yet make specific statements about laterality. It was after eight consecutive case studies in which he noted that all of these lesions had been in the left hemisphere that he commented on the laterality issue. However, as Benton (1981) points out, Broca still hesitated to draw conclusions about the relationship of the left hemisphere to language. Two years later, he had what he considered to be confirming evidence and in 1865 he enunciated his famous dictum, "We speak with the left hemisphere."

Broca's generalization was readily confirmed and led to a major revolution in medical and physiological thinking. From a medical standpoint, aphasia was transformed from a minor curiosity to an important symptom of focal brain disease. From a physiological standpoint, the reality of cerebral localization was established, and this led to a period of intense investigation of functional localization in both animals and human subjects.

More modern approaches to the single case have been undertaken by the Russian neuropsychologist, Aleksandr Luria (1902–1977) who emphasized careful, intensive investigations of brain-damaged individuals. As Lezak (1983) points out, Luria's contribution to neuropsychological assessment consists of a rich store of sensitive qualitative behavioral descriptions, of reproductions of patients' writings and drawings that capture common patterns of distortion while exemplifying the uniqueness of each patient's behavioral product, of techniques for eliciting behaviors that are relevant to the understanding of brain function and the treatment of neuropsychological problems, and of an approach based on the generalization and testing of hypotheses to guide clinical exploration, diagnosis, and treatment (e.g., Luria, 1966, 1970, 1973). Among his many brilliant theoretical insights was his resolution of the localizationist–antilocalizationist debate that had been the focus of much neurobehavioral investigation. His theory of the dynamic localization of function in the cerebral cortex called for a radical revision of the generally held concept that a nonspatial construct could be superimposed on the spatial construction of the brain. Function, as Luria saw it, was not to be considered the business of some particular cerebral organ or group of cells, but as a complex and plastic system performing a particular adaptive task and composed of highly differentiated groups of interchangeable elements (Luria, 1966; Luria & Majovski, 1977). Luria's theoretical formulation influenced the entire field of neuropsychology but this theoretical position grew from his years of intensive study of the single case.

The intensive study of the single case continues to feed into the development of neuropsychological theory but its prominence as a method has been eclipsed by a growing reliance on group research. It is clear, however, from this brief historical overview, that careful investigators did not make theoretical claims on the basis of a single case even though they were engaged in the intensive study of single cases. They noted patterns, formed hypotheses and waited until they had sufficient confirmation of these hypotheses before they made theoretical statements. Although subsequent research has elaborated and sometimes modified their theoretical statements, they formed a solid foundation for modern neuropsychology.

PREMISE THREE

A third criticism of single-subject methodology stems from the fact that case studies tend to be of individuals who have suffered brain damage and show some behavioral impairment as a consequence of that damage. The question is whether a general model of neuropsychological functioning should be built on inferences made from the study of brain-damaged individuals. This argument takes many forms (Segalowitz, 1983). First, brain damage, no matter how it occurs, is rarely tidy, one can never be sure of the extent or the precise nature of the damage. One may observe a loss of function not because of the obvious site of damage but because of some pathway between two other cortical areas has been destroyed. This criticism, however, holds for group data as well. Combining individuals on the basis of lesion site may result in a very heterogeneous group in terms of actual impairment. Second, the damage itself may cause changes in the way undamaged portions of the brain respond. For example, it may be possible for one brain area to have inhibitory effects on some neighboring area and facilitative effects on another (Selnes, 1974). It would then be possible that the performance we attribute to the undamaged portion of the brain may not be representative of that portion of the brain under normal conditions. So although we may be quite adept at predicting loss of function from area of lesion, we still may not know just how the normal brain carries out the functions we wish to study.

THE PROBLEM OF BASELINE DATA IN CLINICAL CASES

A final point of criticism of the use of brain-damaged subjects to build a neuropsychological theory is directly relevant to the single-case study. The gist of the issue is that we never know how the person performed before the damage was done, so it is unclear whether some deficits are the result of the injury or whether the individual was never very good at the task. In most instances, however, the deficit is obvious and very much at odds with the patient's historically validated premorbid functioning. It is the subtle deficits that produce the diagnostic problems.

Another variation on this point is that cases of truly rare or unusual dissociation are not relevant to a theory of normal brain function because they are likely the result of abnormal functional organization prior to injury. So, for example, Nolan and Carramazza's study of the deep dyslexic would be considered somewhat irrelevant to a theory of normal reading because, in order for this patient to have demonstrated such a rare pattern of deficits, she may have had an atypical brain organized for language prior to the damage. Marshall and Newcombe (1984) attack this line of reasoning with vigor. Such objections, they say, rest on a "pernicious equivocation on the meaning of 'normal'." If normal means "appertaining to the norm," then the argument holds, but, they point out, if "normal" means "nonpathological" then clearly it does not (p. 69). If the individual was able to perform some cognitive function (e.g., reading, writing, performing mathematical calculations, speaking fluently) prior to some brain insult, then the loss of that function or the pattern of deficits sustained is important in understanding the mechanisms of the skill that has been lost.

Of course, one can never be certain that a damaged brain was normal, that is, representative of a modal form of brain organization prior to injury. One can, however, be reasonably assured from the literature to date that apparently normal, nonpathological cognitive behavior can be produced by individuals whose brains vary considerably with respect to their physiological organization. The evidence presented by Lorber (Lewin, 1980) about young hydrocephalics is a case in point. These subjects were found to have ventricular enlargement and very little cortical tissue subsequent to early shunt operations to drain excess brain fluid. Despite this gross abnormality in brain structure, evidenced through CT and PET scans, the intellectual performance of these young people was average to above average. Thus, the "normal" brain, in the sense of appertaining to the norm, may not exist, just as the normal learning curve does not exist for any individual learner. It may be that the variability in brain–behavior relationships is much greater than we usually suspect, but we will not know that unless we undertake the intensive investigation of single cases.

The history of memory research attests to the cost of trying to build a theory of normal memory without attending to the ways that memory can break down. Schacter and Tulving (1982) have described how the experimental psychology of memory and the clinical study of amnesia remained quite separate since the late 19th century. They make the point, quite correctly, that the clinical studies of amnesia could have benefited considerably had they incorporated some of the analytical precision that was being developed in experimental psychology. They point out, however, that the progress of the experimentalists was considerably hampered by their failure to attend to the clinical literature.

As evidence, they describe how early experimentalists focused their attention on elucidating the variables that controlled the strength of associations. They believed that if the strength of association between two elements were strong enough, then recall of one in the presence of the other would be assured irrespective of the conditions of retrieval. Schacter and Tulving point out, however, that during this time the clinicians' observations on shrinking retrograde amnesia indicated that retrieval could initially fail and later succeed without any change in the strength of associations. As well, the reports of functional amnesia, such as is seen in fugue states, indicated that the ability to recall information is sensitive to the specific psychological conditions that existed at the time of retrieval. The current interest in the parallels between normal and abnormal memory (see, for example, Baddeley, 1982; Crowder, 1982; Jacoby, 1982) attests to the value of students of normal memory have found in the clinical aspects of memory dysfunction.

PREMISE FOUR

Despite all this concern about the role of clinical research in neuropsychology, the normal–abnormal dimension is theoretically orthogonal to the distinction between the single case and group design. Neurologically damaged individuals are often studied in groups. In practice, however, the converse is rarely done. Brain–behavior relationships in normals are not studied in the single-case paradigm. It is generally felt that the effects observed with normals are so subtle that their significance could only be determined through the use of parametric statistical evaluation and this has always meant using the traditional group experimental design. However, as indicated earlier, statistical analysis of data obtained from a group is a multiple case-study analysis in another guise. In a lateralization study, for example, all subjects are presumed to have identical cerebral specialization of function for the task chosen, which of course necessitates rigorous subject selection. In current practice, the cautious experimenter chooses right-handed males who have no left-handed relatives. If these subjects are identical with respect to brain lateralization, then the only factors influencing the lateralization scores are task variables and experimental measurement error. In practice, of course, there are other variables possibly adding variance to the lateralization scores. Trait and state factors, such as anxiety (Gruzelier, 1984), skill specialization specific to the task (e.g. Bever & Chiarello, 1974), and differences in generalized strategy, cognitive style, or ongoing cognitive activity differences (Bryden & Ley, 1983; Levine, Banich, & Koch-Weser, 1984) may very well influence lateralization scores. It is impossible to control for all such factors,* except by resorting to a series of true case studies.

If instead of testing 30 subjects, we tested the same subject 30 times, we would indeed have controlled for major factors of personality, biochemical, and neurological differences that would contaminate our lateralization scores. These scores would then be "cleaner" in this sense. Also, we would have enough data on which to perform traditional statistics. The cautious experimenter, however, points to several difficulties here. First, a single subject, when retested many times, is not necessarily the same with respect to some personality state variables or relevant cognitive strategy variables. Not only may the subject be bored with the task by the 10th testing, he is also no longer naive with respect to the task demands. As a result, the cognitive strategies used during the 10th testing may be very different when compared with those of the first. For some tasks this problem may be more important than for others. For example, Porter and Berlin (1975) found relatively good consistency on dichotic listening in most individuals over eight test sessions. On the other hand, Segalowitz and Orr (1981) found a visual half-field verbal task produced a steady left hemisphere advantage over six sessions, whereas a spatial task (reading the time on a clockface without numbers) gradually produced an increasing right hemisphere advantage over the six sessions.

There are practical problems as well: getting the subjects to agree to many retestings on the same task is not always easy, and this in itself may produce some subject self-selection that could confound lateralization factors.

The Problem of Generalization

There is a problem of generalization with both group and case-study designs, and the experimenter must make a choice. The experimenter choosing the group design knows that innumerable, but it is hoped minor, factors have been left to vary across the subjects in the study. They can only add unwanted variance to the obtained scores. Thus, as

---

* If the factors can be measured, something requiring much subject testing time, the experimenter can allow them to vary and then remove them from the error term by using a regression model. This has been done with respect to stimulus variables by Graves, Landis, & Goodglass (1981), but we know of no study doing this with subject variables.

indicated earlier, if an effect is found, it is probably quite strong because the error term against which the effects of interest must be tested has been inflated by these factors. Conversely, the traditional difficulty with the case-study approach is that the results may be due to experimental factors of interest in concert with unknown confounds. Some of these confounds can be removed by the use of appropriate control factors, but some cannot. For example, brain-damaged subjects cannot be their own control with respect to side of damage. Despite these limitations, however, a rigorous case-study methodology is possible. We will now outline the requirements for statistical analysis of individual case study in experimental neuropsychology as an attempt to merge the strengths of the experimental paradigm with the case-study approach.

EXPERIMENTAL CASE-STUDY ANALYSIS

For convenience, we will focus on lateralization paradigms and the examples will be from data with non-brain-damaged subjects, but the methods can be generalized to other patients, to other issues in the localization of function, and to other measures, including EEG asymmetries (Segalowitz, 1984).

In any statistical analysis, some effect is measured against an index of its stability. With a single lateralization asymmetry score, such as a score from a dichotic listening task or an EEG alpha asymmetry test, such a stability measure is not immediately apparent. We normally do not have an estimate of the standard deviation associated with such scores, although there are ways of calculating standard deviations in these cases. However, before discussing them we should attend to another serious problem in case-study design alluded to earlier.

Any asymmetry score is a reflection not only of the hemisphere specialization producing it, but also of any other factors promoting asymmetric responding. Some sources are clearly not psychological and are of nuisance value only. Better hearing in one ear, especially in frequency ranges aiding discrimination of consonant sounds, will obviously bias the result if left uncorrected. Undetected visual acuity asymmetries will similarly affect the visual half-field (VHF) paradigm just as an asymmetry in skull thickness will affect the EEG amplitude. Obviously, any such simple asymmetry in the input mechanisms or measuring device will produce an asymmetry in the result that is not of interest.

Similarly, unwanted idiosyncratic asymmetries can be due to non-peripheral factors that are also not related to the hemispheric specialization of interest. For example, right-handed subjects attend to their right ear more than do left-handers independent of which hemisphere is dominant for language (Bryden, 1978). That is, individuals may have lateral attentional biases related to body laterality, such as hand preferences. If performance on the dichotic listening or the VHF task is influenced by this, then the final asymmetry score will reflect this confound.

A more interesting, but nonetheless confounding, factor is the degree of individual difference in relative hemisphere activation during the testing procedure. Some people may have a generally more activated right hemisphere (or left hemisphere) than others. This difference in hemisphericity may be apparent mainly during testing situations or be more general. Whatever the origin, it is clear there are such differences and that they are reliable (Levine et al., 1984).

Thus, an individuals's lateralization score can be seen as a composite of several factors: hemisphere specialization for the task (e.g. speech perception in a dichotic listening test), experimental measurement error, and idiosyncratic lateral biases of various origins. This last factor is a serious confound in case studies, but can be averaged out in group designs (adding variance, of course).

The solution to separating these factors is simply to use each subject as his own control on a neutral task. Instead of administering only a verbal dichotic listening test for speech lateralization, we would give both a verbal test and some neutral or nonverbal dichotic listening test. If the subject has a generalized side-of-body or ear bias that is not simply due to overall acuity differences, then this bias would affect the nonverbal test as well. The difference between these two measures when this bias is controlled for is a cleaner reflection of speech dominance. Considering the variability possible in individual subjects, especially ones that may be interesting for clinical reasons, this control task is a basic precaution. Note that the research question is subtly changed, however. Instead of asking whether there is an absolute asymmetry for language representation, we are now asking for a relative measure: Does the left hemisphere have a greater advantage than the right on the verbal task compared to the nonverbal task? If we have a neutral control task, one that reflects only response or attentional biases and no hemisphere processing asymmetries, then we can have some confidence that our research question speaks directly to the issue of laterality. The critical issue is the neutrality, with respect to functional lateralization, of the control task.

Estimation of Variance

Once we have a measure of lateralization that is relatively unbiased by factors extraneous to our hypothesis, we must calculate an estimate of stability for that measure. A natural way would be to test the subject many times. An estimate of the mean and standard deviation of the subject's lateralization would then be possible.

An alternative has been devised by Bryden & Sprott (1981), who have suggested a metric, lambda, that provides not only a measure of asymmetry but also an estimate of its own stability. The metric is used on the dichotic listening or VHF type of data, where the number of hits and misses for each side are available (see Formula 1).

lambda = Ln [(RIGHT HITS
           × LEFT MISSES)/RIGHT MISSES × LEFT HITS)]   (1)

The variance is the sum of the reciprocals of the hits and misses. Each lambda could thus be tested against its own standard deviation and the result checked against a Z-score table for its probability level. If we want to use a difference score as our indicator of lateralization, this is reflected in the difference between two lambdas. The variance estimate of this difference is the sum of the two variance estimates for each lambda separately. Let us consider some examples for illustrative purposes. Let us say that we have 3 VHF tasks, one of which very consistently produces a right-visual-field (RVF) advantage, one a left-visual-field (LVF) advantage, and one no response asymmetry. Sample LVF and RVF scores for such tests are given in Table 1 with the associated lambdas, variances, standard deviations, and Z-scores. The left hemisphere and right hemisphere tasks are significant at the .02 level. If we want to compare the two tasks, by taking the difference, the associated SD is 1.17 and the Z-score is $(2.023 + 2.023)/1.17 = 3.46$, $p < .001$. If we compare either the left hemisphere or right hemisphere task against the neutral task, though, the difference in lambdas is 2.023, but the Z-score is now $2.023/1.104 = 1.83$, $p > .05$. In a more realistic example, the neutral task would not have a lambda of 0, but the pooled variance would be large just the same. There are, however, few VHF tasks that yield such strong asymmetries. Significance of effects using the difference between two lambdas is, in practice, difficult to obtain.

Let is then return to the multiple testing paradigm. Let us consider

Table 1. Hypothetical VHF Results on a Left Hemisphere (LH), a Neutral, and a Right Hemisphere (RH) Task

Task	Number correct		lambda	Var	SD	Z
	LVF	RVF				
LH	4	11	2.023	.682	.826	2.45
Neutral	8	8	0.0	.536	.732	0.00
RH	11	4	−2.023	.682	.826	−2.45

Note. There are 15 trials in each VHF.

the situation where we have gathered lambda scores for a verbal and a nonverbal task on a number of occasions. Table 2 presents such data for one subject from Segalowitz & Orr (1981).

This subject is fairly consistent in the direction of the asymmetries shown, although the magnitude varies somewhat. If we calculate the Z-scores for each lambda, we find that only the first and last presentation of the clock face condition produced significant results. If we take the difference scores, so that a positive result indicates a relative left hemisphere advantage for the verbal task, we find only the last test session is significant. If we calculate a t-value for each task separately based on the six observations, we find nonsignificant trends in the appropriate direction ($p < .10$ for each condition). However, if we treat each difference score as an observation and calculate a t-value comparing the mean difference to zero, then we find that the differences are significant ($t = 2.37$, $df = 5$, $p < .02$). Thus, this subject shows significant differences in hemisphere specialization for the verbal and nonverbal tasks used, when her six test sessions are taken as a whole. Clearly the difference-score test is more powerful when used in combination with multiple testing.

THE PROBLEM OF SERIAL CORRELATION

When we apply a series of scores from the same subject to a statistical test, we are treating the observations as if they were independent of each

Table 2. Results from Testing a Single Subject Six Times

		Test session					
		1	2	3	4	5	6
CVCs	LVF	18	9	9	6	6	6
	RVF	17	9	12	9	11	10
CFs	LVF	22	19	17	14	17	19
	RVF	16	16	17	14	15	10
CVCs	lambda	− .21	0	.51	.59	.93	.76
	Z	− .32	0	.87	.93	1.49	1.20
CFs	lambda	− 1.71	− .64	0.0	0.0	− .16	− 1.67
	Z	− 1.99*	.97	0.0	0.0	.27	2.57**
Difference	lambda$_d$	1.49	.64	.51	.59	1.10	2.43
	Z$_d$	1.39	.72	.59	.68	1.25	2.68**

Note. Each test session included a verbal task (CVCs) and a nonverbal task (Clock Faces). Report was oral. There were 24 presentations to each VHF. The subject was a right-handed female.
* $p < .05$; ** $p < .01$.

other. If they had been from different people, this would indeed be the case. In our example, however, there is the possibility that the six scores are serially dependent, that is, that the subject's VHF asymmetry given on one day predicts what it would be for the next day, and so on. For example, the subject may be developing a strategy that emphasizes the skills of one hemisphere more than the other. The danger in this is that if the six scores are serially dependent, we do not really have $n - 1 = 5$ degrees of freedom. In fact, in this case we would not easily know what the appropriate degrees of freedom would be, although Cattell (1963) gives a somewhat complex method for calculating them.

If we calculate a serial correlation for the difference scores in the above example, we find that it is .248, clearly not significant. This Pearson correlation coefficient is calculated by pairing the first observation with the second, the second with the third, and so on, until we have $n - 1$ pairs. The nonsignificance of the correlation indicates that we have no reason to suspect that we have anything but six independent measurements of this subject's lateralization pattern. Serial correlations calculated on the two tasks (repeated six times) separately yield quite different results. The six lambdas for the clockfaces are clearly independent $(r = .08)$, but the correlation for the CVCs is .83. This can be seen in the steady rise in the lambdas for the CVCs (a steady shift to a RVF advantage). It may be that this subject's strategy gradually developed and so one score predicts the next to some extent.* This serial dependence is eliminated when the difference scores are used because any strategy shifts will be common to the two conditions.

CONCLUSION

There are inherent limitations to both group and case-study designs in neuropsychological research. Both have merits and drawbacks. Although case-study methods have been applied primarily to clinical settings, this paradigm need not be restricted to a brain-damaged population. With the use of appropriate control tasks and proper statistical methods, rigorous case study research can be done with non-brain-dam-

---

* In discussing the serial correlation problem with respect to time series analysis, Holtzman (1963) suggests that, although exact significance levels are available (Anderson, 1942), the ordinary correlation coefficient tables are approximately correct.

aged subjects in the laboratory. With the increasing interest in relatively subtle neuropsychological factors and in the individual variability possible in brain–behavior relationships, we expect that the case-study approach will become increasingly important in experimental neuropsychology.

REFERENCES

Anderson, R. L. (1942). Distribution of the serial correlation coefficient. *Annals of Mathematical Statistics, 13,* 1–13.

Baddeley, A. (1982). Amnesia: A minimal model and an interpretation. In L. S. Cermak (Ed.), *Human memory and amnesia* (pp. 305–336). Hillsdale, NJ: Erlbaum.

Benton, A. L. (1981). Aphasia: historical perspectives. In M. T. Sarno (Ed.), *Acquired aphasia* (pp. 203–232). New York: Academic Press.

Bever, T. G., & Chiarello, R. J. (1974). Cerebral dominance in musicians and nonmusicians. *Science, 185,* 537–539.

Bryden, M. P. (1978). Strategy effects in the assessment of hemispheric asymmetry. In G. Underwood (Ed.), *Strategies of information processing* (pp. 117–149). London: Academic Press.

Bryden, M. P., & Ley, R. G. (1983). Right hemispheric involvement in imagery and affect. In E. Perecman (Ed.), *Cognitive processing in the right hemisphere* (pp. 111–123). New York: Academic Press.

Bryden, M. P., & Sprott, S. A. (1981). Statistical determination of degree of laterality. *Neuropsychologia, 19,* 571–581.

Bryden, M. P., Hecaen, H., & DeAgostini, M. (1983). Patterns of cerebral organization *Brain and Language, 20* (2), 249–262.

Campbell, D. T. (1975). "Degrees of freedom" and the case study. *Comparative Political Studies, 8,* 178–193.

Cattell, R. B. (1963). The structuring of change by P-technique and incremental R-technique. In C. W. Harris (Ed.), *Problems in measuring change* (pp. 167–198). Madison, WI: University of Wisconsin.

Crowder, R. G. (1982). General forgetting theory and the locus of amnesia. In L. S. Cermak (Ed.), *Human memory and amnesia* (pp. 33–42). Hillsdale, NJ: Erlbaum.

Dukes, W. F. (1965). $N = 1$. *Psychological Bulletin, 64,* 74–79.

Gazzaniga, M. S., Bogen, J. E., & Sperry, R. W. (1962). Some functional effects of sectioning the cerebral commissures in man. *Proceedings of the National Academy of Sciences, 48,* 1765.

Gazzaniga, M. S., Bogen, J. E., & Sperry, R. W. (1963). Laterality effects in somesthesis following cerebral commissurotomy in man. *Neuropsychologia, 1,* 209–221.

Gazzaniga, M. S., Bogen, J. E., & Sperry, R. W. (1965). Observations in visual perception after disconnection of the cerebral hemispheres in man. *Brain, 88,* 221–236.

Graves, R., Landis, T., & Goodglass, H. (1981). Laterality and sex differences for visual recognition of emotional and nonemotional words. *Neuropsychologia, 19,* 95–102.

Gruzelier, J. (1984, October). Individual differences in dynamic process asymmetries in the normal and psychopathological brain. Paper presented at the NATO Advanced Research Workshop, *Individual Differences in Hemispheric Specialization,* Maratea, Italy.

Holtzman, W. H. (1963). Statistical models for the study of change in the single case. In C. W. Harris (Ed.), *Problems in measuring change* (pp. 199–212). Madison, WI: University of Wisconsin.

Inglis, J., & Lawson, J. S. (1981). Sex differences in the effects of unilateral brain damage on intelligence. *Science, 212,* 693–695.

Jacoby, L. L. (1982). Knowing and remembering: Some parallels in the behaviour of Korsakoff patients and normals. In L. S. Cermak (Ed.), *Human memory and amnesia* (pp. 97–122). Hillsdale, NJ: Erlbaum.

Kimura, D. (1961a). Cerebral dominance and the perception of verbal stimuli. *Canadian Journal of Psychology, 15,* 166–171.

Kimura, D. (1961b). Some effects of temporal lobe damage on auditory perception. *Canadian Journal of Psychology, 15,* 156–165.

Kimura, D. (1967). Functional asymmetry of the brain in dichotic listening. *Cortex, 3,* 163–178.

Kolb, B., & Whishaw, I. Q. (1980). *Fundamentals of human neuropsychology.* San Francisco: W. H. Freeman & Co.

Lecours, A. R., Lhermitte, F., & Bryans, B. (1983). *Aphasiology.* Eastbourne, East Sussex: Bailliere Tindall.

Levine, S. C., Banich, M. T., & Koch-Weser, M. (1984). Variations in patterns of lateral asymmetry among dextrals. *Brain and Cognition, 3,* 317–334.

Lewin, R. (1980). Is your brain really necessary? *Science, 210,* 1232–1234.

Lezak, M. D. (1983). *Neuropsychological assessment* (2nd ed.). New York: Oxford University Press.

Luria, A. R. (1966). *Higher cortical functions in man* (2nd ed.). New York: Basic Books.

Luria, A. R. (1970). *Traumatic aphasia* (Translated from the Russian). The Hague/Paris: Mouton.

Luria, A. R. (1973). *The working brain: An introduction to neuropsychology* (B. Haigh, Trans.). New York: Basic Books.

Luria, A. R., & Majovski, L. V. (1977). Basic approaches used in American and Soviet clinical neuropsychology. *American Psychologist, 32* (11), 959–968.

Marshall, J. C., & Newcombe, F. (1984). Putative problems and pure progress in neuropsychological single-case studies. *Journal of Clinical Neuropsychology, 6* (1), 65–70.

Nolan, K. A., & Caramazza, A. (1982). Modality-independent impairments in word processing in a deep dyslexic patient. *Brain and Language, 16,* 232–264.

Nolan, K. A., & Caramazza, A. (1983). An analysis of writing in a case of deep dyslexia. *Brain and Language, 20,* 305–328.

Orgass, B., Hartje, W., Kerchensteiner, M., & Poeck, K. (1972). Aphasie und nichtsprachliche Intelligenz. *Nervenarzt, 43,* 623–627.

Porter, R. J., & Berlin, C. I. (1975). On interpreting developmental changes in the dichotic right-ear advantage. *Brain and Language, 2,* 186–200.

Schachter, D. L., & Tulving, E. (1982). Amnesia and memory research. In L. S. Cermak (Ed.), *Human memory and amnesia* (pp. 1–32). Hillsdale, NJ: Erlbaum.

Segalowitz, S. J. (1983). *Two sides of the brain: Brain lateralization explored.* Englewood Cliffs, NJ: Prentice-Hall.

Segalowitz, S. J. (1984, October). Measuring individual differences in lateralization. Paper presented at the NATO Advanced Research Workshop, *Individual Differences in Hemispheric Specialization,* Maratea, Italy.

Segalowitz, S. J., & Bryden, M. P. (1983). Individual differences in hemispheric representation of language. In S. J. Segalowitz (Ed.), *Language functions and brain organization* (pp. 341–372). New York: Academic Press.

Segalowitz, S. J., & Orr, C. (1981, February). *How to measure individual differences in brain lateralization: Demonstration of a paradigm.* Paper presented to the International Neuropsychological Society, Atlanta, GA.

Selnes, O. A. (1974). The corpus callosum: Some anatomical and functional considerations with reference to language. *Brain and Language, 1,* 111–139.

Sidman, M. (1952). A note on functional relations obtained from group data. *Psychological Bulletin, 49,* 263–269.

Wechsler, D. (1958). *The measurement and appraisal of adult intelligence* (4th ed.). Baltimore, MD: Williams & Wilkins.

Weisenburg, T., & McBride, K. (1935). *Aphasia: A clinical and psychological study.* New York: Hafner.

Young, R. M. (1970). *Mind, brain, and adaptation in the nineteenth century.* Oxford: Oxford University Press (Clarendon).

CHAPTER 12

Psychophysiological Activation Research

An Approach to Assess Individual Stress Reactions?

PETER WALSCHBURGER

SOME GENERAL QUESTIONS ON THE STUDY OF STRESS–STRAIN PROCESSES

Consider the following two questions: (a) Is an uncontrollable aversive event more activating than a controllable aversive event? (b) Will a high anxious person be more activated than a low anxious person when being exposed to conditions of physical danger?

Both questions are related to psychophysiological activation processes but they arise from quite different points of view. In the first case, we are interested in the general difference of *situational* intensity of two stress conditions. We can measure this difference by mean changes of activation variables as sampled over a group of persons. Possible differences in stress reactions of different persons need not be analyzed, except for reasons of statistical inference. This is what I will call a *general psychological* point of view. In the second case, however, we are primarily interested in analyzing possible differences in the reactions of *persons* with certain dispositions, as measured by questionnaires, being exposed

PETER WALSCHBURGER • Institut für Psychologie, Freie Universität Berlin, Habelschwerdter Allee 45, D-1000 Berlin 33, Federal Republic of Germany.

to the same stress condition. The situational intensity and other situational features do not seem to be of primary interest except for reasons of the empirical definition or standardization of the chosen stress condition. I will call this a *differential* point of view. (Note that from an individual-based approach the assumed individual-based definition of action features raises the problem of individual standardization of situational features, which will be discussed later in this chapter.)

The aim of this chapter is to discuss problems of and advances in assessment strategies for stress–strain processes in human beings. The study of such processes is guided by some general questions that are obviously related to our initial questions: Which are the crucial features of typical stress situations for human beings? How do people cope with everyday stressors in their lives? How do different persons react to stress conditions physiologically, behaviorally, and in their subjective or inner feeling? In which way can the patterns and processes of adaptation and maladaptation of persons with certain dispositions, when exposed to stressors with certain features, best be described, analyzed, and predicted?

These questions include both the general psychological and the differential points of view. The main interest, however, is to describe differences in individual stress reactions (differential point of view), or even to give an adequate, comprehensive description of the reaction patterns of any given individual (idiographic approach).

Questions of this kind have stimulated an empirically guided development of psychophysiological reactivity as a personality trait (Eysenck, 1967; Fahrenberg, 1977; Myrtek, 1980) in differential psychophysiology and in biologically oriented personality research. But above all, these questions are relevant for applied research in different fields. In clinical psychology and psychological medicine, for example, stress–strain research is expected to provide better assessment strategies and, ultimately, more efficient treatments for people characterized as being psychovegetatively labile, psychosomatically disturbed, stressed, or as being chronically anxious, irritated, or depressed. In these fields of applied research the idiographic issue is particularly evident.

As will be discussed in the following section, most of the empirical work on stress–strain processes has been derived from concepts like activation or stress. These concepts have been developed in a nomothetic research tradition, that is, they are primarily based on experiments aimed at finding general laws in physiology and behavior rather than at analyzing differences between persons or at describing individual reaction patterns. This may be a major reason for the common but dubious use of concepts like activation, or psychophysiological reactivity in differential psychology and personality research. I will argue in the next two sections that an empirical definition of activation has been only demonstrated convincingly in a general psychological but not in a differential frame of reference. In fact, we are able to predict which one of two stress conditions will be more activating (as measured by group means of psychophysiological stress reactions). But we are usually unable to predict which one of two individuals will exhibit a stronger psychophysiological reactivity (activation) in a given situation. Several approaches have been developed to solve this problem to assess individual stress reactions. As I will point out in the following sections of this chapter, these approaches have not at all been centered on single-subject research. Nevertheless, their advances are closely connected with various attempts to enrich an initially nomothetic research by introducing several idiographic elements.

With respect to the main concern of this volume, the reader might expect that a closed conception of an individual-based methodology in stress research will be offered. In fact, the following discussion resulted from various efforts to overcome the empirical insufficiency of well-established general reactivity concepts, when used to assess individual stress reactions. Taking this context into account the reader may better understand (a) why this chapter demonstrates in length the empirical insufficiency of concepts that seem both conceptually and empirically well established in general psychology, (b) why a combined idiographic-nomothetic approach is favored, and (c) why the empirical definition and applicability are understood as essential features in the evaluation of any conceptual contribution to an individualized research strategy.

ACTIVATION AND STRESS AS MODELS OF UNSPECIFIC REACTIVITY

As was already mentioned, most of the empirical work on stress–strain processes has been derived from (or related to) concepts like activation or stress. These concepts can be characterized as theoretical constructs being originally related to relevant issues in everyday life, where the meanings of these terms have been broad and complex, and relatively variable. Scientific research has to define these constructs in order to improve their precision and their empirical validity.

The basic scientific meanings of the terms *activation* and *stress* have their origins in the field of medicine (Cannon, 1929; Eppinger & Hess, 1910; Selye, 1950). They have been reviewed by several authors (e.g., Cannon, 1929; Duffy, 1962; Eysenck, 1967; Fahrenberg, 1977; Lindsley, 1951; Selye, 1950). Therefore, a few remarks may be sufficient to relate these conceptions to the issue of an assessment of individual stress reactions. Let us first summarize some crucial features of several classical concepts.

Cannon's concept of *emergency reaction* emphasizes adrenergic activation and sympathetic nervous system discharge as an expression of the "wisdom of the body" when being confronted with a fight-or-flight situation. Duffy's *organismic activation* concept covers different consistent changes of most psychophysiological functions, thereby describing a broad psychophysiological intensity dimension. Together with an approach–avoidance dimension of behavior this intensity dimension represents a two-component reconceptualization of traditional research topics (like feelings or drives, which are considered to be unscientific terms). Lindsley's *reticulo-cortical activation* concept was based on Morruzi's and Magoun's work on brain stem reticular activating functions and the corresponding arousal reaction in the EEG (desynchronization, increase in frequency and decrease in amplitude of the EEG). This arousal reaction indicates an intensity dimension that is reflected both in behavioral and experiential changes as well as in the state of consciousness. Selye's *general adaptation syndrome* (GAS) describes an integrated, stereotype response pattern of the organism as stimulus-unspecific reaction to quite different kinds of stressors. In contrast to Cannon's emergency reaction, Selye emphasizes a typical sequence of this GAS over time, including three different stages: a first stage of alarm, a second stage of resistance, and a final stage of exhaustion. The most important physiological mechanism of this GAS is seen in an activation of the pituitary–adreno-cortical system.

These short characterizations of some classical concepts may be sufficient to show that they contain differences and similarities. The differences are rather obvious. They are related to the mechanisms or bodily functions underlying stress reactions. However there is also a more hidden but interesting similarity: all these concepts emphasize a unitary, unspecific, and general process of activation, which has been implicitly understood and used as including very different functions of the total organism. According to this common view, an empirical definition or operationalization of such a general activation construct has to be done by assessing situation-dependent changes in different psychophysiological functions out of a homeostatic resting state. In order to assess individual stress reactions, consistent interindividual differences in general (psychophysiological) reactivity to stress situations should exist, so that different persons can be discriminated as more or less activated, or sensitive to stressors and more or less prone to stress-related disturbances. Obviously, such broadly defined reactivity constructs are of great relevance in different fields of applied or clinical psychology, for example, for purposes of evaluating stress–strain processes in different job situations or of assessment strategies for people with psychosomatic or emotional disturbances.

In empirical research on activation, stress, or anxiety the construct of general psychophysiological reactivity has been used by many authors. They often refer to a construct that covers changes in different physiological functions, as well as in the behavior and in the verbal report of subjects. In order to establish the empirical meaning of that construct, an empirical relation between some indicators of these different psychophysiological responses would have to be demonstrated. But this has almost never been done. Usually, only a single activation variable (like heart rate or GSR-amplitude) is used as an "indicator" of such a reactivity construct which, for its part, is taken as scientific reformulation of terms like anxiety, or stress.

This insufficient situation has stimulated a number of empirical studies aimed to develop a better methodological basis and to examine the empirical relevance of the activation construct. In the following section, I therefore briefly want to outline the main tasks, problems, and results of activation research. This will be done by mainly referring to my own experience with this issue as member of an interdisciplinary research group on psychophysiological problems at the University of Freiburg.

MAIN PROBLEMS AND RESULTS OF PSYCHOPHYSIOLOGICAL ACTIVATION RESEARCH

Starting with a definition of activation as a stimulus- or situation-dependent change (increase) in particular organismic functions, we have to be aware of the fact that individuals may already differ in their basal, or prestimulation, state of activation. Assessment strategies have to take into account these differences in prestimulus activation, because they may influence the amount of stimulus-dependent activation, a problem known as the "problem" (or "law") of "initial values" (Myrtek, 1980).

For assessment purposes one has to distinguish and to evaluate several aspects of activation. As just mentioned, we have to separate state from reaction (stimulus-related) aspects. From an individually oriented view both aspects may be seen as momentary dispositions (Stegmüller, 1974), corresponding to the statement "the individual is activated [i.e., shows activation] at this moment." Furthermore, momentary dispositions have to be differentiated from habitual dispositions, or trait aspects of activation. In order to speak of an activation trait, it is important to demonstrate not only differences between subjects but also the temporal and/or transsituational stability of these differences.

The main task of an empirical study of activation processes is to develop an adequate research strategy in order to analyze more general or more selective aspects of activation as dependent on both situational features and individual dispositions. The difficulties of such an approach may be characterized by four main problems: (a) The first concerns the selection of adequate indicators of activation not only in the physiological but also in the behavioral, and in the experiential domain. (b) The second concerns the measurement of activation processes or the reaction aspect of activation, which is made possible when change scores are independent of differences in baseline scores, or state aspects of activation. (c) The third is the covariation problem—meaning the common observation in most studies that only very small and inconsistent covariations have been found between reactions in different psychophysiological functions and between actual stress reactions and stress-related dispositions, like the "emotional lability" (Eysenck, 1967) of persons. This problem raises the question of the internal validity of psychophysiological reactivity constructs like activation, stress, or anxiety. (d) The fourth and rather comprehensive problem is whether a chosen kind of a typical laboratory study in activation research will be actually suitable to gain meaningful practical knowledge of stress–strain processes outside the laboratory. This problem includes the question of how to control essential features of psychological stress situations adequately.

Among these problems there are some questions that are closely related to our main issue of the role of the individual subject in activation research. Let us begin with discussing the problem of measuring activation processes by change scores.

The problem of measuring situation-dependent aspects of activation by change scores can be solved by simply using difference scores $D = Y - X$ between stress scores Y and prestimulus scores X. In most cases, however, a difference score does not only reflect the reaction aspect but also the state aspect of activation, due in some degree to its dependency on prestimulus activity. Some interesting proposals have been made to solve this problem, the most common and classical solution being the construction of Autonomic Lability Scores (ALS) (Lacey, 1956). An individual ALS corresponds to the deviation of the observed stress score from an expected stress score estimated by regression analysis according to the relations between stress scores and prestimulus scores on basis of a sample of subjects.

The use of more sophisticated change scores like ALS unfortunately brought about some new methodological problems, which are discussed in more detail elsewhere (Fahrenberg, Walschburger, Foerster, Myrtek, & Müller, 1979). With respect to the great amount of interindividual differences in stress reactions it can be argued that such a between-subjects standardization procedure for change scores is not suitable for assessment and has to be replaced by a within-subject procedure. In accordance with this idea some interesting proposals have been made to relate a given response with the individual's prestimulus variability (Malmstrom, 1968) or to the range of scores obtained in one subject when moving from resting states to states of extreme activation of a given function. Lykken (1972) was actually able to demonstrate a reduction of undesired between-persons variance (or error variance) when he corrected heart rate and GSR measures for their intraindividual range. A similar strategy was suggested by Lamiell (1981) as a formal rationale for idiographic measurement in personality psychology.

However, the empirical identification of minimum and maximum values of an intraindividual range of variation in a function that Lykken's and Lamiell's methods require is often deficient. In a recent study, Fahrenberg, Foerster, Schneider, Müller, and Myrtek (1984) compared several ways of transforming change scores, including Lykken's method of range correction. The Lykken scores in the Fahrenberg et al. study were based on the individuals' response range in the lab as well as on the individuals' total reactivity range (from sleeping state to ergometer work). Despite their attention to improving minimum and maximum values, the findings of Fahrenberg et al. cast doubts on Lykken's method of range correction. The Lykken scores used were found to be highly correlated with simple difference scores; they tended to be less stable than difference scores, depended on the respective empirical definition of the individuals' relative range, and did not provide a number of significant correlations with other measures of activation when compared with difference scores (Fahrenberg et al., 1984, p. 113).

On the basis of such findings the conclusion can be drawn that the superiority and practical applicability of such ipsative measures could not be demonstrated convincingly until now. With respect to the reliability problems of Lykken scores, one should be aware of the fact that correcting for intraindividual range normally will result in a reduction of between-subject variance. The problem is to demonstrate that this reduction involves psychologically uninteresting aspects of between-subject variance (e.g., anatomical differences influencing end-organ reactivity) whereas interesting differences (e.g., the central aspects of a certain function) will not be substantially reduced or eliminated. Corrections for intraindividual range will continue to be an attractive idea and a difficult task for future research.

ACTIVATION AND THE PROBLEM OF COVARIATION

Turning now to the problem of insufficient covariation, we find that this classical and central issue for psychophysiological assessment has been the subject of several reviews and of conceptual and empirical discussions and proposals (Duffy, 1972; Fahrenberg & Foerster, 1982; Haynes & Wilson, 1979; Lacey, 1967; Lader, 1975; Myrtek, 1980). This problem is also related to the ongoing debate on the issue of personality and prediction (Epstein, 1983a; West, 1983). The following discussion, however, will be restricted to the impact of the covariation problem on the empirical meaning of activation concepts. One aspect of this problem has been identified under different labels, such as "response fractionation," "discrepancies," or "dissociations" between different psychophysiological response systems. As mentioned earlier, however, another important aspect of the covariation problem still exists—namely, insufficient covariation between actual stress reactions and habitual personality traits, commonly thought to be predictors of such stress reactions.

A number of empirical studies have been performed that deal with the covariation problem. Unfortunately, most of them were based on small samples of subjects, and used only one or very few dependent variables. The following comment on this problem is primarily influenced by findings of a study by Fahrenberg and his colleagues (Fahrenberg et al., 1979). This study was explicitly designed to provide more detailed information about different aspects of the covariation problem using a broad sample of subjects exposed to various stress conditions and considering a broad spectrum of carefully selected activation variables. The crucial finding of this study is consistent with findings in most of the more comprehensive studies on this issue. Between-subjects analyses were not only unable to show consistent and substantial covariations of activation variables across the experiential and the physiological domain. They also failed to establish substantial relations across different bodily functions, like cardiovascular, electrodermal, respiratory, muscular, and central nervous system functions. These results are therefore incompatible with the single-component assumption of the concept of general activation. They do not even support the model of a small number of more limited activation components.

The covariation problem may be illustrated by results of the study of Fahrenberg *et al.* (1979). In this typical activation experiment with 125 male volunteers, activation processes were assessed under five different conditions (rest, mental arithmetic, interview, anticipation of blood taking, and blood taking itself). Conditions of rest and mental arithmetic have been repeated in order to allow reliability measurements. A primary item pool of 167 variables were reduced to 8 final activation variables following an empirically based rational selection procedure. According to this selection procedure, an activation variable has to meet several criteria—to discriminate between subjects as well as between conditions, to show a sufficient amount of reliability, to have an appropriate distribution (skewness, kurtosis), and to show no redundancy to one of the other variables used in the same study.

Table 1 illustrates the between-subject correlations between each pair of these eight selected activation variables: self report of tenseness (T), heart rate (HR), finger pulse volume amplitude (PVA), number of skin conductance reactions per minute (SCR), electroencephalogram 7 to 13 Hz relative power (EEG), respiratory irregularity as measured by relative power of residual spectrum (RESP), eye blink frequency (EB), and electromyogram of forearm extensor (EMG). As it is shown in Table 1, only 5 out of 28 correlation coefficients between resting or state scores, and only 3 out of 28 correlation coefficients between change scores (ALS— see preceding) were significant at the .05 level ($r \geq .18$). None of them, however, demonstrated a relation of any substantial magnitude.

In addition to this finding, the actual stress reactions could not be predicted from conceptually related personality characteristics to any

Table 1. Between-Subject Correlations (N = 125) for 8 Selected Activation Variables under Resting and Stress Conditions in a Typical Activation Experiment

	Activation state (rest)							
	T	HR	PVA	SCR	EEG	RESP	EB	EMG
T	—	.33*	−.14	.13	.12	−.04	−.14	.09
HR	.28*	—	−.22*	.25*	.03	−.01	−.05	.10
PVA	−.07	−.13	—	−.12	−.13	−.01	.00	−.10
SCR	−.01	−.08	.07	—	−.11	−.05	−.07	.22*
EEG	−.17	−.24*	.00	.14	—	−.15	−.09	−.10
RESP	.10	.13	.06	−.05	−.00	—	.14	.21*
EB	.05	.09	−.05	.00	−.07	.06	—	.11
EMG	.09	.22*	.02	.06	.07	.10	.03	—

Activation reaction (mental arithmetic)(stress)

* $p < .05$

substantial degree. In case of the study just mentioned, four dispositional variables were used as predictors of actual stress reactions: (a) questionnaire scores for emotional lability, (b) the tendency to report various physical complaints (like autonomic, muscular, sensory, or "nervous" complaints), (c) physical capacity or fitness based on estimations of maximum oxygen intake during ergometer work, (d) the autonomic (cardiorespiratory) reactivity under conditions of physical stress (cold pressor, breathholding). All these predictors failed to predict actual psychophysiological reactions of subjects under psychological stress conditions in the laboratory (Fahrenberg et al., 1979, 1983).

In order to evaluate these global statements in more detail, one would have to discuss several methodological problems, like the chosen strategy of selecting reliable activation variables, or the fitting of stressors and activation variables. Moreover, the question may be raised whether linear covariation models are appropriate in a field of research in which compensatory regulation processes may play an important role. A polynomial regression analysis (Kerlinger & Pedhazur, 1973) and inspections of scattergrams, however, revealed no substantial amount of nonlinear data relations. This problem and several other methodological problems have been considered to the greatest possible extent. Fahrenberg et al. (1979) discussed these problems at length and no evidence has been found that any of them might have invalidated the main findings mentioned above.

Two main exceptions of the general lack of covariation, however, should not be overlooked. First, there is good evidence for a substantial covariation across different activation variables within the experiential-subjective domain, so that a single or at best a few dimensions (Thayer, 1978) are sufficient to describe the data relations. It is only within this domain that the notion of an activation dimension is supported by data. I suppose that such a tendency to perceive various feelings and bodily reactions as organized around an activation–deactivation continuum also influences our scientific conceptualizations. This could explain to some extent the widespread use of a psychophysiological activation construct, the empirical validity of which has not been studied simultaneously in the experiential and the behavioral-physiological domain. Actually, findings of such a combined approach made clear that verbal reports on bodily reactions cannot simply replace physiological measurements.

Secondly, single-case studies and within-subject analyses (e.g., the use of R.B. Cattell's P-technique in correlational research, and time-series analyses) clearly demonstrated a higher degree of covariation among organismic functions (Fahrenberg et al., 1979; Haynes, & Wilson, 1979; Walschburger, 1976). Walschburger (1976) computed correlation coefficients for each individual subject (in a sample of 61) between pairs of 13 activation measures that were comparable to the measures mentioned above. Thirty data points provided a time series across several stress conditions (including rest). Because there exist special problems in evaluating statistical significance of correlation coefficients in a time series with serial dependency, the coefficients can be used only in a descriptive manner. Average individual coefficients in the sample were found to range from $r = .40$ to $r = .70$. In order to interpret the meaning of these coefficients, it is important to bear in mind that they reflect a process within the person that depends on the situations that the person is in. In the case of bigger differences in intensity between situations the correlation coefficient is also higher. Therefore, a mean within-subject correlation may primarily indicate the *situational* aspect of activation rather than its *differential* aspect of distinguishing persons from one another by their individual reactivity.

Walschburger (1976) found considerable differences between subjects in response covariation. Some subjects demonstrated very high correlational relationships between activation measures, whereas others did not. More detailed analyses revealed high heterogeneity within individuals, and confirmed the impression that different persons have highly variable degrees of covariation between psychophysiological reactions under conditions of stress. This finding may be viewed as a challenge for future research that can evaluate these interindividual differences in response covariation by the use of time-series analysis (e.g., Box & Jenkins, 1976).

Summarizing these findings and evaluating their significance for an empirical meaning or definition of a general activation construct, at least two points of view have to be discriminated.

First, the concept of activation can be evaluated from an experimental, or general psychological (situation-oriented) point of view, thereby neglecting interindividual differences in stress reactions. From this point of view, the term *activation* has indeed an empirical meaning. This meaning consists both in the definition of stress situations by significant mean changes of activation variables (sampled over different organismic functions, and over different persons), as well as in the quantification of situational intensity by the amount of these mean changes.

Second, the concept of activation can also be evaluated from a diagnostic, or differential point of view. This second view is related to the problem of how to assess interindividual differences in stress reactions. This problem is actually of primary interest not only in this chapter but also in the broad field of differential and applied psychophysiology. From this point of view, we have to realize that there is no uniform psychophysiological reactivity. Individuals respond to stress situations with quite heterogeneous activation processes across different organismic functions. It is therefore unwarranted to infer any kind of reactivity of a person in a given situation using a single indicator or even a single composite score of activation.

Such findings are obviously not compatible with any strategy of assessment based on general reactivity constructs. It is therefore important to ask what the consequence of these findings may be for future research.

APPROACHES TO REFINE RESPONSE PARAMETERS IN STRESS–STRAIN RESEARCH

From the point of view of basic research, the answer to this question may be that psychophysiological reactivity concepts are suffering from too high a level of complexity. As a consequence, a number of research groups have resorted to the study of more simple processes of regulation in separate functional subsystems, like the cardiovascular system, or to the analysis of more elementary mechanisms of activation and behavioral plasticity, like the orienting response and its habituation.

Nevertheless, philosophic as well as pragmatic reasons exist to continue research on assessment problems of individual stress reactions.

From a philosophic point of view, it can be argued that preferences for different approaches to psychophysiological problems—among others—are related to the mind–body problem. It seems to be helpful, therefore, when one's position to this problem is made explicit. My own position is in favor of a complementary approach that emphasizes simultaneous assessment of responses both within the frame of reference of physiology or behavioral psychology as well as within the frame of a psychology of conscious experiences. Both of them are understood as being essentially different domains requiring categorically different ways of description and analysis (see Fahrenberg, 1979, for a more detailed discussion of this issue). Before giving up such a basic position we should carefully examine alternative and perhaps more adequate approaches to advance a psychophysiological assessment of stress reactions.

In addition to this philosophic argumentation, there is no doubt that this kind of research will be useful for a scientific evaluation of stress–strain processes in various fields of an applied psychology (e.g., evaluation of stress–strain processes in different work situations or for patients with psychophysiological and emotional disturbances).

If we start from our finding of inhomogeneous activation processes and favor a complementary psychophysiological assessment procedure, the following two lines of thinking seem to be most promising for further developments in stress–strain research: (a) We have to adopt a multi-component model for stress reactions with a correspondent description of reaction profiles or patterns taking into account both person and situation characteristics as determinants of these reaction patterns. (b) Instead of considering response dissociations as a methodological or conceptual problem, we should explicitly ask for the information given by certain dissociations or discrepancies between stress reactions in several response channels. In other words, we should transform the mere methodological question of response dissociations into the more psychological question of specific response patterns. I will shortly outline some central features of three different kinds of approaches based on these lines of thinking.

THE CONCEPT OF DIFFERENT SENSIBILITY OF ACTIVATION VARIABLES

Without leaving the conceptual framework of activation theory we can assume that different organismic functions may be most reactive at different levels of stimulus intensity or "overall activation" (Lader, 1975; Walschburger, 1976). An EEG index of CNS activity, for instance, may be most reactive at low stimulus intensity, whereas measures of blood pressure may reach their optimal sensibility at much higher levels of stimulus intensity. As a consequence of this different sensibility, a weakening of covariations between both variables should be expected.

Such hypothetical differences in sensibility of different activation variables can be examined empirically by relating changes within a single activation variable to changes within a composite activation score (e.g., a sum of standardized T-scores of activation variables). Figure 1 illustrates such empirical sensibility curves for seven physiological activation variables in the study of Fahrenberg et al. (1979). The variables are identical with the ones in Table 1, except for self-report. Each curve is based on 875 data points taken from 125 subjects in each of the following seven situations: two rest phases, two mental arithmetic conditions (in which subjects had to accomplish simple addition tasks as quickly and accurately as possible while being distracted by noise sequences of a realistic content), an interview condition (including a free speech situation and some questions concerning the erotic-sexual sphere, the confrontation with a road accident, and personal experiences during the study), an anticipation of blood taking, followed by an actual blood taking situation. The 875 data points are reduced to 10 intensity classes. The sensibility curve is based on the probability estimation that a particular activation variable will attain a maximum value provided that a certain value of a

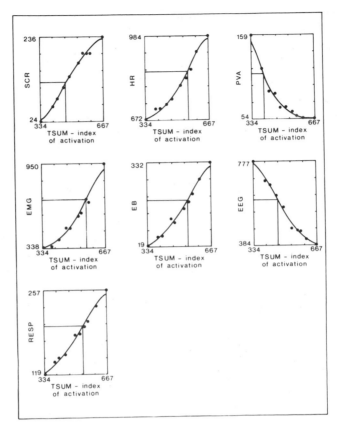

Figure 1. Empirical sensibility curves of seven activation variables as related to an index of overall activation (sum of standard scores TSUM).

standard T-score of overall activation across the sample of all seven activation variables is given (for details see Fahrenberg *et al.,* 1979).

This continuous sensibility curve shows an asymptotic shape for extreme overall activation and for extreme relaxation. It can be discussed with respect to parameters of its distribution, like mean, variance, skewness, kurtosis. For example, the relative power of alpha frequencies in the EEG, the peripheral pulse volume amplitude (PVA), and a measure of phasic electrodermal activity (SCR) show their steepest slope and therefore their greatest sensibility at low levels of overall activation whereas heart rate (HR), forearm EMG, eye blink frequency (EB), or respiratory irregularity (RESP) are most reactive at higher levels of overall activation. Floor and ceiling effects and additional characteristics of certain activation variables under certain conditions may also be discovered by this method.

To sum up, one can say that findings on different sensibility curves may provide interesting information about which response systems or activation variables are most reactive at different situational intensities. Differences in sensibility may represent an important factor of weak co-variations between different response systems. They may also clarify the problem of covariation to some extent and provide a better fitting of situational and response characteristics when planning experiments or assessment strategies. The concept of sensibility, however, does not account for possible qualitative rather than quantitative differences of stress situations. Moreover, this concept is centered on a situation-oriented and general psychological point of view. As we have already seen, this point of view is compatible with the activation construct, but only at the expense of neglecting individual differences by simply averaging reactions of different individuals. Problems of an assessment strategy for individual differences in stress reactions obviously cannot be dealt with adequately from this point of view. We will, therefore, turn to another attempt to modify and refine response parameters in stress–strain research.

THE USE OF RESPONSE DISCREPANCIES AS INDICATORS OF INDIVIDUAL COPING STYLES

Several research groups have emphasized the diagnostic value of discrepancies between measures of verbal report and physiological reactions of a person exposed to stress conditions. Stimulated by psychoanalytic concepts of different defense mechanisms, some researchers have suggested a relation between response discrepancies and certain personality characteristics, such as a more repressive or a more vigilant style of coping with threatening stress situations (e.g., Weinstein, Averill, Opton, & Lazarus, 1968).

This idea is certainly attractive for a number of assessment and treatment purposes, especially in the field of clinical psychology and psychological medicine. There is also some evidence of a possible development of such discrepancies in well-controlled laboratory studies, as the following example may show. By means of an experimentally induced cognitive set we manipulated the expectation of subjects to react to conditions of stress in the laboratory in a more labile or in a more stable manner (Walschburger, 1980). For this purpose, each subject had been informed about his own reactivity following a bicycle ergometer test and some tests of autonomic functions in a realistic medical diagnostic setting. This information was given by a physician who discussed the results with the subject. Actually, the physician did not tell the truth but told half of the subjects that measures of the tests indicated autonomic lability and hyperreactivity (Set A). The other half of the subjects was given the opposite information, suggesting autonomic stability (Set B). Subjects were randomly assigned to Sets A and B. The influence of this explicit cognitive set on both psychological and physiological reactions to some subsequent psychosocial stress situations was analyzed some hours later in another laboratory.

The results demonstrated clear influence of the given cognitive set on stress reactions. This influence, however, depended on the subject's habitual emotional lability, and it was limited to self-reports of perceived activation, bodily changes, attributions, and situational appraisals. No influence of the lability–stability set on physiological stress reactions could be shown.

Studies on relations between personality traits (as measured with questionnaires) and discrepant stress reactions generally used the difference between (standardized) verbal report measures and physiological reactions to stress as a measure of discrepancy. Unfortunately, most of these studies failed to prove a clear superiority (in terms of correlations to the questionnaire score) of their discrepancy measure as compared with the simple reaction components of that measure. Fahrenberg et al. (1979) and Walschburger (1981) examined the methodological and empirical basis of discrepancy concepts and came to the conclusion that some logical and biometrical problems and prerequisites have been overlooked. For instance, a consistent tendency among several discrepancy scores to over- or underestimate the physiological response was not found. Moreover, a minimum covariation of the verbal and physiological components would be an important prerequisite for the superiority of discrepancy scores. This covariation could not be established in most cases and the discrepancy scores then showed undesired high correlations (about ±.70) with both their verbal report and their physiological component.

Due to these and some related problems (Walschburger, 1981), applicability of response discrepancies for assessing characteristic coping styles under conditions of stress could not be demonstrated convincingly. In spite of the methodological shortcomings the idea of using response dissociations as information remains attractive. Among several proposals to overcome the methodological difficulties, two modifications of research strategies seem to be especially promising: (a) to supply or replace the predominant episodic type of studies by a longitudinal perspective including time-series analysis of response measures in the conceptual framework of process models for coping with stress; (b) to take into account discrepancies between more than two reaction components both within and between the physiological, behavioral, and experiential "levels." As we have already seen, there is no consistent empirical evidence for a physiological level. Instead, a rather diverse mixture of different physiological processes was observed. Research on specific response patterns may help us to clarify the problems of multiple response discrepancies.

THE DIFFERENTIATION OF SPECIFIC RESPONSE PATTERNS IN ACTIVATION PROCESSES

Along with research based on unspecific reactivity concepts, there exists a concurrent line of research that examines the inhomogeneous activation processes from the point of view of specific response patterns (Lacey, 1967; Lange, 1887; Mason, 1975; Roessler & Engel, 1974). That direction emphasizes the role of individual, situational, emotional, or motivational factors in determining specific patterns in response processes of certain individuals under certain conditions of stress.

During the last two decades statistical methods have been developed to differentiate and to determine the relative amount of several aspects of specific response patterns occurring in a sample of individuals being confronted with a variety of situations and reacting with various response systems (cf. Fahrenberg *et al.*, 1979; Foerster, Schneider, & Walschburger, 1983). The crucial questions and concepts leading these investigations may be outlined as follows.

Individual-Specific Response Patterns (ISR). The ISR concept refers to the degree to which an individual will maintain a particular idiosyncratic, or stereotypic response pattern (or a maximum response within a specific functional subsystem) when exposed to different stress situations.

The concept of ISR is obviously of special interest for any assessment approach trying to individualize response measures. It is also related to influential concepts in applied fields of stress research, such as to the concept of symptom specificity in psychosomatic medicine (Malmo & Shagass, 1949). Findings of a study by Moos and Engel (1962) may serve as an illustration of symptom specificity. In this study, a group of hypertensive patients—when compared to a group of arthritic patients with respect to their response to a variety of stimuli—exhibited particularly intensive blood pressure reactions to all stimuli, whereas the arthritic patients reacted particularly with muscles surrounding symptomatic joints. Persons exhibiting a strong ISR tendency are generally expected to be more prone to psychophysiological stress disorders than persons with a weak ISR tendency. This assumption, however, has not yet been proved.

Stimulus-Specific Response Patterns (SSR). The SSR concept refers to the degree to which a given stimulus or a stress situation will elicit a particular response pattern in different persons.

The concept of SSR may be viewed as a counterpart to the concept of ISR. It is important to realize that both kinds of response specificities may exist simultaneously. Research and assessment strategies, statistical analyses, and interpretations have to take this possibility into consideration. It is important to note that SSR are not related to physical properties of a situation *per se*, but to the response patterns of a sample of persons in a given situation. For the purpose of an assessment of individual stress reactions it may be helpful to define SSR in a person-related rather than in a stimulus-related way. SSR may then be characterized as the disposition of an individual to exhibit a response pattern to a given situation, that is similar to the other individuals of a reference group.

Motivation-Specific Response Patterns (MSR). The picture gets even more complicated as there are good reasons to assume a basic interaction between characteristics of the individual and the situation (Ax, 1964). There may be characteristic perception and appraisal of a particular situation by a particular person. For example, a woman suffering from a snake phobia may exhibit a specific (habitual) response pattern, differing from response patterns of other persons, but occurring only when exposed to a snake stimulus but not when confronted with other stress situations. It is in this sense that we may differentiate a third type of response pattern, a motivation- (or emotion-) specific response pattern MSR.

Several statistical evaluation procedures utilizing nominal, ordinal, and interval data, as well as various definitions of the similarity between response patterns, have been developed. Methodological problems are elsewhere discussed in detail (Fahrenberg *et al.*, 1979; Foerster *et al.*, 1983). An illustration of a typical finding in the study of Fahrenberg *et al.* (1979) is presented in Table 2. After a common standardization of each activation variable across subjects and situations, a three-factorial ANOVA (subjects, situations, variables) allows to estimate several interesting proportions of variance simultaneously. This ANOVA is based on 125 subjects, four stress conditions (mental arithmetic, interview, anticipation of blood taking, and blood taking itself), seven activation variables of the physiological domain (see Table 1), and four activation variables of the experiential domain (alert, tense, irritated, pulse). The results look rather different in the two data domains. With regard to the physiological domain, individuals differ mostly with respect to their ISR. This is not true for situations that show their greatest variability with respect to their general intensity, but not

Table 2. *Proportions of Variance Related to General Activation and Specific Response Patterns with Respect to 125 Subjects Exposed to 4 Situations and for Activation Variables in the Physiological and Experiential Domain*

Activation variables	Variance component	Interpretation	Proportion of total variance
Physiological	Subjects	Activation (individual)	4 %
functions: 7 ALS	Situations	Activation (situational)	22 %
change scores	Subjects variables	ISR	27 %
	Situations variables	SSR	5 %
Verbal	Subjects	Activation (individual)	23 %
reports: 4 self-scaled	Situations	Activation (situational)	18 %
change scores	Subjects variables	ISR	6 %
	Situations variables	SSR	1 %

with respect to their quality in the sense of SSR. As compared with the physiological domain, activation processes in the experiential domain seem to be more homogeneous, despite a number of efforts to use a representative sample of variables (see Fahrenberg *et al.*, 1979). This study did not facilitate an evaluation of MSR by ANOVA, because the entire study was not repeated (see Foerster *et al.*, 1983, for evaluation of MSR).

Considering these results, the relatively high proportion of ISR on the total variance is of primary interest for the covariation problem of unspecific reactivity concepts and for the problem of assessing individual stress reactions. Similar results were obtained in several studies carried out by different groups, and using different stimuli (Engel, 1960; Knobloch & Knobloch, 1979): about one fourth to one third of the subjects exhibited individual-specific response patterns (ISR). Furthermore, these ISR have been proved to be rather stable over a certain period of time, at least over some weeks (Foerster *et al.* 1983).

It seems possible now to modify the conceptualization of an unspecific reactivity to stress insofar as a substantial proportion of persons may be differentiated qualitatively by means of their idiosyncratic or stereotypic response pattern (ISR) when being exposed to different stress situations.

These results are in accordance with the concept of symptom specificity in psychosomatic medicine. They are also in line with recent modifications of the concept of nonspecificity in Selye's work on stress. Selye's general adaptation syndrome (Selye, 1950) was originally conceived as a nonspecific response to any kind of stressors. Different individuals were thought to exhibit the same response pattern. In a discussion of remaining problems after 40 years of stress research, Selye (1976) refers to "conditioning factors" as responsible for the following two types of observations:

> A. Qualitatively different stimuli of equal stressor potency (judged by their ability to elicit typical stress manifestations—e.g., ACTH, corticoid, or catecholamine production) do not necessarily cause the same stress syndrom in different individuals. B. Even the same degree of stress induced by the same stimulus may provoke different lesions in different individuals. (Selye, 1976, p. 54)

The parallels between these modifications of Selye's concept of a nonspecific stress response and the reported approach to differentiating specific response patterns are quite obvious. The differentiation and simultaneous evaluation of ISR, SSR, and MSR, however, allows one to analyze the amount and quality of response specificities of particular persons exposed to particular situations in a much more precise, systematic, and comprehensive way than was done before (Foerster *et al.*, 1983). These concepts seem to be well suited to test the assumptions of a differential emotion theory approach to stress reactions emphasizing the role of specific emotions instead of unspecific activation in response to psychosocial stress situations (Izard, 1977; Stemmler, 1984; Walschburger, 1982). Furthermore, a number of similarities can be discovered between the response specificity approach in psychophysiology and recent developments of an old interactionist view in personality theory (e.g., Magnusson & Endler, 1977) and anxiety research. These developments in personality and anxiety research have stimulated new kinds of assessment procedures, such as analyses of situation-response (S-R) inventories of anxiety (Endler & Hunt, 1966).

When we keep in mind the rather high proportion of individual-specific response patterns being identified in studies on response specificities, it appears that assessment strategies for individual stress reactions should be crucially improved by taking into account the tendency of each single individual to maintain a stereotyped response pattern in different situations.

It is possible to derive individual indexes of ISR tendency from the procedures mentioned above in a threefold sample of persons, situations, and response variables. This strategy, however, is not very practical, because the costs of gaining the indexes are high. Moreover, studies using such indexes failed to demonstrate their validity convincingly (Fahrenberg *et al.*, 1979). Some proposals were made to use approximations of individual ISR tendency, such as taking the most reactive parameter within the standardized individual reponse pattern (e.g., Fahrenberg *et al.*, 1979; Zahn, 1964), or to correct a given (single) response variable (e.g., skin conductance reaction) for intraindividual range (Lykken score, Walschburger, 1976), which we already discussed as an interesting idiographic change score. Despite these proposals, the practical applicability and validity of all these measures of individual ISR tendency still remains to be demonstrated. The reported findings on specific response patterns suggest that the question of how to individualize assessment strategies for stress reactions in order to keep a better balance between nomothetic and idiographic elements will continue to be an unresolved problem of primary importance in this field of research.

In spite of a number of interesting approaches to the refining and individualizing of response parameters reported in the preceding sections of this chapter, we have to admit that we are not yet able to offer a strategy for an adequate prediction of individual stress reactions. Attempting to provide possible reasons for this unsatisfactory situation, we have to realize that progress in stress–strain research has been centered on response parameters, thereby neglecting to specify and control essential features of psychological stress situations.

A common strategy to induce stress situations in activation research consists of putting several isolated short-time situations together, which are defined or standardized in terms of their physical-objective properties (e.g., noise-level), or in terms of an average response of a reference group. There is little evidence, however, that these types of stress situations allow the experimenter to control central action features of each situation as experienced by individuals and as resulting in different strategies of action. In order to do more justice to an individual's perspective of a stress situation, we have to concentrate more on both his behavioral orientations as well as his expectations, intentions, or (conscious) action goals in this situation.

There is evidence that laboratory situations like those previously mentioned may be quite similar with respect to their impact on the action strategies of persons, or may exhibit a rather similar valence despite their objective differences. It is in this sense that different experimental conditions in studies on response specificities may represent rather homogeneous "action situations" with a consequent underestimation of both stimulus- and motivation-specific response patterns and an overestimation of individual-specific response patterns. In the final section I shall describe our own attempt at standardizing central action features of stress situations for individual subjects. With this approach we tried to explore the empirical meaning of conceptual demands for a scientific reconstruction of stress situations as dynamic interaction processes, or transactions (e.g., Lazarus & Launier, 1978).

TOWARD AN INDIVIDUAL STANDARDIZATION OF STRESS SITUATIONS

At the moment, a sceptical view of any attempts to develop a comprehensive taxonomy of psychological stress situations seems to be justified (e.g., Rotter, 1981). There are, however, good reasons for preferring certain action prototypes of stress situations, which are conceptually more attractive and more ego-involving than others. Conditions of overload, as opposed to conditions of load may be seen as such action prototypes. They are characterized by different relations of situational demands and individual resources, provided that an individual shows a strong motivation, intention, or action goal to cope successfully with that situation. Load conditions can, but overload conditions cannot be coped with successfully—despite a strong interest of the individual to do so. There are interesting contributions on the conceptual and empirical meaning of overload conditions from quite different lines of thinking, such as animal research on experimental psychopathology (Maser & Seligman, 1977), human stress research (Cofer & Appley, 1964; Lazarus & Launier, 1978), and achievement motivation research (Heckhausen, 1980; Kuhl, 1983; Schwarzer & Walschburger, in press).

Our own investigations were based on achievement behavior in success- and failure-type stress situations as models for load and overload conditions. In one of the studies (Walschburger & Kuhmann, 1983), a sequence of stress situations was chosen beginning with a success-type situation (about 80% correct solutions across a series of similar tasks) and—unexpectedly for the subject—ending with 10 successive failure-type situations (about 60% incorrect solutions). According to several process models of stress regulation (Schwarzer & Walschburger, in press), individuals are expected to undergo typical changes in their predominant coping activities when exposed to a series of failures. These changes are assumed to begin normally with initial stages of reactance including positive and negative emotions successively (like interest and anxiety), as well as intensified psychophysiological stress reactions, and an increasing amount of task-irrelevant and self-related cognitive activity (e.g., worries about one's own ability of solving the task). When attempts at coping fail to stop the failure series, people may finally reach a stage of exhaustion or helplessness, including depressive mood, decrease in achievement related behavior, and decrease in activation of various psychophysiological functions.

Empirical evidence for these process models has been found (see Walschburger & Kuhmann, 1983), but we will concentrate now on the more central question of how to construct types of stress situations as defined by a similar amount of success, or failure, for each individual (rather than an average amount of success or failure over a sample of persons).

We used simple tasks of a concentration test (Abels, 1961) as elements of our two types of stress situations. Subjects were asked to decide whether a slide showing 36 two-figure numbers contained two specified numbers. Each of these tasks could be solved within 10 seconds. Subjects indicated their solutions by pressing a button on a keyboard that was attached to the computer. An incorrect solution of a task was indicated by an acoustic signal (designating failure) aimed at ameliorating the subjective perception of the performance outcome. About 60 tasks of this kind may be dealt with by a given subject in 5 minutes. It is within these 5 minutes that our stress situation gradually evolves. On the basis of the sequences of the individual correct or incorrect (salient) solutions, each subject is expected to experience his global quality of performance either as success or as failure depending primarily on the distribution of correct and incorrect solutions. The evidence for this assumption will be presented later on.

The next question is how to standardize such distributions of outcome in order to obtain either success-type or failure-type stress situations.

A success-type situation with a similar high success of about 80% correct task solutions for each single subject can easily be obtained. Subjects were simply asked to handle the tasks at their own optimal speed, an instruction normally eliciting strong achievement motivation and a high quality of performance (Heckhausen, 1980, p. 431).

It is more difficult, however, to obtain an individually standardized failure-type situation in which subjects may expect to be successful when making enough effort, and in which they maintain a high motivation to solve the tasks over a longer period of time.

For this purpose, subjects worked on the same type of tasks as before, but now at a given (variable) speed. An additional instruction made them believe that this was a speed at which a reference group was able to complete the task successfully. So their expectations might be as follows: I can only meet the group norm if I do my best.

While trying to succeed in solving the task, the subject was actually working in a feedback loop aimed at stabilizing the rate of incorrect task solutions at about 60% for each individual. We used a computerized adaptive procedure by which the presentation time of each subsequent task was decreased whenever the subject accomplished the previous task correctly. This resulted in a continuously increasing speed of tasks and difficulty, until the subject made mistakes at two consecutive tasks. Then, the speed (and difficulty) of the task was reduced until the subject did not make any mistakes and the procedure started again. The variations in presentation time came in very short intervals of 250 and 500 ms, and were barely noticed by the subjects (Kuhmann, 1980). The following flow chart (Figure 2) illustrates our procedure.

By means of this procedure, task difficulty can be regulated in a way

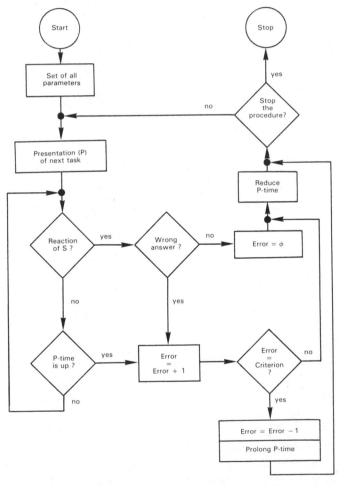

Figure 2. Flow diagram of the procedure aimed at stabilizing the rate of incorrect task solutions at about 60% for each individual.

comparable to typical self-regulatory changes in individual aspiration levels (AL), that is, an increase following correct solutions, and a decrease following incorrect solutions. The dynamics of this procedure support a lasting motivation to solve the tasks. In contrast to self-regulatory changes of AL, however, this procedure actually prevents rather than supports a high amount of success, and enables the experimenter to standardize a failure-type situation individually while the subject is striving for successful task solutions. The procedure may be seen therefore as an attempt to let the subject persue the goal of solving the tasks while simultaneously enabling the experimenter to prevent this goal.

Two studies were carried out in order to test the idea of an individual standardization of success- and failure-type stress situations empirically (Walschburger & Kuhman, 1983; Walschburger, Lachnit, & Meinardus, 1980). We used the percentage of correct solutions (Corr.sol.$_{obj}$) given by a single subject in order to characterize action features of a *stress situation*, but not of a person. For this purpose the parameter Corr.sol $_{obj.}$ should meet three criteria. First, its mean value over each stress situation should indicate predominant success or failure. Second, its variation between persons should be minimal whereas its variation between success- and failure-type situations should be high. Third, action features of situations are understood as including behavioral and experiential aspects. Therefore it should be possible to demonstrate that the actual success of performance (Corr.sol.$_{obj}$) is reflected in the subjective experience (Corr.sol.$_{subj}$).

The results of both studies demonstrated the empirical meaning and replicability of an individual standardization of situational features. Figure 3 shows the average temporal pattern of three parameters, including the percentage of correct solutions (Corr.sol.$_{obj}$) in the study of Walschburger and Kuhmann (1983). Actually, the mean value of Corr.sol.$_{obj}$ is 84% in the success-type situation. Over the series of failure-type situations this mean value shows a rapid decrease that stabilizes at about 42% after the third failure-type situation. These results meet our first criterion. Moreover, they are in line with findings of an earlier study (Walschburger *et al.*, 1980).

Our second criterion gives rise to the question of whether the percentage of correct solutions will only show a substantial variation between situations but not between persons. This question can be answered by conducting an ANOVA on the parameter Corr.sol.$_{obj}$, combined with subsequent estimations of variance components for persons, situations, and unexplained variance. The distribution of these variance components in Table 3 indicates that variations in percentage of correct solutions actually are based on differences between situations rather than between

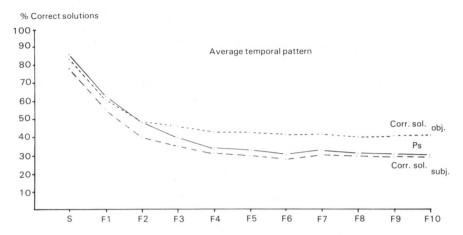

Figure 3. Average temporal pattern (N = 31) of the percentage of correct task solutions (Corr.sol.$_{obj.}$), as well as of its subjective counterpart (Corr.sol.$_{subj.}$) and of the probability of success in the next task (Ps), as reported by the subject. The stress series consists of a success-type (S) and 10 consecutive failure-type (F) situations.

individuals. This is not only true if we select three optimal failure-type conditions (with very similar percentages of correct solutions and complete subject sample size) out of the 10 failure-type situations in the study of Walschburger and Kuhmann (Table 3b). It also holds if all the nine clear-cut failure-type situations are included in the analysis (Table 3c), although the F-value for the subject factor is significant in this case, indicating an increase of reliable differences between persons. In addition, the distribution of variance components is about the same in another study with 40 subjects and 3 situations (Table 3a). All variance components were estimated for the worst case of maximal interaction between the person factor and the situation factor.

Finally, we have to demonstrate that the behavioral rate of success (or failure) is reflected in the subjective experience. There are two kinds of evidence supporting this assumption. First, the average temporal patterns of both the percentage of correct task solutions and its estimation by the subjects are very similar (Figure 3). The reader may be surprised that an initial surplus of hope—as measured by the subjective probability of success (Ps)—gives way to a rather pessimistic expectation of future success after only two failure-type situations. In addition to this observation, Figure 3 shows that subjects not only tend to underestimate their actual success by some 10% (Corr.sol.$_{subj.}$ < Corr.sol.$_{obj.}$), but that this tendency increases over the series of failure-type situations. These findings are contrary to most of the findings in the literature indicating that people tend to overestimate rather than to underestimate their actual success when working on familiar tasks (Heckhausen, 1980, pp. 390–400). In this study, however, a clear distinction in the experience of success and failure, as well as the progressive development of a rather pessimistic view of the actual success rate, was definitely intended: as mentioned earlier, an acoustic signal was given whenever the subject made an error. This failure signal obviously worked well in accentuating the perception of failure. Moreover, the cumulated experience of a rather long series of

Table 3. Results of ANOVA's with Subsequent Estimations of Variance Components for the Percentage of Correct Solutions and for 3 Different Samples of Persons and Situations

(a) Study of Walschburger, Lachnit, & Meinardus (1980) with 40 subjects and 3 success (S)-or failure (F)-type situations (sequence: S-F-S)

	MS	DF	F	p = .05	Estimation of variance components
Situations	22990	2	268.6	sign.	81%
Subjects	105	39	1.2	n.s.	7%
Residual	86	78			12%

(b) Study of Walschburger & Kuhmann (1983) with 30 subjects and 4 situations out of a total of 11 situations (sequence: S-F$_4$-F$_5$-F$_6$)

	MS	DF	F	p = .05	Estimation of variance components
Situations	12334	3	243.7	sign.	86%
Subjects	63	29	1.3	n.s.	4%
Residual	49	87			10%

(c) Study of Walschburger & Kuhmann (1983) with 26 subjects (4 subjects gave up) and 9 situations out of a total of 11 situations (sequence: S-F$_3$-F$_4$-F$_5$-F$_6$-F$_7$-F$_8$-F$_9$-F$_{10}$)

	MS	DF	F	p = .05	Estimation of variance components
Situation	4455	8	144.3	sign.	82%
Subjects	65	25	2.1	sign.	4%
Residual	30	200			14%

failures may have progressively induced a rather pessimistic estimation (and expectation) of the actual results.

There is still a second kind of evidence for our assumption that the actual rate of success (or failure) is reflected in the subjective experience. In the study of Walschburger et al. (1980), a close between-subject correlation ($r = .70$) was found between the percentage of correct task solutions (Corr.sol.$_{obj.}$) and its estimation by the subject (Corr.sol.$_{subj.}$) in the success-type situation. This close correlation could be replicated with results obtained by Walschburger and Kuhmann (1983). In this study, a between-subject correlation of $r = .76$ was found. Note that such a close between-subject correlation cannot be observed in failure-type situations. Here, the reported feedback procedure (Figure 2) prevents any substantial variation of the failure rate between persons, which in turn results in a breakdown of between-subject correlations.

All in all, these results demonstrate the potential usefulness of our procedure for standardizing relevant action features of stress situations by means of a dynamic interaction (or transaction) of task demands and individual resources. We hope that such an approach will do more justice to the individual subject in laboratory studies of stress research than typical studies in activation research have done so far.

REVIEW AND CONCLUSIONS

Psychophysiological activation research was discussed in this chapter as one of the most influential approaches to the study of human reactions to psychosocial stress situations. For assessment of individual stress reactions, the concept of activation is generally assumed to provide a basis for the strategy of distinguishing persons according to their general psychophysiological reactivity, or activation. A number of crucial problems and findings of studies in activation research were discussed, leading to the conclusion that the nomothetic orientation of this approach has to be combined with several idiographic elements. This is not only true for the methodological problem of measuring individual activation processes by change scores. It also applies to the internal validity problem of insufficient covariation of activation parameters and to the external validity problem of how to control essential features of stress–strain processes, for example, particular action features of stress situations.

Several approaches have been used to overcome the problem of insufficient covariation between indicators of different organismic functions and between actual stress reactions and related personality traits as measured by questionnaires. Among these approaches, a simultaneous analysis of individual-specific, stimulus-specific, and motivation-specific response patterns seems to be most promising. There is consistent evidence from different laboratories that subjects differ most with respect to their individual-specific response pattern (ISR) maintained in different stress situations, but not with respect to their unspecific reactivity, or activation. According to this finding, assessment strategies for individual stress reactions should be improved by basing them on individual indexes of ISR tendency. Two problems, however, have to be considered with this proposal. First, it is a very costly and difficult task to obtain such indexes from an appropriate threefold sample of individuals, situations, and response variables. In addition, neither such indexes, nor some approximating strategies (e.g., correcting a single response variable for intraindividual range), are sufficient to improve assessment strategies. This raises another and even more challenging problem of the response-specificity approach: strategies to develop stress–strain research were centered on refining response parameters, thereby neglecting an adequate control of situational features. The validity of ISR indexes, however, and of other results of the response specificity approach, and, finally, the external validity of any laboratory approach to assess stress–strain processes heavily depends on an adequate control of psychological stress situations.

Although a comprehensive taxonomy of stress situations is not to be constructed in the near future, control of some basic features of psychological stress situations in laboratory studies seems to be a promising strategy. We therefore discussed our own approach to standardize individually success and failure as important action features of stress situations. In addition, this research strategy is theoretically and empirically related to processes of coping with persistent failure as opposed to the episodic and isolated type of stress conditions used in most studies in activation research.

What direction should future research on the assessment of individual stress reactions take? In general, I think we should advance a combined idiographic-nomothetic approach. In order to do more justice to the individual subject in a nomothetic research tradition, it may be an attractive strategy to begin with single-subject designs (Haynes & Wilson, 1979) or with other individualized assessment procedures (see this chapter, or Lohaus, 1983). This kind of single-subject research, however, should be followed by between-subjects analyses. It is only this combination of idiographic and nomothetic strategies that will enable us to interpret individual stress responses in the frame of reference of responses given by other persons.

In our own research on individual stress reactions we have combined a general psychological stage of conceptualization and data analysis with a person-oriented stage. We will continue to follow a psychophysiological approach, taking into account patterns of physiological, behavioral, and verbal responses, and relating them to situational and individual determinants. The approach should be guided by a particular kind of a combined nomothetic-idiographic conceptualization that may be characterized as a process model of interindividual differences in stress regulation. Two sets of assumptions are made in this model. A first set of assumptions refers to the average temporal pattern in the reactions of a sample of persons being exposed to a particular series of stress conditions. Complementary to these general psychological and situation-oriented assumptions, a second set of differential and person-oriented assumptions should explicitly deal with distinguishing temporal patterns in the reactions of persons supposed to be more or less vulnerable (or resistant) to the particular series of stress conditions. A first stage of data analysis—applying to our general psychological assumptions—may help us to describe psychologically relevant action features of more or less extended stress conditions. In other words, this kind of analysis provides a nomothetic frame of reference for a second stage of differential (or even idiographic) analyses, referring to our person-oriented assumptions. I think it is important to relate individual dispositions and psychologically relevant action features of stress conditions closely to each other within a consistent approach.

The study of Walschburger and Kuhmann (1983)—partly reported in the preceding section—illustrates this kind of approach. The study was based on a particular process model of interindividual differences in the regulation of overload conditions, and two groups of persons were selected by means of a questionnaire (Kuhl, 1981, 1983). These two groups of persons are assumed to exhibit a more "state-oriented" or a more "action-oriented" style of coping with persistent failure (Kuhl, 1981, 1983). Each person in our sample was confronted with the same failure series (as standardized individually). According to our general assumptions we expected a particular average temporal pattern when analyzing the central tendency in the reactions of all persons over the failure series (e.g., changes from positive to negative moods, a tendency to give up, etc.). According to our differential assumptions, we expected different temporal patterns for state-oriented versus action-oriented persons. State-oriented persons are assumed to be more vulnerable to persistent failure (e.g., to exhibit a progressively more pessimistic, and helpless mood) than action-oriented persons.

A first stage of data analysis revealed an average temporal pattern that confirmed our general assumptions on the motivational and emotional effects of persistent failure and improved our knowledge of these effects. In addition to this main effect of the situation factor, we found significant interactions between the person (group) factor and the situation factor. For example, self-reports of state-oriented subjects indicated a more pronounced and more rapid increase of helplessness over the series of failure-type situations. Such different temporal patterns in current coping reactions of persons differing in certain dispositions underline the importance of process models of interindividual differences in the regulation of particular stress conditions. With respect to the problem of predicting actual stress reactions from personality dispositions, these results draw our attention to the fact that an expected relation between a dispositional measure and a current stress reaction may only be established after a certain duration, or accumulation, of stress episodes.

As for our own approach, the reported problems of insufficient covariation and of individual specific response patterns made us conclude, at least for the moment, that group designs instead of single-subject designs may be more promising. Group designs allow more economic tests of differences in group means of few response parameters instead of confronting us with the laborious task of selecting indexes of ISR tendency for each single subject. In our next studies we will therefore continue to select extreme groups characterized by different styles of action control (Kuhl, 1981), and by different resistance or vulnerability to ego-involving overload situations. The action features of those situations will be carefully controlled (or individually standardized). We should give, however, just as much attention to an individual standardization of response characteristics as to that of situational features. Although this chapter mainly referred to physiological responses, several interesting proposals were made to individualize also the assessment of verbal responses (e.g., Lohaus, 1983). We further agree with Bem (1983), who pointed out that we have to construct triple typologies or equivalent classes of persons, of situations, and of response modes in a conceptual framework and to relate them to one another. Moreover, we will carefully expand the scope of action for subjects in future studies by discriminating several subtasks (goals) within the frame of a more comprehensive task (goal). We hope to progress from the mere study of processes of regulation over a failure series within a single type of task to the study of both situational and individual determinants of accentuating different aspects of a given task and of actively selecting one out of several possible subtasks or goals of action. If we want to improve our predictions for individual stress reactions, we will have to take into account that people normally have to make choices between competing goals of actions and strategies to reach these goals. From this point of view it seems to be promising to induce cases of conflict between different tasks, and to weaken situational constraints without losing too much control in order to improve the predictive value of such studies.

Recently, Epstein (1983a, b) presented an alternative approach to combine both idiographic and nomothetic research by examining responses of multiple individuals on multiple measures over multiple occasions. Epstein emphasizes the fact that behavior is often highly dependent on situational aspects. As a consequence, it is very difficult to establish replicable findings and generalizations from typical laboratory experiments. Epstein shows that aggregating data, for example, pooling them over situations and occasions, can reduce variance associated with individual situations, or occasions. He agrees, however, with our view that some types of laboratory and field settings exist that are clearly more appropriate than others in predicting behavior. Among them, Epstein (1983a) points out the examination of ego-involving stimuli or events, the examination of situations with relatively few or weak constraints, and the examination of prototypical stimuli. Such settings may have the additional advantage of not only improving prediction but also refining psychological analysis of behavior.

Despite some clear differences between Epstein's approach and our own priorities for future studies, it should be realized that both proposals are complementary rather than contradictory because they are based on a common idea: the combination of idiographic and nomothetic research.

REFERENCES

Abels, D. (1961). *K-V-T Konzentrationsverlaufstest*. Göttingen: Hogrefe.

Ax, A. F. (1964). Goals and methods of psychophysiology. *Psychophysiology, 1*, 8–25.

Bem, D. J. (1983). Constructing a theory of the triple typology: Some (second) thoughts on nomothetic and idiographic approaches to personality. *Journal of Personality, 51*, 566–577.

Box, G. E. P., & Jenkins, G. M. (1976). *Time series analysis, Forecasting and control*. San Francisco: Holden-Day.

Cannon, W. B. (1929). *Bodily changes in pain, hunger, fear, and rage*. New York: Appleton-Century-Crofts.

Cofer, C. N., & Appley, M. H. (1964). *Motivation: Theory and research*. New York: Wiley.

Duffy, E. (1962). *Activation and behavior*. New York: Columbia University Press.

Duffy, E. (1972). Activation. In N. S. Greenfield & R. A. Sternbach (Eds.), *Handbook of psychophysiology* (pp. 577–622). New York: Holt.

Endler, N. S., & Hunt, J. McV. (1966). Sources of behavioral variance as measured by the S-R-inventory of anxiousness. *Psychological Bulletin, 65*, 336–346.

Engel, B. T. (1960). Stimulus-response and individual-response specificity. *Archives of General Psychiatry, 2*, 305–313.

Eppinger, H., & Heβ, L. (1910). *Die Vagotonie*. Berlin: Hirschwald.

Epstein, S. (1983a). Aggregation and beyond: Some basic issues in the prediction of behavior. *Journal of Personality, 51,* 360–392.

Epstein, S. (1983b). A research paradigm for the study of personality and emotions. In M. M. Page (Ed.), *Personality—Current theory and research: 1982 Nebraska Symposium on motivation* (pp. 91–154). Lincoln, NE: University of Nebraska Press.

Eysenck, H. J. (1967). *The biological basis of personality.* Springfield, IL: Charles C Thomas.

Fahrenberg, J., (1977). Physiological concepts in personality research. In R. B. Cattell & R. M. Dreger (Eds.), *Handbook of modern personality theory* (pp. 585–611). Washington DC: Hemisphere.

Fahrenberg, J., (1979). Das Komplementaritätsprinzip in der psychophysiologischen Forschung und psychosomatischen Medizin. *Zeitschrift für Klinische Psychologie und Psychotherapie, 27,* 151–167.

Fahrenberg, J., & Foerster, F. (1982). Covariation and consistency of activation parameters. *Biological Psychology, 15,* 151–169.

Fahrenberg, J., Walschburger, P., Foerster, F., Myrtek, M., & Müller, W. (1979). *Psychophysiologische Aktivierungsforschung. Ein Beitrag zu den Grundlagen der multivariaten Emotions- und Stress-Theorie.* München: Minerva.

Fahrenberg, J., Walschburger, P., Foerster, F., Myrtek, M., & Müller, W. (1983). An evaluation of trait, state, and reaction aspects of activation processes. *Psychophysiology, 20,* 188–195.

Fahrenberg, J., Foerster, F., Schneider, H. J., Müller, W., & Myrtek, M. (1984). *Aktivierungsforschung im Labor-Feld-Vergleich. Zur Vorhersage von Intensität und Mustern psychophysiologischer Aktivierungsprozesse während wiederholter psychischer und körperlicher Belastung.* München: Minvera.

Foerster, F., Schneider, H. J., & Walschburger, P. (1983). The differentiation of individual-specific, stimulus-specific, and motivation-specific response patterns in activation processes: An inquiry investigating their stability and possible importance in psychophysiology. *Biological Psychology, 17,* 1–26.

Haynes, S. N., & Wilson, C. C. (1979). *Behavioral assessment. Recent advances in methods, concepts, and applications.* San Francisco: Jossey-Bass.

Heckhausen, H. (1980). *Motivation und Handeln. Lehrbuch der Motivationspsychologie.* Berlin: Springer.

Izard, C. E. (1977). *Human emotions.* London: Plenum Press.

Kerlinger, F. N., & Pedhazur, E. J. (1973). *Multiple regression in behavioral research.* New York: Holt.

Knobloch, H., & Knobloch, J. (1979). Zum Problem physiologischer und psychologischer Reaktionsspezifität. *Psychologische Beiträge, 21,* 522–539.

Kuhl, J. (1981). Motivational and functional helplessness: The moderating effect of state versus action orientation. *Journal of Personality and Social Psychology, 40,* 155–170.

Kuhl, J. (1983). *Motivation, Konflikt und Handlungskontrolle.* New York: Springer.

Kuhmann, W. (1980). *Ein Konzept zur experimentellen Induktion individuell standardisierter Belastungen.* Referat, 9. München: Arbeitstagung "Psychophysiologische Methodik."

Lacey, J. I. (1956). The evaluation of autonomic responses: Toward a general solution. *Annals of the New York Academy of Sciences, 67,* 123–164.

Lacey, J. I. (1967). Somatic response patterning and stress: Some revisions of activation theory. In M. H. Appley & R. Trumbull (Eds.), *Psychological stress: Issues in research* (pp. 14–37). New York: Appleton-Century-Crofts.

Lader, M. (1975). Psychophysiological parameters and methods. In L. Levi (Ed.), *Emotions: Their parameters and measurement* (pp. 360–365). New York: Raven.

Lamiell, J. T. (1981). Toward an idiothetic psychology of personality. *American Psychologist, 36,* 276–289.

Lange, C. (1887). *Ueber Gemüthsbewegungen. Eine psycho-physiologische Studie.* Leipzig: Thomas.

Lazarus, R. S., & Launier, R. (1978). Stress-related transactions between person and environment. In L. A. Pervin & M. Lewis (Eds.), *Perspectives in interactional psychology* (pp. 287–327). New York: Plenum Press.

Lindsley, D. B. (1951). Emotion. In S. S. Stevens (Ed.), *Handbook of experimental psychology* (473–516). New York: Wiley.

Lohaus, A. (1983). *Möglichkeiten individuumzentrierter Datenerhebung.* Münster: Aschendorff.

Lykken, D. T. (1972). Range correction applied to heart rate and GSR data. *Psychophysiology,* 9, 373–379.

Magnusson, D., & Endler, N. S. (Eds.). (1977). *Personality at the crossroads: Current issues in interactional psychology.* Hillsdale, NJ: Erlbaum.

Malmo, R. B., & Shagass, C. (1949). Physiologic study of symptom mechanisms in psychiatric patients under stress. *Psychosomatic Medicine,* 11, 25–29.

Malmstrom, E. J. (1968). The effect of prestimulus variability upon physiological reactivity scores. *Psychophysiology,* 5, 149–165.

Maser, J. D., & Seligman, M. E. P. (Eds.). (1977). *Psychopathology: Experimental models.* San Francisco: Freeman.

Mason, J. W. (1975). Emotion as reflected in patterns of endocrine integration. In L. Levi (Ed.), *Emotions: Their parameters and measurement* (pp. 143–181). New York: Raven.

Moos, R. H., & Engel, B. T. (1962). Psychophysiological reactions in hypertensive and arthritic patients. *Journal of Psychosomatic Research,* 6, 227–241.

Myrtek, M. (1980). *Psychophysiologische Konstitutionsforschung. Ein Beitrag zur Psychosomatik.* Zürich: Hogrefe.

Roessler, R., & Engel, B. T. (1974). The current status of the concepts of physiological response specificity and activation. *International Journal of Psychiatry in Medicine,* 5, 359–366.

Rotter, J. B. (1981). The psychological situation in social-learning theory. In D. Magnusson (Ed.), *Toward a psychology of situations. An interactional perspective* (pp. 169–178). Hillsdale, NJ: Erlbaum.

Schwarzer, R., & Walschburger, P. (in press). Stress, Angst und Hilflosigkeit. In H. Häcker, H. D. Schmalt, & P. Schwenkmezger (Eds.), *Persönlichkeitssysteme und Persönlichkeitskonstrukte.* Weinheim: Beltz.

Selye, H. (1950). *The physiology and pathology of exposure to stress.* Montreal: Acta.

Selye, H. (1976). Forty years of stress research: Principal remaining problems and misconceptions. *CMA Journal,* 115, 53–56.

Stegmüller, W. (1974). *Probleme und Resultate der Wissenschaftstheorie und Analytischen Philosophie. Bank II/1. Theorie und Erfahrung.* Berlin: Springer.

Stemmler, G. (1984). *Psychophysiologische Reaktionsmuster. Ein empirischer und methodologischer Beitrag zur inter- und intraindividuellen Begründbarkeit spezifischer Profile bei Angst, Ärger und Freude.* New York: Lang.

Thayer, R. E. (1978). Factor analytic and reliability studies on the activation–deactivation adjective check list. *Psychological Reports,* 42, 747–756.

Walschburger, P. (1976). *Zur Beschreibung von Aktivierungsprozessen.* Doctoral dissertation, University of Freiburg, FRG.

Walschburger, P. (1980). Beeinflußt die Erwartung, labil zu reagieren, die Bewältigung nachfolgender Belastungen? *Archiv für Psychologie,* 132, 207–220.

Walschburger, P. (1981). Die Diskrepanz zwischen subjektiven und physiologischen Belastungsreaktionen: Ein informativer Indikator des individuellen Bewältigungsstils? *Schweizerische Zeitschrift für Psychologie und ihre Anwendungen,* 1, 55–67.

Walschburger, P. (1982). Emotionsforschung und Klinische Psychologie. In U. Baumann, H. Berbalk, & G. Seidenstücker (Eds.), *Klinische Psychologie. Trends in Forschung und Praxis: 5* (pp. 18–55). Bern: Huber.

Walschburger, P., & Kuhmann, W. (1983, March). *Verhaltensleistung und emotional-motivationale Bewältigungsprozesse in einer individuell standardisierten Belastungsserie mit fortlaufendem Mißerfolg.* Paper presented at the 25. Tagung experimentell arbeitender Psychologen, Hamburg.

Walschburger, P., Lachnit, H., & Meinardus, B. (1980). Anforderung und Überforderung. Ein Ansatz zur Diagnostik von Belastungs-Beanspruchungsprozessen. *Archiv für Psychologie, 133,* 293–321.

Weinstein, J., Averill, J. R., Opton, E. M., & Lazarus, R. S. (1968). Defensive style and discrepancy between self-report and physiological indexes of stress. *Journal of Personality and Social Psychology, 10,* 406–413.

West, S. G. (1983). Personality and prediction: An introduction. *Journal of Personality, 51,* 275–285.

Zahn, T. P. (1964). Autonomic reactivity and behavior in schizophrenia. *Psychiatric Research Reports, 19,* 156–173.

CHAPTER 13

Sequence–Structure Analysis
Study of Serial Order within Unique Sequences of Psychological Phenomena

JAAN VALSINER

INTRODUCTION

Psychological processes that underlie individual subjects' acting and thinking can be studied in different ways. Outcomes (products) of these processes have usually served as bases for generalizations about these processes. These outcomes—be these responses to test questions or recorded behavior in experiments or observations—constitute static representations of the processes that have generated them. Therefore, a conceptual difficulty in much of psychology has been the contradiction between the dynamic character of psychological processes and the static nature of empirical data used in the analysis of these processes. From time to time, calls for inquiring into these processes directly, rather than through their products, have been made (Bertalanffy, 1952; London, 1949; Vygotsky, 1962, 1978; Werner, 1957). Nevertheless, the majority of psychologists have remained faithful to the tradition of studying static products of their subjects' actions and of making inferences to the dynamic processes that may have generated these outcomes.

An alternative to this strategy would be to study psychological processes during the time period when they are generating their outcomes.

JAAN VALSINER • Developmental Psychology Program, Department of Psychology, University of North Carolina at Chapel Hill, Chapel Hill, North Carolina 27514.

In this case, direct access to the processes has to be empirically available. This may be the case only in some areas of psychological research. In this chapter, it is assumed that in one domain—the study of goal-directed actions—psychological processes that produce outcomes can be studied directly. This possibility has been demonstrated in the case of cognitive processes that involve concurrent verbalization and action (Ericsson & Simon, 1980; Luria, 1932, 1976), and on other occasions where verbal reports of subjects are unavailable in principle as in the case of animal experimentation (Köhler, 1925, 1973).

The most important issue in the effort to study psychological processes is to take their temporal order into account. Very rarely has this issue been addressed by psychologists. Psychology has often eliminated temporal unfolding of its phenomena from consideration. Sequentially patterned research material is usually reduced to elements of data, which are subsequently aggregated into summary indexes. Such strategy eliminates temporal relationships between different aspects of the original phenomena. The goal of this chapter is to analyze human goal-directed action, both theoretically and in terms of different empirical applications, from the perspective of its sequential organization. A research strategy that preserves information about a succession of phenomena when the latter are transformed into data is outlined, and its philosophical and practical implications analyzed. Sequence–structure analysis makes it possible to analyze the full range of variability of an individual's sequentially organized action patterns. Even though aggregation of data from sequence–structure analysis across individual subjects is technically possible, it is scientifically unfeasible.

THEORETICAL BACKGROUND OF SEQUENCE–STRUCTURE ANALYSIS

The Nature of Organisms: Open Systems

Definition and Its Application

All living organisms are open systems, because exchange of material with their environments is the ultimate condition of their existence. A system is called *isolated* if it exchanges neither matter nor energy with environment; *closed* if it exchanges energy but not matter; and *open* if it exchanges both (Bertalanffy, 1960, p. 144). All open systems—and that means all living organisms—are characterized by equifinality in their functioning. *Equifinality* refers to two aspects in the existence of open systems: (a) the same intermediate (or end) result of some process can be reached through different routes; and (b) the outcome of a process involving open systems cannot be predicted from the initial state of the system (Bertalanffy, 1950).

It is easy to apply these abstract characteristics of open systems to particular phenomenologies of living organisms. In the course of the evolution of species, a similar functional outcome has emerged that helps both whales and fish to move around in aquatic environment. While adapting to their surroundings, they have developed bodies similar in form. In human lives, different individuals can reach the same end state (e.g., school graduation, marriage, divorce, childbirth, death, etc.) via different courses of life. In psychology, a similar outcome (e.g., a self-rating of "very high self-esteem" by different subjects) can be reached through different psychological processes—some subjects rate themselves this way because the rating seems to apply to them, others would arrive at the same rating on the basis of compensation for what is actually low self-esteem, for instance. In childhood, different children reach certain developmental milestones by different pathways. For example, some infants start walking independently after crawling and creeping, whereas others may bypass the crawling stage. Almost all children learn to speak their mother tongue, but some reach that capability earlier, others later; some progress by a linear increase in their vocabularies, others take a long time to start and then progress with catastrophic rapidity. Still some other children may have ups and downs in their development, but they end up in the same end state—being able to speak the language of their culture. The progress of particular children's development cannot be predicted from their beginning state—for example, it is not possible to make accurate predictions of a normal child's language development on the basis of its behavior as a newborn. Examples that reveal the open-systems nature of psychological phenomena can also be taken from the realm of social interaction. Some parents carefully censor their children's TV-viewing habits, with the aim that this can prevent these children from becoming aggressive later. Other parents may not bother to do that, or some of them may even believe that TV viewing helps their children to develop aggression that is useful in the given society. Despite all these measures, some children from each of these subgroups of parents end up becoming aggressive criminals, and others will not. The open-systems nature of psychological phenomena guarantees variability in the forms in which these phenomena may occur. For an individual person, it also grants the presence of plasticity in adapting to environmental demands, which change over time.

Meaning for Psychology

Conceptualization of psychological phenomena in terms of open systems undermines the traditional world view of the majority of psychologists. First, a widespread definition of psychology as a science of the prediction and control of behavior is inapplicable in principle if a psychologist wants to build a theory in accordance with the notion of open systems. Prediction would be an impossible task for a psychologist, because outcomes (behavior) in open systems are inherently unpredictable from a previous state of the system. It is often assumed that prediction is equivalent to explanation in science—if we cannot predict, then we cannot explain the phenomenon scientifically. This rule of thumb, which has been based on the dominant status of physical sciences in the social history of the meaning of what qualifies as scientific, reflects a misunderstanding of the phenomena studied by biological and social sciences (Scriven, 1959). Theories similar to those in conventional physics (which deal with closed systems) are inadequate for accounting for biological and social phenomena (Bertalanffy, 1960).

The other aspect mentioned in the definition—control of behavior— also has its peculiarity in the case of open systems. Control of the functioning of the system can be established through alterations in the exchange process between the system and its environment. The system, however, does not react to such alterations in a simple, linear fashion. Rather, when the system–environment exchange process is modified by some experimental manipulation, the system can actively modify its functioning, turning it into a state that cannot be predicted from the experimental manipulations of the independent variable(s). Again, different systems behave differently when the system–environment exchange process is altered. In psychology, examples of such variable coping by human beings under conditions where their relationships with their environments are altered, are abundant. In experimental psychology, issues of subjects' noncooperation, or alternatively, overwhelming cooperation, illustrate the fact that subjects are active participants in psychological research, rather than passive and subordinate creatures who are run by experimenters.

Finally, the theme of this volume—the role of the individual subject in scientific psychology—follows directly from the assumption that psychology deals with open-systems phenomena. It is the case of open systems where every individual is unique where classes of individuals are heterogeneous as a consequence. The open-systems nature of organisms is a precondition for interindividual variability within a class. As such, this variability need not justify the claims made in humanistic and similar fields of psychology, which deny in principle the possibility of the existence of general laws in psychology, given individuals' uniqueness. Instead, the open-systems nature of psychological phenomena can lead us to the development of theoretical models that tolerate individuals' uniqueness as a concrete example of self-constructive activity of individuals within their environments. Thus, relative to the particular environment and its (unique) history, every individual is an organized, systemic whole—rather than a deviation from a prototypic or average ("ideal") person. As such, a study of every unique individual can provide us with information about general laws that have made it possible for the individual's concrete uniqueness to emerge. In this sense, universal general laws could be sought in diversity. Traditionally, psychologists have had great difficulty understanding this unity of diversity and universality. The reasons for that difficulty can be traced, again, to the transfer of the paradigm of classical physics (that deals with closed systems and homogeneous classes of phenomena) to the young science of psychology, whose phenomena are qualitatively different from those studied by macrophysicists (Lewin, 1931).

The Time Dimension and Open Systems

In classical physics and psychology, time has often been eliminated as the fourth dimension of the existence of phenomena, or reduced to a status similar to that of the three dimensions of space. However, the dimension of time is qualitatively different from the dimensions of space—whereas space is symmetrical and reversible, time is asymmetrical and irreversible, as it flows from past to future. Reduction of the time dimension into a state similar to that of space has made it necessary for psychologists to infer from psychological outcomes to psychological processes that possibly may have produced these outcomes. Similar elimination of the time dimension from consideration, or amnesia about its irreversibility, has plagued other human sciences as well (e.g., archaeology—cf. Bailey, 1983). In the case of open systems, the irreversible nature of time makes the emergence of qualitatively new, structured systems (organisms) possible—through catastrophic amplification of intrasystemic fluctuations (variability *in* time) that no longer can be attenuated to reach the equilibrium of a steady state (Prigogine, 1973, 1978; Prigogine & Nicolis, 1971). In contemporary psychology, Gottman (1982) has called for reintroduction of the time dimension into psychologists' world views through the introduction of the concept of *temporal form*—a structure of relationships (between people, or—we may extend it—between a person and his environment) that is constructed, step by step, over time. In the individual-cognitive domain, the sequential organization of phenomena is taken into account when a person attempts to estimate some of the static characteristics of the phenomena (e.g., their variability—cf. Lathrop, 1967). In their acting and their thinking, human beings create relationships between phenomena that are based on the organization of sequences of these phenomena in time. Psychology has not taken sequential organization of its phenomena into account. Instead, data are necessarily collected over a certain time period (e.g., observation period, or the course of an experimental session), and the results are subsequently treated as if they had occurred simultaneously—each data unit viewed independently of every other, and of their sequential organization.

The transformation of sequentially organized information into its static, time-free representation may well be a widespread adaptation process in organisms, one that is built into their physiological organization. The shape of an object pressed onto the skin of a subject is difficult to recognize from such simultaneous pressure. However, if that shape is traced on the skin with a moving point (or explored actively by the subject), the object can easily be distinguished. Such adaptation of our sensory systems that helps us to take in static information about the world (shape) through the reception of a sequence of input information units, which is later translated back to a static representation (of the shape) illustrates the difference between time-related process and time-free outcome. Concentrating our research efforts on studying outcomes with the hope that some miraculous new method could lead us to valid inferences about their underlying processes can serve as a blinder in our science. Lashley (1951, p. 134) reminded psychologists and physiologists of that danger.

> The problems of the syntax of action are far removed from anything which we can study by direct physiological methods today, yet in attempting to formulate a physiology of the cerebral cortex we cannot ignore them. Serial order is typical of the problems raised by cerebral activity; few, if any, of the problems are simpler or promise easier solution. We can, perhaps, postpone the fatal day when we must face them, by saying that they are too complex for present analysis, but there is danger here of constructing a false picture of those processes that we believe to be simpler.

The "fatal day" for psychologists can indeed be postponed. However, as soon as a psychologist rejects the idea of closed-systems nature of psychological phenomena and accepts an open-systems viewpoint, his treatment of the time dimension in psychological research would change. It is not the case that sequential order of phenomena is "too complex for present analysis," but the axioms onto which psychology has been built have been inadequate in their oversight of the open-systems nature of the phenomena.

Theoretical Background

Determinacy and Indeterminacy. As in any other science, psycho logical phenomena can be viewed from a deterministic or a nondeterministic perspective. In the first case, events in a sequence (e.g., a hypothetical behavioral sequence A-B-C-D-A-B) are assumed to follow one another with rigid certainty. Historically, such a deterministic assumption was prominent in early ethology, when C. H. Whitman and O. Heinroth described patterns of rigid courtship behavior among pigeons and waterfowl. Such rigid sequences of behavior have been labeled in ethology fixed motor patterns or fixed action patterns (FAPs) (Lorenz, 1981). At first, ethologists were interested in discovering behavior patterns that would characterize a particular species, that is, would be performed similarly by every member of the species, with minimal interindividual variation. Explanation for the existence of rigid fixed motor patterns took the form of attributing causality to innate physiological programs that can be triggered by some specific sign stimuli (e.g., Tinbergen, 1972), and executed without modification by environmental conditions. More thorough analysis of behavioral sequences by ethologists, however, has revealed that these sequences display remarkable plasticity to environmental and organismic conditions—the same organism on different occasions may produce a different behavioral sequence, dependent on the intensity of stimulation and the state of readiness of the organism. This has guided ethologists to question the deterministic status given in their discipline to the fixed action pattern (Dawkins & Dawkins, 1973; Schleidt, 1974), and to suggest alternative conceptualizations (e.g., modal action pattern—see Barlow, 1968, 1977). As a result of increasing attention to the variability within FAPs, statistical methods have been suggested to eliminate the noise from the "true" modal description of action patterns, typical for all specimens within the given species. Theoretically, the operation of getting rid of interspecimens variability in their behavioral sequences is based on assuming qualitative similarity of all specimens in the given species, and quantitative dissimilarity in their behavioral sequences, which is due to different strengths of environmental stimulation and organismic state (Lorenz, 1981, pp. 112–113).

Both inter- and intraindividual (temporal) variability in animals' behavior sequences can also be conceptualized from the perspective of indeterminacy of the sequential organization of behavior. This perspective has led investigators to study probabilistic relationships between different behaviors in a sequence of FAP, in an effort to find some traces of statistical regularity in highly variable behavior sequences that fulfill the same function.

Concomitants of the Probabilistic Approach. Ethologists and psychologists who have studied behavioral sequences as stochastic phenomena have usually followed a frequency-based interpretation of probability. This interpretation is most easily available for empirically oriented scientists because their observations can produce frequency counts of behaviors. Likewise, frequencies of transitions from one to another behavioral event across an observation period can easily be counted. The tradition of frequency interpretation of probability provides a basis for treating empirical frequencies of behaviors and their transitions as probabilities, and from that point onward the scientific discourse turns from descriptive statistical analysis into a predictive probabilistic endeavor.

The adoption of the frequency interpretation of probability as a theoretical framework for analysis of behavior sequences requires acceptance of a number of assumptions. First, the system of behavior categories has to be defined, so that the categories are mutually exclusive, and that all behaviors in a category are treated as equivalent. This assumption can be unwarranted if behaviors in a category are expected (or shown) to differ greatly in form (Slater, 1973). Second, it must be assumed that the behavioral sequence is stationary if accumulated frequencies of behavior categories are to be treated as bases for probability estimation. This assumption can be valid only for special cases of open systems—when these systems are in steady state and any occasional fluctuation from that state is compensated for homeostatically. However, if a sequence of behaviors of an organism functions to bring the behaving organism into a new state, the assumption that the sequence is stationary is unwarranted.

Dependence on the Accumulative Past. Frequency interpretation of probability relies on the analysis of aggregated frequencies of behavior, which have been taken out of their temporal context. So, conditional probability $p_{B/A} = 0.45$ can result from a frequency count of different transitions from A to any *observed* subsequent code, over the whole sampling period. All *particular* transitions $A \rightarrow B$ in the sequence analyzed become lost in the process of counting the frequency of occurrence of cases in different classes of transitions from A to some other behavior. The resulting frequency count is treated as a basis for probability calculation with the expectation that in the future, the past is likely to repeat itself. This prediction of future from the past follows from the assumption of the stationariness of the probabilities. Given that, it is not possible to determine the probability of the emergence of new qualitative behaviors, because what was not observed during the observation in the past does not participate in the estimation of probability of novel events in the future.

Elimination of the Context of Sequence. Probabilistic approach to sequence structure of action is context free—all frequencies (probabilities) are treated as if they represent phenomena that do not depend on their immediate environmental contexts. When an investigator characterizes an action sequence through analysis of transitional probabilities, then the organism's assessment of consequences and its intelligent production of alternatives to fit expected consequences are not taken into account (Reynolds, 1982a, p. 345). Even relatively simple motor activities— like children's digging with a spade (see Reynolds, 1982b)—cannot be analyzed as examples of context-free behavior. Digging includes varia tions in the particular action sequences that are related to the particular contextual conditions of that activity. For example, the process of digging depends on the nature of the tool (spade), and to the object of action (substance in which the child is digging). The motor programs that organize children's digging activity are not comparable to ethological probability trees because a tool user does not flip a coin at each behavioral choice point. Rather, he selects the behavior that is relevant to the particular situation. Probabilistic accounts of action sequences reduce complex, context-related action into decontextualized behavior strings, and aggregate only quantitative occurrence–transition information from these sequences. Although empirical records of behavior sequences can be described through conditional probabilities, such descriptions cannot provide an explanation for the observed transitions from one behavior to another. Neither can such records predict and explain the occurrence of novel behaviors, because the latter are likely to emerge under new task conditions set by environmentally organized needs (Reynolds, 1982a).

Elimination of Hierarchical Organization. Reduction of behavior sequences into matrices of transitional probabilities eliminates the structure of hierarchical organization of action, which is an essential characteristic of any organismic phenomenon. Usually, empirical efforts to analyze sequential data have taken place at a molecular level of analysis. However, human action proceeds simultaneously at different levels of organization (Cranach & Kalbermatten, 1982), and its reduction of the most elementary level inevitably leaves out of consideration more molar levels. Such "molecular reductionism" (Petrinovich, 1976) obscures psychology's understanding of the reality of its phenomena. Instead of analyzing probabilistic processes that participate in generation of some action sequence investigators have resorted to the study of frequencies of elements of the outcome as if those were probabilities, forgetting that these elements are interdependent with one another in time and across levels of the hierarchical organization of the phenomenon.

Practical Methods Used to Analyze Sequences

Construction of Sequential Data. Sequences of naturally occurring psychological events are usually highly complex. Multiple events, which cannot easily be delineated from one another, can take place in parallel. Besides, a transition from one event in a sequence to the next need not be strictly organized—there can be intervals of time during which it is not clear whether event A has ceased to exist and B (the next event) is already in existence. Rather, during the observation of sequences of psychological events, there can be periods of time where both the previous (outgoing) and the subsequent (emerging) phenomena are present in parallel, and are intrinsically related to each other. Such organization of reality makes it possible for investigators to construct sequential data on the basis of the phenomena in accordance with the conceptual framework of the researcher. Each operation the investigator performs to turn the fuzzy raw material (observed phenomena) into data illustrates the constructive nature of data creation process. A researcher looking at sequences of behavior between interactants in social interaction may decide to treat the phenomena either as a *continuous* or *discrete* series (Bakeman, Cairns, & Appelbaum, 1979). In the first case, the data are then constructed by the researcher with the assumption of gradual change taking place in the quantitative amount of variable(s) over time (time-series or multiple time-series data). In the second case, the sequence is assumed to consist of discrete (qualitatively different and mutually exclusive) events that follow one another. Empirical data are then constructed with that defining condition in mind, either in the form of *event sequence data* (that capture only the succession of qualitatively different events, irrespective of their duration), or *time–event sequence* data (which preserves the duration of events by application of a time unit to the ongoing event, and recording the number of time units that the event lasted). Finally, time–event sequences that take place in parallel can be captured by extending the time and even coding to *multiple time–event* sequence data.

All these strategies of constructing data from observations necessarily eliminate some aspects of the original phenomena in order to concentrate on others. For example, once an investigator considers his observations to represent continuous fluctuation in some quantitative parameter, then possible qualitative change in the observed phenomena over the period of study is excluded. When the phenomena are considered to be discrete, then the possibility of gradual change from one event to the next in the sequence is lost from the investigative process.

   Treatment of Individual Subjects. Very often, psychologists who analyze event sequences in their data are insensitive to the problem of differences between individual subjects' original event sequences and the average sequence for a group of individual subjects. The practice of lumping sequence data from individual subjects into group data and subsequently interpreting the results of group analyses as pertaining to individuals is widespread. Different investigators have voiced concerns about such data aggregation practices, because it is evident that pooling sequential data from individual subjects into group data is an effective way of distorting the phenomena under study (Barlow, 1977; Bobbitt, Gourevich, Miller, & Jensen, 1969; Castellan, 1979).

   Treatment of Nonstationariness in Sequences. The question of actual nonstationariness in time-series data and conditional probability estimates has been of concern for investigators who have invented sequential analysis techniques. The solutions offered have usually concentrated on how to get rid of nonstationariness in a way that still renders sequential analysis methods adequate. For example, Gottman and Ringland (1981, pp. 398–399) suggested that the data be divided into successive chunks, which are reasonably stationary in themselves, even if the whole sequence is nonstationary. The general tendency in this respect seems to be to transform the data from a nonstationary state into a stationary one by some technical manipulation, and once this is done, continue with the procedure of sequential analysis. The question of whether the transformation of the data, although technically possible, is theoretically feasible, is seldom asked. That question may require reformulation of the theoretical background of our efforts to analyze behavior sequences. For example, the empirically demonstrated nonstationariness of conditional probabilities could provide the basis for abandoning the assumption of stationariness in our theoretical framework. As a consequence methods that are based on the assumption of stationariness will become unfeasible to use. An investigator may prefer to proceed along the lines suggested by MacKay (1972), assuming from the outset that the whole probability distribution changes as a result of every new event that unfolds in time. Such an approach may require a different view of probabilities than that based on frequencies of occurrence of elements in the sequence, or of transitions between them.

Probabilistic Lag-Sequential Analysis.   The majority of sequential dependency studies in psychology have concentrated on probabilistic relationships between adjacent events in sequential data. Therefore, Markovian assumptions that postulate the existence of relationships no more back in time than one event (or time) period are often used. However, sequences of psychological events can be expected to have temporal organization that covers a longer time span than that between periods t and t-1. Furthermore, sequential data may include cyclical regularity, which would remain undetected in a Markovian analysis. These considerations have led to development of lag-sequential analysis techniques (Gottman & Bakeman, 1979; Sackett, 1979, 1980). These techniques involve calculation of conditional probabilities (frequencies of transition) across different time–event lags, for example, between t and t-2; t and t-3; t and t-4, and so forth. (t = time or event unit in the sequence). The requirement of stationary probabilities across the sequence observed remains an important issue in lag-sequential analysis. The other dificulty of this analysis is the artificial extraction of a certain longer-lag relation between units from the actual sequential context. Thus, a calculation of t-4 conditional probability in the following sequence, A-B-C-D-E-A-C-D- ..., involves counting the frequency of occurrence of E/given A; A/given B; C/given C; D/given D—all of the given units occurring across a "gap" of 3 units. Such calculation excludes the events that are in between the specified units at t, and at t-4. So, for example, considering E/given A at t-4, the subsequence -B-C-D- between A and E in the original sequence is left out of the analysis. This is a case of the artificial decontextualization involved in lag-sequential analysis—the original subsequences in data are not preserved; instead, elements of the sequence are taken out of their temporal context and examined from the point of frequencies of co-occurrence over different event–time lags. It is interesting to note that this artificial decontextualization has been avoided in some other efforts of analyzing sequences (Bakeman & Brown, 1980; Bobbitt *et al.*, 1969; Chatfield & Lemon, 1970; Souček & Venzl, 1975).

SUMMARY

In traditional research in psychology, the time dimension has been treated as similar to the dimensions of space. It has necessarily been important as a dimension in which all data collection efforts are arranged. However, once the material has been collected, psychologists have paid little or no attention to sequential organization in this material. Instead, they have reduced sequentially organized material into time-free accumulated data, and proceeded further in an effort to explain the data.

This research practice in psychology—decontextualizing phenomena from their temporal contexts—has rendered the discipline helpless if it tries to deal with dynamic aspects of its phenomena. Methods of sequential analysis that would preserve the temporal (sequential) contexts of events would make it possible to retain information about dynamic aspects of the phenomena when these are translated into data. The remaining part of this chapter outlines some concrete research efforts by the author to develop methods of sequential analysis of time structure ("temporal form" in Gottman's, 1982, terminology) that can preserve the variability of psychological phenomena in individual subjects.

DESCRIPTION OF SEQUENCE–STRUCTURE ANALYSIS

BASIS OF THE METHOD

Sequence–structure analysis is based on efforts by a number of psychologists (Bakeman & Brown, 1980; Bobbitt et al., 1969; Keats et al., 1983; Patterson & Moore, 1979; Simon, 1979), anthropologists (Métraux, 1943) and biologists (Schleidt & Crawley, 1980; Souček & Venzl, 1975) who have attempted to look at regularities in structured sequential phenomena. Sequence–structure analysis breaks a sequentially organized phenomenon into subsequences and describes the whole action sequence of an individual subject as a distribution of such patterns of different lengths. The analysis keeps original intrinsic relationships between events in a sequence intact within the subsequences that make up a subject's set of action patterns. Furthermore, frequency interpretation of probability need not be used in the analysis. Or—if an investigator insists on accepting that interpretation—different manipulations of the empirically established frequencies of subsequences in the repertoire can be performed on the basis of the results of sequence–structure analysis. If additional assumptions are used (e.g., see Baldwin, 1940), it becomes possible to make sense of the frequencies of recurrence of different subsequences in the repertoire. It is also possible to use sequence–structure analysis results for further analysis of individual subjects' action repertoires with the help of information theory concepts and calculations (Chatfield & Lemon, 1970; Gottman & Bakeman, 1979). This latter directon of thinking, however, may lead an investigator back into the labyrinth of closed-systems assumptions about the statistical nature of the phenomena under study (see Frick, 1959).

THE METHOD: GENERAL CASE

The Importance of the Action Context

In principle, sequence–structure analysis can be applied to any string of sequences of symbols, whatever the latter denote. However, meaningful use of the method depends on the consideration of the environmental context in which the observed action sequences were generated by individual subjects. The method can be useful for interpretation of the temporal structure of actions by individual subjects when these actions are explicitly goal-directed.

General Principle of Analysis

Consider the following string of symbols where each letter (A . . . E) denotes a particular event in an individual string:

$$-A-B-C-D-A-C-A-D-A-D-E-A-C-D-E-C-E-A-B-C-D-A-C-D- \qquad (1)$$

This string is the *sequence* that is subjected to sequence–structure analysis. It is unique, because it cannot recur as this particular sequence (because it, as this sequence, has already occurred). Given that uniqueness of the string, our analysis must proceed within the sequence, so that its constituent subsequences can be recognized. Metaphorically speaking, this activity is similar to somebody's efforts to go through a computer program with the aim of understanding how the program works. There is, however, one important difference between a computer program and organisms' action sequences—the former remains the same over time (e.g., can be rerun, with similar results, many times), whereas the latter is amenable to change every time the subject has to solve a problem—even if it looks similar to previously solved ones. Besides, a computer program is an end result of the programmer's work, which need not include the programmer's intermediate (later abandoned) units that were generated in his search for a good solution for the program. In contrast, individuals' action sequences are phenomena where these corrected dead ends remain an important part of the analysis of the ongoing process.

Sequence–structure analysis of the original sequence of events proceeds as follows. If starts from the minimal length of the possible subsequences (length 0, which equals the frequency count of constituents, as 0 designates subsequences of 1 unit). Going through the original sequence unit by unit, previously unobserved units are recognized and their frequency count started. After the analysis of length-0 subsequences is finished the procedure is repeated at greater subsequence lengths (1, 2, etc.). So, at length 1, the subsequence -A-B- is extracted from the sequence in the beginning, followed by -B-C-, -C-D-, and so forth. Similarly, length-2 subsequences that are extracted from the beginning include (a) -A-B-C-,(b) -B-C-D-, and (c) -C-D-E-. The recognizing system always (at all subsequence lengths) advances through the original sequence by steps of 1 unit. This creates a situation where adjacent subsequences in the list have partial overlap. For example, the first and second length-2 recognized subsequences had in common the link -B-C-. Likewise, greater-length subsequences include into them some of the shorter-length subsequences. For instance, the first length-2 subsequence contains two previously recognized units at length-1 (-A-B-C- = -A-B- & -B-C-).

This partial overlap of subsequences within and between length levels is the reason why the application of frequency-based probability concepts to the results of sequence–structure analysis is unwarranted. The extracted subsequences are not independent from one another, which makes it theoretically invalid to convert frequency distributions of subsequences of any length into probabilities. Likewise, statistical comparison of frequency distributions at different subsequence lengths (e.g., through use of information measures) is a conceptionally questionable procedure for the same reason.

Table 1 contains the results of sequence–structure analysis of the original sequence(s).

Because of partial overlap of adjacent recognized subsequences, the minimal frequency at which a subsequence of a given length is listed as belonging to the set of subsequences is 2. In other words, a subsequence of length-X has to *recur* at least once in the sequence to be included. Among the subsequences that were found to recur (and therefore were included in the repertoire), those adjacent to one another can still have partial overlap. For example, consider length-1 subsequences -D-E- and -E-A- from Table 1. Both of these were found to recur ($f = 2$), and if we refer back to the original sequence we can observe that these two subsequences were originally extracted from adjacent positions (. . . -A-*D-E*-A- . . .), but then recurred later in the sequence in nonadjacent positions (-D-E- and -E-A-, separated by -C-). For that reason, no subsequence of length 2 of the form D-E-A- was recognized by the analysis program (as its $f = 1$, i.e., it did not recur). On the other hand, the second occurrence of the length-1 subsequence -E-A- is partially overlapping with the beginning of the longest recurrent subsequence that the analysis produced

Table 1. Results of Sequence–Structure Analysis (Example 1 in the Text)

	Subsequence lengths				
0	1	2	3	4	5
-A- ($f = 7$)	-A-B- ($f = 2$)	-A-B-C- ($f = 2$)	-A-B-C-D- ($f = 2$)	-A-B-C-D-A- ($f = 2$)	-A-B-C-D-A-C- ($f = 2$)
-B- ($f = 2$)	-B-C- ($f = 2$)	-B-C-D- ($f = 2$)	-B-C-D-A- ($f = 2$)	-B-C-D-A-C- ($f = 2$)	
-C- ($f = 6$)	-C-D- ($f = 4$)	-C-D-A- ($f = 2$)	-C-D-A-C- ($f = 2$)		
-D- ($f = 6$)	-D-A- ($f = 3$)	-D-A-C- ($f = 2$)			
-E- ($f = 3$)	-A-C- ($f = 3$)	-A-C-D- ($f = 2$)			
	-A-D- ($f = 2$)				
	-D-E- ($f = 2$)				
	-E-A- ($f = 2$)				

(length 5: -A-B-C-D-A-C-), through the second unit in the -E-A- subsequence.

Auxiliary Analytic Terminology

Extraction of subsequences of different lengths is not the final step in sequence–structure analysis. Given partial overlap between adjacent subsequences, and potential partial membership of shorter subsequences in longer ones, further steps are necessary to reconstruct the wholistic structure of the sequential process.

Cyclicity Analysis. It is easy to observe in the case of subsequences length ≥ 2 that cyclicity can occur. *Cyclicity* is defined here operationally as recurrence of a certain unit–event within a recurrent subsequence. As an example of *loose cyclicity*, consider the longest subsequence in Table 1 (length 5: -A-B-C-D-A-C-). This subsequence includes two repetitive units (-A- and -C-), observed over 3 and 2 other units in the subsequence, respectively.

A more stringent version of cyclicity is present in the case where units repeat themselves at fixed regular intervals in a subsequence. For example, a hypothetical subsequence of length 9: -F-G-F-G-F-G-F-G-F-G- is *strictly cyclic*, because both events (F and G) are repeated intermittently within the extracted subsequence over an interval of 1 unit.

Decline Function. Gradual decline in the number of different recurrent subsequences at increasing lengths is called the *decline function* (Hill & Valsiner, 1984). It provides general information about the organization of a given sequence. The cutoff point of the decline function, that is the length after which no recurrent subsequences are detected in the sequence, illustrates depth of sequential organization of a serially ordered phenomenon. Table 2 provides the decline function for sequence(s) analyzed in Table 1.

Apart from information about the depth of sequential organization of a series, the decline function can also characterize other aspects of the sequential organization. For example, a relatively low decline function over all subsequence lengths can reveal the existence of short-but-frequent subsequences. Consider for an example another hypothetical sequence:

$$-A-B-C-D-A-B-C-D-A-B-C-E-A-B-C-F-A- \qquad (2)$$
$$B-C-G-A-B-C-H-A-B-C-I-A-B-C-K-A-B-C-$$

It is easy to see that this sequence of 35 events consists of a repetitive subroutine (-A-B-C-), separated by uniquely occurring events (D, E, F, G, H, I, K). The sequence–structure analysis starts from extracting 3 sub-

Table 2. Examples of Decline Functions of Two Sets of Subsequences of Hypothetical Sequences (Examples 1 and 2 in the Text)

A. Example 1

Sequence length	0	1	2	3	4	5	6
Number of different $f \geq 2$ subsequences	5	8	5	3	2	1	0
Total number of different subsequences	5	1	17	18	18	18	18
Percentage different/total	100	72.7	29.4	16.7	11.1	5.5	0

B. Example 2

Sequence length	0	1	2	3	4
Number of different $f \geq 2$ subsequences	3	2	1	0	0
Total number of different subsequences	10	16	22	28	27
Percentage different/total	30	12.5	4.5	0	0

sequences of length 0 that are recurrent (A, B, C, each recurring 8 times), then proceeds to length 1 and recognizes 2 recurrent subsequences (-A-B- and -B-C-) each of which occurs with the same frequency ($f = 8$). Furthermore, at length 2 the analysis recognizes the single subsequence -A-B-C- ($f = 8$). The decline function for sequence (2) is presented in Table 2B.

In comparison with the decline function of sequence (1) (Table 2A), the decline function in Table 2B is relatively lower in percentage value at every subsequence length. This is directly dependent on higher frequencies of occurrence of the recurrent units in the sequence, and their combinations in subsequences. It also depends on the presence of non-recurrent units of the sequence in positions in between the recurrent -A-B-C- subsequences.

Purified Sequence Structure. As was seen in the examples above, certain subsequences of greater length include other subsequences of shorter lengths as their components. Returning to example (2), it is clear that at length 2, the -A-B-C- subsequence includes all recognized length-1 subsequences (-A-B- and -B-C-) as its components. Similarity in the frequencies of length-2 and length-1 subsequences certifies that all -A-B- and -B-C- subsequences are components of the -A-B-C- subsequence (all three having equal frequency, $f = 8$). Thus, sequence–structure analysis has to proceed to eliminate these subsequences, recognized at lower lengths, that are actually components of longer strings.

In order to accomplish this goal, a procedure of elimination of detected shorter patterns that are components of longer recurrent subsequences can be adopted. Table 3 describes the procedure for such purification.

The process starts from the longest detected subsequence(s) in the list (length x). It then moves to the sets of subsequences of shorter lengths (x–k), where k = 1 ... n, so that x − k > 0. Each shorter subsequence is compared to the longer subsequences that have been found to be unique and retained in the list. The example in Table 3 illustrates the procedure. The procedure starts from the only length-5 subsequence (a) which is retained in the list as unique. Comparison of shorter subsequences (b–j) with the first target (a) reveals that these subsequences are components of (a), and are therefore eliminated from the list. The comparison of (k) with (a) results in retaining (k) because (k) is not a component of (a). As the analysis proceeds, subsequences (1 ... p) are found to be components of either (a) or (k), and therefore eliminated, whereas (q), (r), and (s) are

Table 3. Reconstruction of Nonoverlapping Subsequence Lists from Overlapping Subsequence Lists of Different Lengths (Example 1 in the Text)

Procedure	Illustration	
1. Begin with the longest extracted subsequence (length x) and include it in the list.	(a) -A-B-C-D-A-C-	Retain (a)
2. Consider subsequences at lengths x–k (where k = 1, 2, ... n, so that x − k > 0). Eliminate every subsequence that is a component of a longer, retained subsequence.	(b) -A-B-C-D-A-	
	(c) -B-C-D-A-C-	
	(d) -A-B-C-D-	
	(e) -B-C-D-A-	Eliminate all subsequences
	(f) -C-D-A-C-	(b) ... (j) because they
	(g) -A-B-C-	are components of (a)
	(h) -B-C-D-	
	(i) -C-D-A-	
	(j) -D-A-C-	
	(k) -A-C-D-	Retain (k)
	(l) -A-B-	eliminate (l) ... (p) because
	(m) -B-C-	these are components of
	(n) -C-D-	(a) and (k)
	(o) -D-A-	
	(p) -A-C-	
	(q) -A-D-	Retain (q)
	(r) -D-E-	Retain (r)
	(s) -E-A-	Retain (s)

found not to be components of any other (longer) subsequence and therefore retained in the list of unique subsequences. Instead of the 19 partially overlapping subsequences in example (1) the elimination of components has left only 5 subsequences in the list. The list of recurrent subsequences has thus been purified to include only those items that are independent of one another in the sequence.

Qualitative Analysis. The list of purified subsequences detected in particular original sequence (e.g., action sequence on some task, observed in the case of a single subject) can be further analyzed qualitatively. The particular ways in which such qualitative analysis is worthwhile depend on the content of the symbols in the sequence. In our example (1) it is possible to analyze both invariant and unique content of the purified subsequences. For example (see Table 3), event A is represented in 4 out of 5 of the nonoverlapping subsequences, and may be a candidate for the invariant event status for the list. Feasibility of the search for invariant and unique events in the subsequences depends on the particular research questions asked, on the basis of the content of serially ordered phenomena that are studied by sequence–structure analysis.

Limits on Applicability of the Method

No method in any science is of universal applicability. This general specificity of methods makes it necesssary to specify general conditions and aspects of phenomena to which the method might be applied.

Assumption of Nonrandomness. The present method is applicable to phenomena which can be *a priori* considered to be organized as complex serial phenomena, and in which case the symbols that make up the original sequence denote units which are structurally complex. Therefore, the occurrence of such units in a sequence cannot be attributed to chance or randomness. In the case of sequences consisting of elementary (atomistic or nonstructured) units, where chance or randomness may constitute valid agents for causal attribution, traditional probability (and lag-sequential) analysis methods may be sufficient, and on occasions even more adequate, than the present method.

Assumption of Recurrence. The present method can be applicable if the serially ordered phenomenon under study can be expected, *a priori*, to include recurrences of some lengths. Phenomena where such recurrences are not expected, or where they are in principle impossible, render it unfeasible to apply the present method. For example, consider an original sequence:

$$-A-B-C-D-E-F-G-H-I-J- \qquad (3)$$

This sequence includes *no* recurrences of any subsequences that can be extracted from it, of any possible length. However, this need not mean that the string is not sequentially organized. Rather, this sequence may be an example of a strictly specified algorithm for solving some problem, where the particular event sequence is strictly determined in its serial order to arrive at the solution. For some tasks that organisms solve by an action sequence that is narrowly determined (with no possibilities of different ways of reaching the same goal, or repetition of previous efforts), the present method is blind because it misses all unique nonrecurrent serial organization in the sequence.

EMPIRICAL EXAMPLES

STUDY 1— ADULTS' STRATEGIES FOR CALMING A CRYING INFANT

The Research Problem

Background. In adults' lives, encounters with infants necessarily involve situations where the latter may be crying. Newborn and infant crying has powerful effects on adults. In the context of adult–infant interaction, an infant's crying usually leads to efforts to calm the baby by the caregiver(s). This process—calming down a crying infant—takes place in individual caregiver–infant dyads. Therefore, the phenomenon should be studied in individual dyads in order to examine how the process of infant calming is organized. Theoretically it is possible to expect to find variety in the ways in which babies are calmed, as the infant's relationship with its environment is a good example of an open-systems phenomenon—the baby's state depends on exchange with its environment and certain conditions of that exchange (e.g., state of hunger) can lead to signals informing about that state (cry), which result in adaptations in the exchange (feeding). Empirical support for the theoretical notion of individual uniqueness of baby-calming activities has also been obtained (Valsiner, 1982).

Within an individual adult–infant dyad, baby calming can vary from one instance to the next, given differences in the particular conditions of the setting, and of the interactants. Furthermore, once the infant cries, there is no ready-made solution to the problem of calming it down. Rather, on each particular occasion, the caregiver has to construct a means that will eventually help the infant to calm down. That strategy is constructed by trying out different ways of acting, one after another, and shifting back and forth between them. On some occasions, the baby can be calmed easily and by a simple (and even unitary) strategy—for example, taking the baby onto the mother's lap may suffice. On other occasions, the same baby may keep on crying, even as the mother puts it into different positions (on lap, on shoulder, in arms) in a sequence, and tries simultaneously in other ways (by rocking, singing, redirecting visual attention, etc.) to calm the baby. Thus, the task of *the* mother to calm down *her* baby varies in difficulty from one situation to another, and in every single situation the mother has to work towards her goal (a calm state of the child) by creatively constructing baby-calming action sequences that may lead to the goal. There cannot be, in principle, a single set of efficient strategies of baby calming that would be of guaranteed success for all babies all of the time. Instead, there can be some strategies (action sequences) that a mother develops in her experience with her baby, that seem to work some of the time, still leaving open the uncertainty that these strategies might not work the next time her baby cries.

Applicability of Sequence–Structure Analysis.   The inherent uncertainty of the baby-calming task in the life of every single infant–adult dyad makes it feasible to apply sequence–structure analysis to observational data of baby calming. It is *a priori* possible to expect that an adult, in his or her efforts to calm a baby, will try out different recurrent subroutines of action that may lead to the goal. These actions are inherently complex—for example, while holding the baby on the knees, the adult may rock the baby, sing to the baby, and show the baby some object to get its attention. Such behavioral complexity need not be attributed to chance or random causes, but is a necessary part of purposeful activity of a goal-directed caregiver.

The Experiment

Rationale and Procedure.   The present experiment used an artificial task to trigger baby-calming efforts in adult subjects. In an effort to eliminate unpredictability of infant's cry from the study, subjects were given a cloth doll and asked to imagine that the doll is a 3-month-old infant. Tape-recorded infant cry (recorded during circumcision) was presented in the laboratory room through loudspeakers. Each subject (tested individually) was asked to do anything with the "baby" (doll) that the subject thought would help to calm down a 3-month-old infant when the cry could be heard through the loudspeaker. The task was presented as a test of social intelligence, where there are correct solutions available, and the subject was expected to find the correct solution as quickly as possible. The laboratory room (3 × 3 meters in size) was furnished with a small table and two chairs. The loudspeaker was placed under the table, and the doll on the table (as its starting position). The doll was 30 cm in height, made of soft cloth. A number of toys (a dog, building blocks) were also on the table and available to the subject.

The subject was explicitly told that the correct solution to the problem would result in the cessation of the cry heard over the loudspeaker. When that happened, the subject was to put the doll back into the starting position on the table, and wait until the cry was heard again to start the next trial.

The rationale of this procedure was to create an experimental condition in which the subject is led to believe that the problem can be solved, and to search for a solution. In reality, during the actual trials (lasting for about 3 minutes) the problem could not be solved. The first of the actual trials was followed by one or more filler trials, lasting 20 to 60 seconds, during which the cry sound was switched on, the subject acted with the doll, and the cry was switched off again. Care was taken not to switch the cry off in the filler trials if the subject behaved in the way that had ostensibly stopped the cry during the previous trial. If the subject started from that previously successful behavior during a filler trial, then the trial was extended until the subject switched to some other behavior, after which the cry was switched off. The sequence of experiment included 3 actual trials and a varied number of filler trials (1 to 5), which were organized in the following sequence for all subjects: ACTUAL 1—Filler trials (1 or more)—ACTUAL 2—ACTUAL 3. The function of the filler trials was to confuse the subject as to what behavior seemed to be effective, and also to strengthen the subject's belief that the problem could be solved (after the long first actual trial, it was expected that the subject could begin to doubt that a solution is possible, or persist on trying out the behavior that worked at the first trial).

The whole experimental session (12–18 minutes in duration) was videotaped, but only action sequences from the actual trials (the last 2 1/2 to 4 1/2 minutes) were analyzed. After the session, each subject was interviewed about his or her understanding of "which methods seemed to work" in the study. This was a control condition to verify the success of the instruction and the effectiveness of the filler trials in confusing the subject. None of the subjects stated that they had found the solution to the problem during the session. They were also asked in the interview about their beliefs in the presence of some successful ways of solving the problem. All subjects reported that they had continued to believe in the solubility of the task, although behaviors that worked were perceived as changing through the session. In addition to that, some subjects reported that the realistic infant cry had facilitated their imagination that the doll was a baby, and thus helped them to play the role of the baby calmer.

Subjects.   Undergraduate students with explicit interest in children participated in the study. The sample consisted of 48 subjects (39 female, 9 male), whose average age was 23.6 years (range 20–43 years). Six out of the 39 female subjects had children of their own (within an age range of 14 months to 11 years). The subjects reported varied experience with children and infants in their past life history, for example, babysitting, and working in a nursery. A very small number of subjects (2 out of 9 males, 3 out of 39 females) reported no former experience with children and infants. All subjects volunteered to participate in the study.

Coding of Videotaped Material.   Analysis of the videotapes was performed in terms of complex behavioral units. The basic coding unit of this study was the doll position. The subject could keep the doll only in one position at a time. Therefore, coding the doll positions from videotapes was an unambiguous task. Different behaviors could occur within the framework of each doll position in complexes (e.g., patting, looking at, or talking to the doll could occur in one doll position, such as being held on an arm, but not in combination with another, such as being held against the shoulder). Thus, doll positions constituted molar categories of action, within which different combinations of behaviors could, and did, occur. Changes in doll positions were easily identifiable by coders. Coders registered time spent in each of the doll positions, behaviors that co-occurred within the positions, and the position of the subject (sitting, standing, walking). The list of doll position categories used for coding videotapes is presented in Table 4.

Interobserver agreement on the sequences of doll positions (without taking the exact timing into consideration) was measured on the basis of comparing the coding of the 2 coders who had separately and independently coded 13 different trials from the videotapes. The average percentage of agreement was 96.1% (range was 78.5% to 100%, with 10 out of 13 trials coded at 100% agreement). In trials where less than 100% agreement was found between observers, this was due to the omission by one of them of some short-duration, transient doll positions (e.g., subject holding the doll on shoulder, than taking it into hands for 2 seconds, and then placing the doll on the knees for a longer period). These disagreements on short-duration doll positions were easily correctable in reviewing the tapes.

Analyses of Data and Results

Nonsequential Analysis within the Sample.   The data obtained from videotaped materials—information about the position in which the doll

Table 4. Categories of Behavior Complexes (Doll Positions) Used in the Study for Coding Videotaped Adult–Doll Interaction

Code	Label	Description
SH/L	Shoulder	The doll is put on the shoulder (left or right).
SH/R	(L = left, R = right)	Behaviors that can occur: pat, talk to, sing to, rock
A/L	Arm	The doll is put on the arm; may be looked at,
A/R	(L = left, R = right)	talked to, patted, rocked, shown an object
KN/V	Knee	Doll in a sitting position on the knees of the
	(V = visual	subject, in face-to-face orientation. Behaviors:
	orientation)	talk to, show object, sing, bounce, pat
KN/U	Knee/up	The doll is laying on the knees, facing upward
KN/D	Knee/down	The doll is on the knees, facing downward
KN/O	Knee/outward	The doll is in a sitting position on the knees,
KN/L or K		facing outward
H	Hands	The doll is held by both hands, in a vertical, face-to-face position
T	Table	The doll is positioned on the table:
	T/S	—sitting, either face-to-face or facing away position
	T/D	—laying down, facing downward
	T/U	—laying, facing upward

was held—can be analyzed in many different ways. In this section, non-sequential frequency analysis of data aggregated over the whole sample is presented (for greater detail, see Valsiner, 1982).

Table 5 reports the frequencies of different doll positions observed in the whole sample.

At the level of data analysis for the whole sample, the present experiment revealed a general statistical finding that left-side positioning of the doll occurred more frequently than right-side positioning. If only the subgroup of codes referring to left–right symmetry in doll position is concerned (i.e., A/L vs. A/R and SH/L vs. SH/R), then 60% of this group of positions occurred on the left side, and 40% on the right ($p < 0.001$). This finding is similar to data in the literature (Brüser, 1981; DeChateau & Andersson, 1976; DeChateau, Holmberg, & Winberg, 1978; Rheingold & Keene, 1965; Salk, 1960, 1973), where the left side has been found to be the side on which infants are more frequently held.

Limitations of Frequency Analysis of Aggregated Data. The traditional analysis of frequencies of doll positions that has just been provided is of very limited usefulness for understanding how the baby-calming process is organized by individual subjects in the sample. First, by aggregating data over all individuals the analysis proceeded immediately to the populational level of scientific discourse. The results—a statistical

Table 5. Frequency Distribution of Different Doll Positions Used by the Sample (N = 47) in the Baby-Calming Task

Code	Description	Frequency	Percentage	p^a
SH/L	Shoulder/left	406	21.6	0.001
SH/R	Shoulder/right	279	14.8	0.001
A/L	Arm/left	293	15.6	0.001
A/R	Arm/right	187	10.0	0.001
H	Hands	204	10.8	0.001
KN/LR	Knees/facing left or right	168	14.3	0.001
KN/V	Knees/vis-à-vis	65	3.5	0.05
KN/O	Knees/facing outward	43	2.3	n.s.
KN/U	Knees/upward	54	2.8	n.s.
KN/D	Knees/downward	11	0.5	n.s.
T/U	Table/upward	29	1.5	n.s.
T/D	Table/downward	4	0.2	n.s.
T/S	Table/sit	33	1.8	n.s.
	Total	1876		

^{a} Significance of comparison of the percentage with O-level (t test).

dominance of the left-side doll positions over the right-side positions— is a fact that may be interpretable in that it may perhaps tell us something about the sample (= group of individuals), and at best about some population for which the sample represents. However, interpretations based on such aggregated data can be applicable to individuals in the sample only if certain additional *a priori* assumptions are made. Two of these are (a) homogeneity, the assumption that all individual subjects in the sample are considered qualitatively similar to one another in the function under study; and (b) that the frequency within the sample equals individual preference, that is, items that were found to occur more frequently in the sample are preferred to other items (that occurred less frequently) by individual subjects.

In other words, according to these assumptions every subject in the sample is similar to every other, and their similarity takes the form of independently preferring different items in a similar manner. In our example, if these two assumptions were true, every subject in the sample should consider putting the infant onto the left shoulder (SH/L) the most preferable way of calming the infant, followed by left arm (A/L), right shoulder (SH/R), and so forth, in a descending order of frequency (percentage) for the occurrence of these positions in the study.

Sequence–Structure Analysis. In a previous paper (Valsiner, 1982), I reported that all individual subjects in this doll-calming task demonstrated unique but self-consistent sequential strategies in their efforts to solve the problem. All trials of every individual subject in the sample were entered into sequence–structure analysis in order to find out what kinds of different and recurrent subsequences of doll positions occurred in their behavior. First, repertoires of overlapping recurrent subsequences were found for each subject individually. Second, these repertoires were further purified to exclude excessive overlap. Final results of sequence-structure analysis for five selected subjects are presented in Table 6.

These data illustrate the existence of qualitatively unique individual repertoires of action subsequences. Some of the subjects were observed to produce very few (e.g., B, with one length-3 subsequence), others produced a multitude of sequence–structure units (e.g., D, with 15 different, but short, recurrent subsequences). Subject D attempted to solve the problem by repeatedly using a variety (14) of different length-1 subsequences, and was observed to use only one longer (length-2) pattern. The latter, however, is also cyclically organized (switching from left shoulder to on-lap position, and then back to the left shoulder). In contrast, Subject B's single recurrent subsequence illustrates the subject's avoidance of wider-scale experimentation with a variety of doll positions so that the problem could be solved.

The other subjects in Table 6 (A, C, E) demonstrate neither strictly limited nor excessively variable action strategies (the number of different subsequences is 5, 7, and 7 respectively). However, the quality of the sets of subsequences of these subjects differs from one another. Subject A's repertoire jumps after four length-1 subsequences to include a length-5 pattern, which also demonstrates loose cyclicity (recurrent units: SH/L and KN/LR). In contrast, Subject E's repertoire includes a greater variety of length-1 subsequences (five out of seven: four at length 2, one at length 3). Finally, Subject C is somewhat similar in the distribution of subsequences at different lengths to A, but qualitatively they differ in the relevance that positioning the doll on knees plays for A, which is not the case for C (in a comparison of the longest subsequences).

Discussion

Application of sequence–structure analysis to sequences of baby-calming actions observed in the present laboratory task made it possible to go beyond the limits of frequency-based analysis of behavior that is taken out of the context of time (serial order). Every subject was found to attempt to solve the problem differently, as was found from the interindividual comparison of the repertoires of subsequences. The temporal form of their solution efforts was retained in the sequence–structure

Table 6. Repertoires of Nonoverlapping Subsequences of Baby-Calming Actions of Individual Subjects

Subject	Length	Frequency	Content
		Subsequence	
A. 2DL-F	1	2	-SH/L-A/L-
		2	-H-A/R-
		2	-SH/L-A/R-
		2	-A/R-SH/R-
	5	2	-SH/L-KN/LR-SH/L-KN/LR-A/L-SH/L-
B. 9JM-F	3	2	-SH/L-A/L-SH/L-H-
C. 12HB-F	1	2	-KN/V-SH/L-
		2	-SH/L-KN/V-
		2	-SH/L-H-
		3	-A/L-KN/LR-
		2	-SH/L-A/R-
	2	2	-KN/V-SH/R-A/L-
	5	2	-SH/L-A/L-SH/L-A/L-SH/L-A/L-
D. 19JF-F	1	2	-SH/R-A/R-
		2	-A/R-H-
		2	-KN/V-SH/R-
		2	-KN/O-KN/V-
		2	-SH/L-A/L-
		2	-A/L-T/S-
		3	-KN/LR-A/L-
		2	-A/L-KN/LR-
		2	-SH/R-KN/V-
		3	-SH/L-SH/R-
		2	-SH/R-KN/LR-
		2	-SH/L-H-
		2	-SH/R-SH/L-
		2	-A/R-SH/L-
	2	2	-SH/L-KN/LR-SH/L-
E. 31PP-F	1	2	-SH/R-SH/L-
		2	-H-SH/R-
	2	2	-A/L-SH/R-A/L-
		2	-SH/L-H-KN/LR-
		2	-SH/R-KN/LR-SH/R-
		2	-KN/LR-SH/R-KN/LR-
	3	2	-SH/L-KN/LR-H-KN/LR-

analysis. As another important benefit of the analysis, within-individual variability of temporal forms constructed by each subject in his or her effort to solve the problem was also preserved. Because one of the main features of open-systems phenomena is their variability, its preservation in the individual subjects' baby-calming data brings us closer to the possibility of analyzing the functioning of the person as an open system that generates that variability.

A note of caution is in order here to remind the reader that the way sequence–structure analysis has been applied to serially organized data in this example remains purely descriptive. By arriving at any single subject's repertoire of baby-calming subsequences we have not accomplished more than preserving some important characteristics inherent in the data (recurrent serial organization, variability within subject) that require further theoretical explanation. Furthermore, it is not possible to treat the results of this application of sequence–structure analysis of baby calming as if these represent the static (immutable) repertoire that the particular individual subjects would use without modification when the next opportunity arises. In this sense, to talk about the validity of individual subsequence repertoires in ways that this issue has traditionally been tackled in psychology would be counterproductive for efforts to find ways in which the open-systems nature of psychological phenomena could be captured.

STUDY 2—TODDLERS' STRATEGIES IN CONTACTING A STRANGER

The Research Problem

Background. The next example of the application of sequence–structure analysis to sequential data comes from a study on toddlers' social interaction strategies (Hill, 1984; Hill & Valsiner, 1984). Starting from birth, human beings are surrounded by social environments that change over time. For every individual infant and child, it is important to learn to establish new social relationships. Such social relationships are constructed by the developing child and the child's interaction partner(s). In its turn, the possibility for social interaction depends on the rules by which the developing child is exposed to other people (and other people to him). Across cultures, the extent to which infants are exposed to other people shows remarkable variability—from limiting the exposure to some subgroup of the kinship network, to exposing it to everybody around. Frequent arrival of visitors may provide the basis of exposure of the infant to previously unfamiliar people on a regular basis. Among the Baganda, propriety requires that the parents give their infant to any visitor to hold, which makes the infant a potential center of attention and conversation (Kilbride & Kilbride, 1974). On the other hand, the persons who are exposed to the developing children are expected to know the cultural rules of acting with other people's children. The same applies to persons in the developing child's kinship network. In cultures where sibling caregiving is practiced, parents gradually delegate their caregiving functions to older siblings in accordance with their understanding of whether or to what extent the particular older sibling can be trusted with a particular caregiving task. In the majority of cultures, older siblings are required to perform a variety of caregiving functions (Weisner & Gallimore, 1977).

The developing child has to learn to establish and maintain new social relationships and adapt to the dissipation of the relationships. The setting in which this learning takes place are varied. For example, in the case of toddlers in contemporary American conditions, the children may encounter new people inside their familiar settings (homes, baby-sitters' house), in public situations (shopping malls, etc., usually separated from the home by a car ride), or semiprivate, possibly unfamiliar settings (e.g., parents take the toddler with them while visiting the house of a friend). Each of these settings provides a different framework for the child's establishment of new social contacts. For example, the appearance of an unfamiliar visitor in the home of the toddler (where the whole environment is highly familiar) provides the child with a very different reference frame for encountering the stranger and establishing contact with him, than the arrival of a stranger to the home of the parents' friends, where the particular child is visiting for the evening. Different settings define the outer limits of the child's possible actions, but need not determine the particular actions of the child in any strict way (Valsiner, 1984). Thus, a toddler who encounters an unfamiliar adult can pursue different routes in his or her efforts to establish contact with the stranger. Observational material on how the social interaction process between toddler and stranger proceeds toward the establishment of a relationship can be productively analyzed from the perspective of recurrent action sequences that the child displays.

Goals of the Study. The aim of the study was to analyze the process of social interaction of toddlers with unfamiliar persons who visit the home territory of the target children in the presence of at least one of the parents, for a brief time. The appearance of the stranger in the home environment creates the possibility for the child to encounter the unfamiliar newcomer, and to attempt to involve the stranger in the child's activities. Given the open-systems nature of the phenomenon under study, every child can be expected to construct action strategies that are unique for the given child in the given setting. The uniqueness of these strategies need not move them beyond the realm of scientific study.

The Experiment

Subjects and Procedure. Sixteen families with toddlers (ranging in age from 12 1/2 to 15 3/4 months at the beginning of the study) participated in the study. Eleven girls and five boys were the subjects. All were single children in their families.

The procedure consisted of repeated visits by two experimenters (S1—"Stranger 1"—with videotaping equipment, S2—"Stranger 2"—who was unfamiliar for the child and his or her parents prior to the session) to the home at a prearranged time. The parent(s) were instructed previously to try to distract their child's attention from S1, and to try to facilitate his or her interaction with S2.

Each experimental session (visit to the subject's home) was conducted according to the following constant script. In the beginning of a session, the first stranger (S1) was admitted to the home by the parent(s). She began to videotape the child in whatever activity the child was engaged. Approximately 10 minutes later, a person unfamiliar to the family, S2, knocked and entered. S2 engaged the parent(s) in conversation for approximately 5 minutes, or less if the infant made attempts at social contact sooner. S2 responded to any overture on the part of the child. S2 conversed with the parents on such topics as the child's health and favorite toys. Then S2 asked the child to show him or her the favorite toy if the child had not yet attempted to make contact. For 5 to 10 minutes, S2 attempted to play with the infant. Then S2 left, waving and saying "bye-bye"; the child was encouraged by S2 and the parents to respond in the same way. After S2's departure, S1 continued to videotape for 10 minutes longer.

The videotaping sessions were repeated at different intervals for different subsamples of the subjects. Seven children (3 girls, 4 boys) were videotaped again within a 1 1/4-month interval after the first session. Nine children (3 boys, 6 girls) were videotaped again 6 to 8 months later.

Different persons (all but one were college students) served as S2s in the course of the study. On every occasion in the course of repeated videotaping sessions, the child had never met the S2 before. The S2s were recruited mainly from basic courses in developmental psychology.

Transformation of the Videotaped Observations into Data. The videotapes obtained in home settings were subjected to a two-step analysis. First, observers transcribed observable actions of the interactants from the videotape onto transcription sheets, using everyday language descriptions of the child's and adults' (parents', S2's, and S1's, if relevant) actions in the videotaped setting. Different observers transcribed the same videotapes, and care was taken to eliminate differences in the transcription results through comparison of the transcripts. Because the transcripts are in the narrative form, it is not possible to use quantitative methodology to ascertain the interobserver similarity in transcription of the tapes.

The second step in the analysis involved the coding of the target child's actions into a more generalized form, amenable to sequence–structure analysis. The coding system used for that purpose consisted of action codes aimed at capturing the basic structure of the child's actions, rather than describing in detail the whole complexity of the concurrent behavior that could be observed. Table 7 includes the codes and their descriptions. Reliability of the coding procedure was established by comparing coding results of the same (single) coder for two different occasions of coding of the same transcript. Reliability was found

Table 7. Coding System for Describing Toddlers' Contacting of Strangers

Code	Description
TOS1 or TOS2	Touches stranger (S1 or S2) directly
GVS1 or GVS2	Gives stranger (S1 or S2) an object
SS10 or SS20	Shows stranger (S1 or S2) an object
OFS1 or OFS2	Offers stranger (S1 or S2) an object
UOS1 or UOS2	Uses other person to give the stranger (S1 or S2) an object, or to take an object from a stranger
VOC	Vocalizes
APS1 or APS2	Approaches the specified stranger (S1 or S2)
APM or APF or APO	Approaches mother, or father, or another adult (who is not S1 or S2)
IMS1 or IMS2	Imitates the specified stranger
LAS1 or LAS2	Lifts arm towards the specified stranger
MAS1 or MAS2	Moves away from the specified stranger
SNS1 or SNS2	Sits near the specified stranger
SOS1 or SOS2	Sits on the lap of the stranger (S1 or S2)
INTS1 or INTS2	Interacts with the stranger (S1 or S2)—via book, toy, or game
AOS1 or AOS2	Accepts object from the specified stranger
CRY	Cries
FUSS	Fusses
OOS	Goes out of sight of S1
OBS1 or OBS2	Follows instructions (obeys stranger)
INTM or INTF	Interacts (via toy, book, or game) with mother or father
TM or TF or TO	Touches mother, or father, or other adult (not S1 or S2)
THS1 or THS2	Throws object towards the specified stranger
OTH	Other action (individual play within the visual field of S1, S2, and parent)
LS1 or LS2	Looks at the specified stranger (S1 or S2)
SMS1 or SMS2	Smiles at the specified stranger
LM or LF or LO	Looks at mother, or father, or other adult (not S1 or S2)
MBYE	Mother tells child to say bye-bye and wave
S2BYE	S2 asks the child to wave and say bye-bye
OBYE	Other adult (not S1 or S2) asks the child to wave and say bye-bye

to be sufficient (intercoder agreement in recoding sequences ranged from 72% to 92%).

Results

Decline Functions. The cutoff subsequence lengths of decline functions illustrate the temporal depth of the child's contact-making strategies. All 32 full sequences of actions (from the moment of S2's entry to the signal given by S2 indicating his or her intention to leave) were analyzed by the sequence–structure analysis program. The whole range of action codes in the sequences varied from 41 to 174.

Some subjects in the study demonstrated consistency from the first to the second experimental session in the cutoff points in the decline functions of their subsequence distributions. Nine toddlers revealed similar (±1 unit) cutoff points in the second session with those they had in the first. The remaining seven subjects showed difference in the cutoff points, where the maximum variation was from length 5 (1st session) to length 9 (2nd session). Within the sample of 32 sequences, the range of variation in decline function parameters (their cutoff lengths) was remarkable—from 2 to 9. It is possible to argue on the basis of these data that the process of toddlers' entry into contact with unfamiliar persons is organized through flexible subsequences of action that extend over time (and action) to a remarkable extent.

An Individual Subject's Subsequences. The content of subsequences that result from sequence–structure analysis provides the investigator with data for understanding the particular child's ways of contacting an unfamiliar person. It is expected (on the grounds that establishing relationships with other human beings is an open-systems phenomenon) that different occasions (experimental sessions) result in different ways of making contact.

As an illustration of a case of single-subject data. Table 8 includes the sets of nonoverlapping subsequences discovered by sequence– structure analysis in the behavior of a male toddler (C.S.) at two ages, with exposure to a male and a female stranger at both ages. It should be remembered that the particular subsequences in Table 8 are only descriptions of recurrent sequences of behavior, which are not assumed to be a part of permanent means that the child would select from when meeting a new person. Instead, the data in Table 8 are examples of the particular means that the same child constructed in four experimental sessions. In an analogy with speech, the examples in Table 8 constitute different sentences generated from an alphabet of events (given by the coding system used here), by certain generation rules with certain goals

Table 8. Recurrent Subsequences in a Toddler's (C.S.) Action in Repeated
Experimental Stranger–Exposure Situations

A. At age 13–14 months

Session 1 - male stranger		Session 2 - female stranger	
Number	Subsequence	Number	Subsequence
1	-OTH-LM-LS2-OTH-	1	-LS2-OTH-LM-LS2-OTH-
2	-VOC-AOS2-INTS2-	2	-OTH-LS2-AOS2-OTH-
3	-TM-OTH-	3	-OTH-AOS2-OTH-
4	-CRY-TM-	4	-INTS2-AOS2-OTH-
5	-VOC-OTH-	5	-AOS2-OTH-VOC-
6	-INTS2-VOC-	6	-OTH-VOC-LAS2-
7	-OBS2-INTS2-	7	-LS2-OTH-VOC-
8	-INTS2-IMS2-	8	-OTH-LS2-OTH-
9	-OTH-VOC-	9	-VOC-LM-LS2-
10	-OTH-LS2-	10	-LM-LS2-LS1-
		11	-LM-OTH-LM-
		12	-LAS2-OTH-
		13	-OTH-INTS2-
		14	-LS2-MAS2-
		15	-APS2-LAS2-
		16	-TM-OTH-
		17	-VOC-OTH-
		18	-LS1-OTH-
		19	-MAS2-LS2-

B. At age $17\frac{1}{2}$–18 months

Session 3 - female stranger		Session 4 - male stranger	
Number	Subsequence	Number	Subsequence
1	-LS2-LS1-LS2-LS1-LS2-	1	-VOC-OTH-VOC-LS2-LM-OTH-
2	-OTH-LS2-LS1-LS2-	2	-LS2-APS2-LS2-MAS2-
3	-LS1-SMS1-LM-OTH-	3	-LS2-SMS2-LM-LS2-
4	-OTH-LS1-VOC-	4	-SMS2-LS2-SMS2-
5	-VOC-LS2-OTH-	5	-VOC-OTH-APS2-
6	-OTH-LS2-LM-	6	-LS2-MAS2-SMS2-
7	-LM-OTH-LS2-	7	-APS2-SNS2-LS2-
8	-OTH-LS2-OTH-	8	-LS2-MAS2-OTH-
9	-TM-VOC-TM-	9	-LS2-SMS2-LS2-
10	-LS2-SMS2-OTH-	10	-APM-TM-TM-
11	-CRY-OTH-	11	-LM-LS2-SMS2-
12	-TM-FUSS-	12	-MAS2-APS1-VOC-
13	-FUSS-TM-	13	-SMS2-OOS-
14	-OTH-FUSS-	j14	-THS2-LS2-

(continued)

Table 8. (**Continued**)

B. At age $17\frac{1}{2}$–18 months			
Session 3 - female stranger		Session 4 - male stranger	
Number	Subsequence	Number	Subsequence
15	-OTH-TM-	15	-GVS2-LS2-
16	-LS2-VOC-	16	VOC-SMS2-
17	-LM-LS2-	17	-GVS2-VOC-
		18	-SMS2-GVS2-
		19	-INTS2-SMS2-
		20	-SMS2-AOS2-
		21	-OTH-APS2-
		22	-LS1-LS2-
		23	-LS2-VOC-
		24	-SMS2-VOC-
		25	APS2-OTH-
		26	-OTH-LM-
		27	-APM-TM-
		28	-LS2-OTH-
		29	-OTH-LS2-
		30	-FUSS-OTH-
		31	-LS1-VOC-
		32	-OTH-LS1-

in sight. Which set of goals may be used by the child in our experimental sessions can only be assumed from the nature of the setting, but not verified. Let us assume that when a toddler has the possibility of establishing a social relationship with an unfamiliar person in the home setting, the toddler will use it and set the goal of making friends with the stranger. This goal, however, cannot be taken for granted because the experimental setting is sufficiently open to make it possible for the child to set other goals (e.g., playing actively with a toy, and excluding the stranger from that activity). In other words, we as experimenters (who have set up a naturalistic setting that we expect would facilitate contacting a stranger by the child) cannot expect that the child would interpret the situationally given task in the way that we have in mind.

The open-systems nature of the establishment of social contact is also captured in the account that the mother of the toddler (C.S.) gave in a questionnaire before the experimental sessions took place:

> His behavior depends on that of the adult. He is generally friendly and responds with a smile and giggle to anyone who plays with him. If he senses apprehension on the part of the adult he will cry and usually seek out his mother or father. His reaction to people doesn't seem dependent on their

sex. The difference may be that most men he has met have not been as attentive
to him as most women were. But if he meets a man who shows interest in
him and feels at ease with him he responds the same as if it were a woman.
If a woman is apprehensive and ill at ease with C., he will ignore the person
or go to a parent.

The conditional nature of the child's behavior is clearly reflected in
the mother's statement. Psychologists have often dismissed narrative
statements about their subjects as anecdotal, and have preferred to rely
on observations or experimental results as hard data. However, the mother's description of *her* child's behavior is not an occasional narrative, but
a condensed account of her observations of her child across many situations. In contrast to that, psychologists' observations of her child can
sample only a very limited subset of situations in which the child encounters unfamiliar persons. Bearing this in mind, we should conduct
an analysis of individual subject's psychological phenomena by combining information from different sources, recognizing the difference in the
intent and perspective that these different sources may have in characterizing the target subject.

Table 8 reflects further the conditional dependence of the contact-
making efforts of the child. The table includes all nonoverlapping subsequences that were found to recur in C.S.'s behavior during the time
that a target stranger was in the room (i.e., from the time of stranger's
entry, until the time the stranger initiated leave-taking). The sets of subsequences from one session to the next, and (not surprisingly) from those
recorded at 13 to 14 months of age and 17 to 18 months of age are different
from one another. However, if we examine the contents of the subsequences more closely, some similarities in their sequential organization
can be observed.

First, in the case of all four sessions, the majority of recurrent subsequences involved some action relating to the stranger. Only in a minority of cases (Session 1, subsequences 3, 4, 5, 9; Session 2, subsequences
11, 16, 17; Session 3, subsequences 9, 11, 12, 13, 15; Session 4, subsequences
10, 26, 27, 30) were the child's actions entirely unrelated to the strangers.
At least 60% (Session 1) and up to 87.5% (Session 4) of the different
recurrent subsequences included actions directed toward the stranger(s)
(S1 or S2).

Second, if we consider subsets of length-2 subsequences, it is easy
to observe the presence of transitions between the child's individual
action within the visual field of the strangers and the child's direct efforts
to contact the stranger (accepting objects from S2, smiling toward S2,
interacting with S2). Visual referencing (looking toward S2, S1, and the
mother) is another prominent aspect of the subsequences. In all 4 ses-

sions, this visual reference appears in the longest recurrent subsequences.

> No. 1. → ACTS INDIVIDUALLY → LOOKS AT MOTHER → LOOKS AT S2 → ACTS INDIVIDUALLY → ; No. 2. → LOOKS AT S2 → ACTS INDIVIDUALLY → LOOKS AT MOTHER → LOOKS AT S2 → ACTS INDIVIDUALLY → ; No. 3. → LOOKS AT S2 → LOOKS AT S1 → LOOKS AT S2 → LOOKS AT S1 → LOOKS AT S2; No. 4, → VOCALIZES → ACTS INDIVIDUALLY → VOCALIZES → LOOKS AT S2 → LOOKS AT MOTHER → ACTS INDIVIDUALLY →.

It is easy to observe that these (and other length ≥ 2 subsequences) often are cyclical in nature—including both strict (e.g., Session 2, No. 3; Session 3, No. 1 and 8; Session 4, No. 4 and 9) and loose forms of cyclicity. Examples of the latter are Session 2, No. 1 (LOOKS AT S2 recurs in the subsequence, but other constituents do not); Session 3, No. 2 (similar LS2 recurrence); Session 4, No. 1, 2, 3. Subsequence No. 2 of Session 4 illustrates cycle in the child's approach pattern: LOOKS AT S2 → APPROACHES S2 → LOOKS AT S2 → MOVES AWAY FROM S2 →. In the same session, other recurrent subsequences illustrate similar approach patterns:

> No. 6. → LOOKS AT S2 → MOVES AWAY FROM S2 → SMILES AT S2; →; No. 8. → LOOKS AT S2 → MOVES AWAY FROM S2 → ACTS INDIVIDUALLY → .

Such cyclicity in the process of toddler's contact making with a stranger can be observed only if the original action sequence is retained in the process of data analysis. Frequency counts of behavior that extract elements from the sequential context do not allow any adequate analysis of the cyclicity characteristic of the actual sequences. Table 9 includes frequency counts of C.S.'s recurrent $(f \geq 2)$ units in each of the four experimental sessions. It is evident that none of the recurrent subsequences in Table 8 can be recovered from that frequency count. At most, frequency data in Table 9, if treated as a basis for probability estimation, can allow making predictions of which units occur with higher probability in general, in comparison with others. The actual subsequences of units by which the child constructed his relationship with the strangers— that is, the data about the process of contact making—have been lost as the sequential aspect of the phenomena has been lost in case of data in Table 9.

Discussion

Application of sequence–structure analysis to an individual toddler's action sequences in situations where an unfamiliar person was available for social interaction illustrates the potential usefulness of the analysis of the temporal form of a social interaction process. Establishing new

Table 9. Frequency Count of Different Recurrent Action Units in C.S.'s Contact-Making Behavior during Different Experimental Sessions

		Age 13–14 months		Age 17 1/2–18 months	
Code	Contents	Session 1	Session 2	Session 3	Session 4
LS2	Looks at S2	6	20	21	39
LS1	Looks at S1	0	4	12	6
APS2	Approaches S2	0	4	0	10
MAS2	Moves away from S2	0	7	0	10
AOS2	Accepts object from S2	6	13	0	3
INTS2	Interacts with S2	7	8	0	2
IMS2	Imitates S2	3	2	0	0
OBS2	Obeys S2	2	3	0	0
LAS2	Lifts arm towards S2	0	6	0	0
SNS2	Sits near S2	0	0	0	3
GVS2	Gives object to S2	0	0	0	5
THS2	Throws object towards S2	0	0	0	4
APS1	Approaches S1	0	3	0	3
OTH	Acts individually	13	37	23	22
VOC	Vocalizes	11	13	8	20
LM	Looks at mother	3	11	8	13
APM	Approaches mother	0	4	0	6
INTM	Interacts with mother	0	2	0	0
TM	Touches mother	3	4	7	8
FUSS	Fusses	0	2	5	0
CRY	Cries	2	0	3	0
OOS	Out of sight	0	0	0	3
	Total	56	143	87	157

social relationships—be those between a toddler and an unfamiliar adult, or between two adults who had not met before—is a process for which the assumption of stationariness that is the basis for sequential analysis methods (that are constructed around frequency interpretations of probability) is unwarranted. In this process, interaction partners are expected to move, step by step, from a state of lack of relationship between them to a state of an established relationship of some kind. Sequence–structure analysis makes it possible to save the original sequential structure of recurrent action subsequences that are observed in that dynamic process. However, our method does not automatically build a theoretical model of the processes that generate the observed action subsequences. That task cannot be accomplished inductively, by mere analysis of the sequential data, but only in conjunction with particular theoretical concepts that take time as an essential and qualitatively unique (irreversible) dimension of things into account.

CONCLUSION: SEQUENCE–STRUCTURE ANALYSIS OF INDIVIDUAL SUBJECTS' DATA

Two themes have been prominent in this chapter—the relevance of data analysis within individual subjects, and the importance of preserving original sequential relationships between events in the data. Both of these themes follow from an open-systems conceptualization of the nature of psychological phenomena. Because each individual subject whom psychologists may study develops along an individual, unique life course, high interindividual variability within a sample (population) is to be expected. Traditional research practices in psychology have dealt with that variability in ways that would reduce it to abstract notions of the average subject or type. Such traditions have treated an individual's uniqueness as if it were an obstacle to generalizations about universal psychological laws, applicable to all subjects. Following a static view on psychological phenomena, and supported by the belief in the law of error, traditional psychology has overlooked the possibility that *each individual subject is lawful in his or her particular relationships with the concrete environment.* This lawfulness can be studied only if psychologists begin to investigate individual subjects' relationships with their environments, and changes in both these relationships and the environments over time. Universality in psychology is embedded in interindividual diversity, rather than in superficial similarity between individuals.

Furthermore, an individual's relationship with the environment is a dynamic process, and should be studied in psychology as such. In other words, not only are individual subjects different from one another in the forms of their static being, but they also differ in their dynamic change, or becoming (developing into new state, starting from a previous one). It is in this respect that psychologists' data analysis methods must preserve the original serial organization inherent in the phenomena. Psychological phenomena of becoming cannot be studied by methods that assume stationariness as a basic axiom. Such methods are adequate for those phenomena that can be considered to be in a steady state and are not expected to undergo development. For all developing phenomena, however, acceptance of the assumption of stationariness of the processes under study leads to the elimination of the most important aspect of the phenomena—their developmental dynamics—from consideration. The method of sequence–structure analysis described in this chapter affords preservation of some aspects of original sequentially ordered phenomena in the course of data analysis. This method can be useful in the study of processes that are characterized by equifinality, meaning that a certain outcome can be reached through different pathways. Sequence–structure analysis can help to extract recurrent sequential patterns of events in an individual subject's efforts to reach a certain goal by constructing different action routes in the pursuit of the goal. Application of sequence–structure analysis, however, is no substitute for theoretical analysis of the phenomena under study. The primary means by which psychology can learn to understand lawfulness in individual subjects' psychological phenomena is the construction of explicit theoretical models of the phenomena. These models have to recognize the open-systems nature of the phenomena. Usefulness of sequence–structure (or any other) analysis methods that can be applied to sequential data can be determined only relative to the theoretical models with which the investigator operates. Contemporary psychology does not lack valid methods so much as it lacks adequate theoretical understanding of the nature of the phenomena it studies. That understanding may begin to emerge if individual subjects are considered to be lawfully organized open systems that can develop through different routes in their life courses. For that purpose, the time dimension should not be eliminated from psychological data. Instead, information about serial order of psychological phenomena helps in our efforts to understand lawfulness in individual subjects' psychological development.

ACKNOWLEDGMENTS

The empirical studies used in this chapter were conducted with the help of a number of people, whose assistance is gratefully acknowledged here. Study 1 was possible thanks to the facilities of the University of Minnesota that were made available to the present author by Herbert L. Pick and Pat Broen. Vicki Turnquist assisted the author in the analysis of data in Study 1. The empirical research in Study 2 was conducted by the author together with Paula E. Hill, and benefited greatly from the help in transcription and coding of data by Nina E. Tracy, Claudia Mackie, Rebecca Peterson, and Pamela D. Fleming.

REFERENCES

Bailey, G. N. (1983). Concepts of time in quarternary prehistory. *Annual Review of Anthropology, 12,* 165–192.
Bakeman, R., & Brown, J. V. (1980). Analyzing behavioral sequences: Differences between-preterm and full-term mother–infant dyads during the first months of life. In D. B. Sawin, R. C. Hawkins, II, L. O. Walker, & J. H. Penticuff (Eds.), *Exceptional infant: Vol. 4. Psychological risks in infant–environment transactions* (pp. 271–299). New York: Brunner/Mazel.

Bakeman, R., Cairns, R. B., & Appelbaum, M. (1979). Note on describing and analyzing interactional data: Some first steps and common pitfalls. In R. B. Cairns (Ed.), *The analysis of social interactions: Methods, issues, and illustrations* (pp. 227–234). Hillsdale, NJ: Erlbaum.

Baldwin, A. L. (1940). The statistical analysis of the structure of a single personality. *Psychological Bulletin, 37,* 518–519.

Barlow, G. W. (1968). Ethological units of behavior. In D. Ingle (Ed.), *The central nervous system and fish behavior* (pp. 217–232). Chicago, IL: University of Chicago Press.

Barlow, G. W. (1977). Modal action patterns. In T. A. Sebeok (Ed.), *How animals communicate* (pp. 98–134). Bloomington, IN: Indiana University Press.

Bertalanffy, L. von. (1950). The theory of open systems in physics and biology. *Science, 111,* 23–29.

Bertalanffy, L. von. (1952). Theoretical models in biology and psychology. In D. Krech & G. S. Klein (Eds.), *Theoretical models and personality theory* (pp. 24–38). Durham, NC: Duke University Press.

Bertalanffy, L. von. (1960). Principles and theory of growth. In W. W. Nowinski (Ed.), *Fundamental aspects of normal and malignant growth* (pp. 137–259). Amsterdam: Elsevier.

Bobbitt, R. A., Gourevitch, V. P., Miller, L. D., & Jensen, G. D. (1969). Dynamics of social interactive behavior: A computerized procedure for analyzing trends, patterns, and sequences *Psychological Bulletin, 71*(2), 110–121.

Brüser, E. (1981). Child transport in Sri Lanka. *Current Anthropology, 22*(3), 288–290.

Castellan, N. J. (1979). The analysis of behavior sequences. In R. B. Cairns (Ed.), *The analysis of social interactions: Methods, issues, and illustrations* (pp. 81–116). Hillsdale, NJ: Erlbaum.

Chatfield, C., & Lemon, R. E. (1970). Analysing sequences of behavioural events. *Journal of Theoretical Biology, 29,* 427–445.

Cranach, M. von, & Kalbermatten, U. (1982). Ordinary interactive action: Theory, methods, and some empirical findings. In M. von Cranach & R. Harre (Eds.), *The analysis of action: Recent theoretical and empirical advances* (pp. 115–160). Cambridge: Cambridge University Press.

Dawkins, R., & Dawkins, M. (1973). Decisions and the uncertainty of behaviour. *Behaviour, 45,* 83–103.

DeChateau, P., & Andersson, Y. (1976). Left-side preference for holding and carrying newborn infants. II. Doll-holding and carrying from 2 to 16 years. *Developmental Medicine and Child Neurology, 18,* 738–744.

DeChateau, P., Holmberg, H., & Winberg, J. (1978). Left-side preference in holding and carrying newborn infants. I. Mothers holding and carrying during the first week of life. *Acta Paediatrica Scandinavica, 67,* 169–175.

Ericsson, K. A., & Simon, H. (1980). Verbal reports as data. *Psychological Review, 87*(3), 215–251.

Frick, F. C. (1959). Information theory. In S. Koch (Ed.), *Psychology: A study of science* (Vol. 2, pp. 611–636). New York: McGraw-Hill.

Gottman, J. M. (1982). Temporal form: Toward a new language for describing relationships. *Journal of Marriage and the Family, 44*(4), 943–964.

Gottman, J. M., & Bakeman, R. (1979). The sequential analysis of observational data. In M. Lamb, S. Suomi, & G. Stephenson (Eds.), *Social interaction analysis: Methodological issues* (pp. 185–206). Madison, WI: University of Wisconsin Press.

Gottman, J. M., & Ringland, J. T. (1981). The analysis of dominance and bidirectionality in social development. *Child Development, 52,* 393–412.

Hill, P. E. (1984). *The structuring of toddlers' strategies for contacting unfamiliar men and women.* Unpublished honors thesis, Department of Psychology, University of North Carolina at Chapel Hill.

Hill, P. E., & Valsiner, J. (1984, April 6). *Contacting a visitor in home settings: Toddlers' strategies of entry into social contacts with unfamiliar adults.* Paper presented at the International Conference on Infant Studies, New York.

Keats, J. A., Keats, D. M., Biddle, B. J., Bank, B. J., Hauge, R., Wan-Rafaei, & Valantin, S. (1983). Parents, friends, siblings and adults: Unfolding referent other importance data for adolescents. *International Journal of Psychology, 18*(3/4), 239–262.

Kilbride, P. L., & Kilbride, J. E. (1974). Sociocultural factors and the early manifestation of sociability among Baganda infants. *Ethos, 2,* 296–314.

Köhler, W. (1973). *The mentality of apes.* New York: Liveright.

Lashley, K. S. (1951). The problem of serial order in behavior. In L. A. Jeffress (Ed.), *Cerebral mechanisms in behavior* (pp. 112–146). New York: Wiley.

Lathrop, R. G. (1967). Perceived variability. *Journal of Experimental Psychology, 73*(4), 498–502.

Lewin, K. (1931). The conflict between Aristotelian and Galileian modes of thought in contemporary psychology. *Journal of General Psychology, 5,* 141–177.

London, I. D. (1949). The concept of the behavioral spectrum. *Journal of Genetic Psychology, 74,* 177–184.

Lorenz, K. (1981). *The foundations of ethology.* New York: Springer.

Luria, A. (1976). *The nature of human conflicts.* New York: Liveright.

MacKay, D. M. (1972). Formal analysis of communicative process. In R. Hinde (Ed.), *Nonverbal communication* (pp. 3–25). Cambridge: Cambridge University Press.

Métraux, R. (1943). Qualitative attitude analysis. *Bulletin of the National Research Council, 108,* 86–94.

Patterson, G. R., & Moore, D. (1979). Interactive patterns as units of behavior. In M. Lamb, S. J. Suomi, & G. R. Stephenson (Eds.), *Social interaction analysis: Methodological issues* (pp. 79–96). Madison, WI: University of Wisconsin Press.

Petrinovich, L. (1976). Molar reductionism. In L. Petrinovich & J. L. McGaugh (Eds.), *Knowing, thinking, and believing: Festschrift for Professor David Krech* (pp. 11–27). New York: Plenum Press.

Prigogine, I. (1973). Irreversibility as a symmetry-breaking process. *Nature, 246,* 67–71.

Prigogine, I. (1978). Time, structure, and fluctuations. *Science, 201,* 777–785.

Prigogine, I., & Nicolis, G., (1971). Biological order, structure, and instabilities. *Quarterly Reviews of Biphysics, 4*(2 & 3), 107–148.

Reynolds, P. C. (1982a). The primate constructional system: The theory and description of instrumental object use in humans and chimpanzees. In M. von Cranach & R. Harré (Eds.), *The analysis of action: Recent theoretical and empirical advances* (pp. 343–385). Cambridge: Cambridge University Press.

Reynolds, P. C. (1982b). Affect and instrumentality: An alternative view of Eibl-Eibesfeldt's human ethology. *Behavioral and Brain Sciences, 5,* 267–274.

Rheingold, H. L., & Keene, G. C. (1965). Transport of the human young. In B. M. Foss (Ed.), *Determinants of infant behavior* (Vol. 3, pp. 87–109). London: Methuen.

Sackett, G. P. (1979). The lag sequential analysis of contingency and cyclicity in behavioral interaction research. In J. Osofsky (Ed.), *Handbook of infant development* (pp. 623–649). New York: Wiley.

Sackett, G. P. (1980). Lag sequential analysis as a data reduction technique in social interaction research. In D. B. Sawin, R. C. Hawkins, II, L. O. Walker, & J. H. Penticuff (Eds.), *Exceptional infant: Vol. 4. Psychosocial risks in infant–environment transactions* (pp. 300–340). New York: Brunner/Mazel.

Salk, L. (1960). The effects of the normal heartbeat sound on the behaviour of the newborn infant. *World Mental Health Journal, 12,* 168–175.

Salk, L. (1973). The role of the heartbeat in relation between mother and infant. *Scientific American, 228,* 24–29.

Schleidt, W. W. (1974). How "fixed" is the Fixed Action Pattern? *Zeitschrift für Tierpsychologie, 36,* 184–211.

Schleidt, W. M., & Crawley, J. N. (1980). Patterns in the behaviour of organisms. *Journal of Social and Biological Structures, 3,* 1–15.

Scriven, M. (1959). Explanation and prediction in evolutionary theory. *Science, 130*(3374), 477–482.

Simon, H. (1979). *Models of thought.* New Haven, CT: Yale University Press.

Slater, P. J. B. (1973). Describing sequences of behavior. In P. P. G. Bateson & P. H. Klopfer (Eds.), *Perspectives in ethology* (Vol. 1, pp. 131–153). New York: Plenum Press.

Souček, B., & Venzl, F. (1975). Bird communication study using digital computer. *Journal of Theoretical Biology, 49,* 147–172.

Tinbergen, N. (1972). *The animal in its world: Vol. 1. Field studies.* Cambridge, MA: Harvard University Press.

Valsiner, J. (1982, March). *Strategies of dyadic problem-solving with infants in a simulated laboratory situation.* Paper presented at the International Conference on Infant Studies, Austin, TX.

Valsiner, J. (1984). Construction of the Zone of Proximal Development in adult–child joint action: The socialization of meals. In B. Rogoff & J. V. Wertsch (Eds.), *Children's learning in the "Zone of Proximal Development"* (pp. 67–76). San Francisco: Jossey-Bass.

Vygotsky, L. S. (1962). *Thought and Language.* Cambridge, MA: M.I.T. Press.

Vygotsky, L. S. (1978). *Mind in society.* Cambridge, MA: Harvard University Press.

Weisner, T., & Gallimore, R. (1977). My brother's keeper: Child and sibling caretaking. *Current Anthropology, 18*(2), 169–180.

Werner, H. (1957). The concept of development from a comparative and organismic point of view. In D.B. Harris (Ed.), *The concept of development: An issue in the study of human behavior* (pp.125–148). Minneapolis, MN: University of Minnesota Press.

Different Perspectives on Individual-Based Generalizations in Psychology

JAAN VALSINER

The role of the individual subject in scientific psychology has always been a controversial issue in psychology. Throughout the aim of the present volume was to analyze some of the theoretical and methodological sides of that issue, and to bring together psychologists from different fields who have attempted to work out research techniques that are based on individual subjects. The contributors to this book emphasized different aspects of the role of the individual subject, and suggested various ways of constructing basic knowledge in our discipline on the basis of individual subjects' data. These ways ranged from an emphasis on the integration of idiographic and nomothetic research approaches (Walschburger, Grossmann) to the separation of the two approaches, depending on whether the given field is aimed at explanation of the generalized (individual) system, or a population of such systems (chapters by Cairns, Valsiner, Thorngate). Furthermore, the contributors expressed widely different opinions on the role of statistical methodologies in the study of individual subjects—ranging from the innovative use of traditional methods (Walschburger, Thoman, and Rogoff & Gauvain) to the need for devising novel methodology (Dywan & Segalowitz and Mace & Kratochwill),

JAAN VALSINER • Developmental Psychology Program, Department of Psychology, University of North Carolina at Chapel Hill, Chapel Hill, North Carolina 27514.

and further to the need for preserving the integrity of the psychological phenomena, whatever methods are being used (Cairns, Ginsburg, Valsiner). Going beyond ordinary single-case statistical methodology, Thorngate and Carroll outlined a strategy for comparison of hypotheses that can be applied to individual subjects. Grossmann and Franck outlined the historical side of the issue of inference from individual subjects. The issue that is of concern to the contributors to the present volume has been in the center of attention of psychologists and philosophers in the past. It has been discussed; psychologists have fought over it to prove their claim for the scientific nature of their points of view (e.g., Allport, 1940, 1946; Holt, 1962; Skaggs, 1945, and others, discussed in Chapters 1 and 2 in this volume)—and after a while the whole issue was abandoned as a topic, until another generation of psychologists picked it up again. Such episodic interest in the role of the individual subject in psychology illustrates how psychologists' social environment guides scientists in their efforts to explain psychological phenomena (Buss, 1978; Flanagan, 1981; Gergen, 1973, 1982). Unfortunately, the issue itself has remained unsolved, and much dispute around it has facilitated selective forgetting of what had been actually said by our predecessors (see Grossmann's chapter for details).

AGGREGATION OF SYSTEMIC KNOWLEDGE ABOUT INDIVIDUAL SUBJECTS

It is hoped that this book will contribute to reawakening the interest of psychologists and other scientists in the issues of scientifically valid inference from individual subjects. Very often, psychologists' everyday activities make them rush ahead in their research without allowing them sufficient opportunity to reflect on the nature of the knowledge their research actually constructs. Obviously, all the authors represented in this volume emphasize the need for greater attention to the systemic integrity of generalized knowledge about individual subjects in psychology. Again, the old truth that an organized system is more than a simple sum of its constituent components surfaces in our renewed interest in the individual subject. As Thorngate emphasizes, it is impossible to preserve the systemic nature of the individual phenomenon if we break it into its elements and aggregate those into some summary accounts of these elements, in the hope that this activity will afford the reconstruction of the generalized, abstract, or ideal system. If we are really concerned with explaining how the system functions in general, we should analyze its particular (individual) examples intrasystemically, and then proceed to aggregates of knowledge about the particular examples of a system across other examples of a similar kind. Or, to use Vygotsky's terminology, analysis in psychological research should proceed to reduce the complexity of the phenomena not to their elementary constitutents, but to *units*—which, unlike elements, retain all the basic properties of the whole and which cannot be further divided without losing them (Vygotsky, 1962, pp. 4–5).

Knowledge about how individual subjects function as systems cannot be generated by mere summation of the data about individual cases with the aim of finding the average or typical among a sample of systems. Instead, the functioning of the system should be explained first in the individual case. Subsequently, the model of how the individual system works can be abstracted from that particular case by testing whether the model is applicable to other individual cases. Thus, analysis of a new individual subject constitutes an independent effort to replicate the adequacy of the systemic model. If that effort succeeds, then the new model explains the functioning of the individual subjects *in general,* and is thus part of general scientific knowledge in psychology. Many important discoveries and theoretical inventions in physiology and psychology have been based on the systemic analysis of individual subjects (see Dukes, 1965). Subsequent addition of other individual subjects need not guarantee any further improvement of the validity of the systemic knowledge of the individual case, unless it disproves our model of the generic system of the phenomenon. In other words, the investigators' belief that the generality of their empirical research findings can be automatically guaranteed by studying a large number of subjects may be a convenient illusion. It substantiates the hard work of "getting the sufficient number of subjects" (an organizational task), and often legitimizes neglect of the systemic functioning of the individual subjects as systems.

In quite a few areas of psychology the individual subject has preserved its relevance as the inferential basis for generalizations. For example, experimentation in psychophysics is often accomplished on the basis of testing individual subjects and using individual data for revealing general laws of the phenomena studied. Likewise, in neuropsychology (see the chapter by Dywan & Segalowitz) the focus has historically been, and remains, in the systemic organization of psychological functions within the working brain. The systemic nature of neuropsychological knowledge is evident in Luria's emphasis on the role of syndrome analysis in granting the validity of psychological investigation (Luria & Artemyeva, 1970), and in his response to the standardization of his neuropsychological methods on large samples of patients in the United States (cf. Luria & Majovski, 1977). In order to explain the systemic nature of a

psychological phenomenon, the separation of intrasystemic and intersystemic views of the phenomenon is useful to clarify to ourselves what
kind of knowledge we are trying to obtain to answer the particular question we have in mind.

GROUPS AS INDIVIDUAL SUBJECTS

The strategy of generalization in psychology that emphasizes intrasystemic analysis (with subsequent testing of the emerging models on
other individual cases) need not be considered necessarily alien to the
traditional practice of accumulation of a sufficient number of subjects in
a sample, aggregating the data into group data (e.g., averages, standard
deviations), and arriving at generalized inductive knowledge as a result
of this manipulation. In its essence, this strategy is similar to the intrasystemic one—but *only in the special case where the population (to which
generalization is made from a sample) acts as the individual subject* in
investigators' thinking. Thus, for scientific disciplines that make populations their objects of investigation—as is the case, for example, in much
of sociology—inference from samples to populations may fit the theoretical goals of the discipline. Group data, statistically aggregated, may
allow us to make generalizations about populations as collective individual subjects. However, inference from the populational system to its parts
(individual persons in the sample) is epistemologically questionable, even
if practical predictions from group data to individual subjects sometimes
happen to be accurate (e.g., Dymond, 1953; Sarbin, 1942).

What psychologists often do is to confuse the levels of the system
that is to be explained with its parts that are related with one another
and make that system work. The pervasiveness of that confusion is evident
in the overwhelming majority of empirical research reports published
(and unpublished) in psychology. Two domains of psychological discourse where the data are consistently misinterpreted by psychologists
are the interpretation of validity and reliability indexes of their measures,
and statements about individual subjects based on average group norms.

Many psychological methods emphasize the relevance of reliability
and validity estimates (in terms of correlations) as a proof for the scientific
applicability of these methods. These correlation coefficients describe
the consistency of the *method*, based on some sample of subjects. However, quite often a reasonably high reliability coefficient obtained for a
method gets interpreted as if it reflects the consistency of the particular
psychological characteristic within the individual persons within the sample

studied. Lamiell (1982, pp. 7–8) has made explicit the inferential fallacy involved in such interpretation:

> Unfortunately, this rationale (that high group-based reliability equals high intraindividual consistency) rests on the faulty premise that the reliability and validity coefficients obtained in studies of individual differences provide adequate empirical grounds for inferring the degree of consistency with which *individuals* in a sample have manifested a particular attribute or characteristic over time or across situations. The premise is faulty simply because such coefficients are *group statistics*, computed on the basis of data summed *across persons.* It is the aggregate nature of those coefficients which virtually precludes their appropriateness as grounds on which to infer anything about (in)consistency in the levels at which any *one* individual has manifested a particular attribute or characteristic. Indeed, a reliability or validity coefficient cannot be unequivocally interpreted in this manner unless it is perfect (i.e., unity), because it is *only* then that the degree of consistency (in relative position with respect to some attribute) manifested by the individuals *as a group* unambiguously reflects the degree of consistency (in relative position) manifested by *each* individual *in* the group.

A confusion similar to the one criticized by Lamiell is often evident when psychologists attempt to interpret averaged group data in terms of their meaning for individual persons in the group. An average result, obtained on the basis of a sample of subjects, characterizes the sample (and perhaps the population to where the sample belongs), and need not represent any individual subject in that sample (or in the population). The sample is the individual subject in this context, and the particular individual persons in the sample constitute parts of that system. For example, an empirically found average height of some sample drawn from a population is an abstract description of that population, and may adequately describe only a small proportion of individual persons (e.g., if the average height of a sample is 175 cm, then the average adequately represents only those individual persons who happen to be exactly 175 −/+ 0.5 cm tall). The only case where the average adequately represents every individual in the population is the situation where every person is similar to every other person (e.g., if every person in the population is 175 cm tall, which obviously renders the average of 175 cm as well, although the whole rationale for calculating the average in such a case is obscure).

The confusion of populational and individual levels of discourse in psychology is not just an error in the thinking of psychologists. As I attempted to show in Chapter 5, it is supported by some aspects of the language we use, both in everyday life and in psychology. Our language use guides our common sense, which in turn has a profound influence on psychologists' thinking. As a result, the overwhelming confusion of the levels of discourse, and social consensus about the acceptability of that confusion, are natural outcomes of our own cognitive processes.

In order to overcome the confusion surrounding the role of the individual subject in scientific psychology, different perspectives from which the individual subject can be studied may prove useful. These perspectives constitute frames of reference in which the phenomena are conceptualized. These perspectives—intellectual angles from which the investigator may view the phenomena—belong to the world view of an investigator. Four different frames of reference are outlined in the following, each of which sets the question of the individual subject in a different light.

FOUR FRAMES OF REFERENCE IN PSYCHOLOGICAL RESEARCH

Every target in a science is looked on from the perspective of a certain background reference system of axiomatic knowledge. Here I outline four frames of reference that either have been used in psychology, or that can be used in the discipline.

THE INTERINDIVIDUAL FRAME OF REFERENCE

The interindividual frame of reference is by far the most widely used reference framework in psychology. It involves comparison of an individual subject (or samples of subjects) with other individuals (samples) in order to determine the standing of these subjects relative to one another. For example, any comparison of two (or more) persons with one another, resulting in statements like "Jimmy is better than Johnny and Mary in reading and writing, but worse than the other kids in arithmetic"—involves thinking within the interindividual frame of reference. A quick introspective scanning of our everyday life activities and thinking may reveal that this frame of reference is very often used, particularly in situations (and societies) where interindividual competition is required, and emphasized. This reference frame is also used to compare groups of individual persons—because the groups are individual subjects. For instance, a comparison of two samples of subjects—boys and girls, as an example—on some measure that leads to a statement about a difference (or lack of it) between the samples—uses this frame of reference. In psychology's research literature, we often come across statements like "the experimental group was *found to do better* than the control group" in an experiment. In such comparisons, the samples—boys and girls, and also the experimental and the control group—are dealt with as if they were individuals, who are compared with each other. This is similar to a comparison between Johnny and Jimmy, where a parent of one of them may arrive at the statement "Jimmy does better than Johnny in X, Y, or Z."

In psychology, the interindividual frame of reference has guided the *normative* ("populometric" or "parametric") tradition of psychological measurement (Cattell, 1944). In the case of that reference frame, the environmental contexts in which the individuals function is excluded from consideration as a part of the particular issue. For example, finding out that Jimmy's IQ score is 115 can lead a psychologist to the comparison of Jimmy to the average norm for a population (100), and to relative statements, such as "Jimmy has above-average IQ." However, the particular environmental context with which Jimmy interacted in the process of his IQ being tested—the test materials, Jimmy's conceptualization of the tasks involved, and his motivation to pass the test at the level of his maximum performance at the testing time—all these aspects are fully and irreversibly eliminated from the psychologist's information base concerning Jimmy's "intelligence." The interindividual frame of reference leads to *decontextualized knowledge,* which in its turn leads to an attribution-based explanation of the psychological phenomena in question (see Valsiner, 1984, for further analysis of that aspect). Psychology's theoretical mainstream emphasizes decontextualization of individual psychological phenomena (see Super & Harkness, 1981). From that perspective, the use of the interindividual frame of reference is very natural. However, its effect on the advancement of psychological knowledge may divert psychologists from the theoretical pursuit of explaining the issues that the empirical data represent. This danger was explicitly emphasized by Cattell (1944, p. 300):

> If individuals can be given a score simply from putting them in rank order—and people can be put in rank order for anything under the sun—there is very little incentive to find the exact nature of the thing with respect to which they are being put in rank order. The facility with which IQ or percentile scores can be used in educational and placement problems has apparently obscured interest, for example, in the problem of the nature of intelligence at different age levels ... while the readiness with which interests can be ranked ... seems to have made superfluous to ask "What is interest?"

This quote from Cattell explicates the conceptual problems inherent in the frequent confusion of the interindividual and intraindividual frames of reference.

THE INTRAINDIVIDUAL FRAME OF REFERENCE

The intraindividual frame of reference is widespread in some areas of psychology that either by the nature of their objects of research or theoretical convictions have preferred psychological explanations derived from the study of the individual. These areas include clinical case analyses of patients in clinical psychology, psychophysical or cognitive experimentation that is confined to the study of phenomena within a subject, neuropsychological syndrome analysis, and analyses of unique psychological events or phenomena. The defining characteristic of this reference frame is the emphasis on *processes that take place within the individual and that are believed to be explainable as aspects of the individual.* One the one hand, this reference frame comes close to recognizing the analysis of individual subjects. On the other hand, however, it is similar to interindividual methods in its discounting of the context of the individual. Even in the case of analysis of psychological phenomena within an individual—in Cattell's terms, through "ipsative" measurement (see Cattell, 1944, p. 294; also Baldwin, 1940, 1942, 1946)—the knowledge about the individual subject is decontextualized. The decontextualization of psychological phenomena violates their integrity as open systems, which are necessarily interdependent with their environments. Only those frames of reference that retain the systemic unity of an individual and the environment will advance psychological theorizing in directions that fit the reality more adequately. Cattell (1944) called for (but did not develop) ways of looking at psychological phenomena that would emphasize individual–environment interaction. For that purpose, our thinking about the individual must preserve the interdependence with the environment, as is indicated in the remaining two reference frames.

THE INDIVIDUAL-ECOLOGICAL FRAME OF REFERENCE

The individual-ecological frame of reference considers an individual person (or a social group, a collective individual) as it acts on its environment to solve some problem, created at the given time by the given structure of the environment and by the individual's goals. The individual's actions are viewed in the context of problem-solving situations that emerge when the individual acts in the environment. The questions asked about these actions concentrate on the issue of *how* an individual solves the given problem. Whether an individual is better or worse at that than other individuals are is unimportant. In psychology, any study of cognitive processes that an individual uses in solving a particular problem (e.g., Anzai & Simon, 1979; Duncker, 1945; Köhler, 1976), or Piaget's description of children's explanations of physical phenomena (e.g., Piaget, 1960), are examples of psychological research within that frame of reference. The individual-ecological framework suits cognitive psychology well because it deals with the general explanation of the thinking processes and uses the analogies of computer programs to try to explain how people think and solve problems. Likewise, the individual-ecological frame of reference has become more prominent in the thinking of personality psychologists in the last couple of decades, as it was facilitated by the interactionist movement in the study of person–situation relationships (Magnusson, & Endler, 1977; Mischel, 1968, 1979) and related to the problem of constancy versus change in personality (Bem, 1983; Bem, & Allen, 1974; Bem, & Funder, 1978). However, the emphasis in the contemporary interactive approach to personality has largely remained on intrapersonal psychological phenomena, which vary from one situation to another. This emphasis harms the interactionist approach because the description of the processes that mediate an individual's goal-directed activity, interdependently with the structure of the environment, often remain unstudied. Only few efforts to understand the processes that mediate a person's interaction with the environment are notable in the interactionist research tradition (Magnusson, & Allen, 1982).

The issue of mediating processes in person–environment relationships brings the interactionist psychology of personality together with research interests of cognitive psychology (e.g., Lamiell, 1982; Lamiell, Foss, Larsen, & Hempel, 1983; Mischel, 1979). Traditionally, cognitive psychology has taken the cognition of organismic (or technical) systems for granted. The question of how these systems have emerged is rarely of interest to cognitive psychologists. Psychologists of that persuasion may study how the cognitive processes of individual persons are organized, and not worry too much about the ontogeny of these processes. Likewise, cognitive anthropologists may analyze the presence of certain cognitive operations in the culture, or in individual carriers of the culture and disregard the history of these cognitive phenomena in the culture. The developmental perspective on individuals' transactions with their environments requires the adoption of a frame of reference, however, in which the complexity of developmental reality can be captured. In the case of the development of psychological phenomena, the social nature of the developmental process is often de-emphasized. The fourth frame of reference described here is outlined with the purpose of conceptualizing the cooperative nature of development.

The Individual-Socioecological Frame of Reference

The individual-socioecological frame of reference differs from the individual-ecological frame by the attention given to the *assistance from another individual* (or individuals, or groups, etc.) in the process of individual–environment transaction. In this reference frame, an individual's actions and thinking to solve a problem that has emerged in the person–environment transaction is not a solitary, but a social event. A person who is confronted with a problem may ask for help from somebody else, who may be more experienced in solving that kind of problem. For example, for quite a long time during a child's development, the child depends on other people who help the child to acquire culturally appropriate and successful ways of solving problems. Help can also be sought even from another person who has less experience with the given problem, but who, by being related to the problem solver, can be used as a social other whose presence helps the problem solver to deal with the problem. An example of the latter case is a young mother with a two-year-old child, whose husband has suddenly died. The child, and the mother's feeling of responsibility toward the child, may help that young woman to cope with the psychologically traumatic event. In psychology, Vygotsky's cultural-historical approach to psychological phenomena is a good example of the individual-socioecological reference frame (see Van IJzendoorn & Van der Veer, 1984; Vygotsky, 1962, 1978). Particularly, Vygotsky's concept of the "zone of proximal development" (see Rogoff & Wertsch, 1984; Vygotsky, 1978, pp. 84–91) is a direct example of how the individual-socioecological reference frame is used in some traditions in psychology.

It must be admitted that the four reference frames outlined in this chapter are quite uneven in their popularity among psychologists. The interindividual framework is the framework that the majority of psychologists continue to use, either consciously or implicitly. In contrast, only some areas in psychology, particularly in the cognitive domain, have adopted the individual-ecological reference frame. The individual-socioecological frame of reference is extremely rare in contemporary psychology, although the renewed interest in Vygotsky's psychological heritage may perhaps lead more investigators to attempt its adoption. It can be adequate for many problems in developmental psychology, where psychological phenomena undergo changes that are guided by "social others." For example, a psychological analysis of children's accidents for the purposes of prevention may benefit from the individual-socioecological reference framework (Gärling & Valsiner, 1985). Again, it is necessary to stress here that whichever of the four reference systems is adopted by a psychologist depends upon her or his general view of the phenomena under study. The scientific value of adopting any of the four is determined by the adequacy of the reference frame to the phenomena under study. If, for example, a psychologist's goal is to select certain individual persons, social groups or business firms, for particular applied purposes (e.g., so as to satisfy a "consumer": an employer, a politician, or a businessperson who wants to select the best investment option out of the ones present), then the interindividual frame of reference is most suitable. If a psychologist wants to help a computer programmer to construct a program that could accomplish a certain task (e.g., play chess, recognize visual or acoustic patterns, etc.), then the individual-ecological reference frame is adequate. The constructed computer system—be this a chess player or pattern recognizer—has to succeed in the task for which it is constructed. Whether or not it is better than other computer systems according to some criteria (a question in the interindividual reference frame) need not be relevant, as long as the program accomplishes its task. The interindividual frame of reference includes the maximizing (or minimizing) strategies (e.g., selecting a candidate for a job, who is "the most qualified" out of the pool of applicants), whereas the individual-ecological reference frame is centered around the strategy of *satisficing* (see Simon, 1957, pp. 204–205, 241–260)—accomplishing the given objective in ways that are "good enough."

Finally, the individual-socioecological frame of reference emphasizes the role of social coordination of actions and knowledge of individuals (persons, or collective individuals—social groups, institutions, cultures, countries) in solving different problems that a particular individual is confronted with. In this reference frame, problems are solved by individuals who face them, interdependently with other individuals (who may know how to solve such problems, or be interested in the other's solving, or failing to solve, the problem).

There is, of course, always the possibility that any of the four frames of reference can be combined with others. The designers of the computer systems bear in mind not only the task that the new problem must perform, but also its chances of success in the interindividual reference frame of the competitive marketplace. Likewise, parents of children who teach the children new skills (that the children develop within their individual-socioecological reference frameworks), are eager to find out from a psychologist of "how my child is doing" in comparison with other children in the given age group. The parents also hope that knowledge and skills that they have helped their children to develop will be used by the children individually in situations where parental presence and guidance are not available (within individual-ecological contexts). Educating children for facing different life situations in their future illustrates that combination of the individual-socioecological and individual-ecological frames.

The *possibility* of combining the four reference frames need not make it a viable or desirable strategy for psychologists to mix these four in their research and practice. Although every psychological phenomenon can be simultaneously considered within each of these frames, different research goals of psychologists may make it necessary to consider them separately. A conscious and explicit decision by a psychologist not to use a particular frame of reference on theoretical grounds (although its use is practially possible) may help in reaching the particular scientific goals that the psychologist has. The frames of reference, although combinable, provide science with distinctly different kinds of knowledge, each of which may have its place somewhere in the knowledge structure of psychology.

CONCLUSION: THE ROLE OF THE INDIVIDUAL SUBJECT IN SCIENTIFIC PSYCHOLOGY

It may be sufficient to end this epilogue by reiterating the basic theme that has united the contributors to this book—that inference about the systemic functioning of psychological processes, based on the study of individual subjects, is of decisive importance for scientific psychology. This message of the book stands on solid grounds, as long as the adequacy-to-phenomena criterion of science is accepted, and the organized (systemic) nature of psychology's individual subjects is recognized.

By reinstating the role of the individual subject in psychology, the authors in this volume do not intend to declare themselves to be soft scientists in the eyes of their peers or the lay public. Instead, the contributors tend to be united in their world views at the highly abstract level of looking at their subject matter from a systemic perspective. That systemic perspective sets the stage for the study of how individual subjects (persons or collective individuals) function as organized wholes, rather than mere summary aggregates of their elements. Despite the obvious complexity of the systemic functioning of individual subjects in psychology, it is possible, and necessary, to disentangle that complexity in ways that preserve the systemic organization of the phenomena under study. There can be a variety of ways of accomplishing that task that may be adequate for reaching different goals of research.

Psychologists' interest in the individual subject has always been present somewhere in their minds, as we can learn from the history of our discipline. A psychoanalytically inclined (and systemically minded) outside observer of our discipline may be quite amused by psychologists' thinking and acting. As is the case with any difficult or ambivalent subject matter in our thinking and feeling, the interest in the individual subject in psychologists' minds has sometimes been repressed, denied, or substituted by the study of populations. From time to time, our professionally socialized scientific superego becomes weakened, and our interest in the individual subject bursts out to the surface. The difficult task during such periods is to guide that outburst away from the replacement of scientific interests by emotional declarations about the importance of understanding others as human beings (which has always been a nonscientific everyday task for human beings), and to continue our search for abstract knowledge about the psychological functioning of individual subjects as self-organized systems. It is hoped that the present volume has clarified some issues of the scientific role of individual subjects in psychology. The question closest to the hearts of empirically minded psychologists—how to study the individual subjects in detail—remains largely (and purposefully) unanswered in this volume. It can be answered only on the basis of serious theoretical reorganization of much of psychology—to which, we hope, the contributions to this volume may prove to be relevant.

REFERENCES

Allport, G. W. (1940). The psychologist's frame of reference. *Psychological Bulletin, 37,* 1–28.

Allport, G. W. (1946). Personalistic psychology as science: A reply. *Psychological Review, 53,* 132–135.

Anzai, Y., & Simon, H. (1979). The theory of learning by doing. *Psychological Review, 86,* 124–140.

Baldwin, A. L. (1940). The statistical analysis of the structure of a single personality. *Psychological Bulletin, 37,* 518–519.

Baldwin, A. L. (1942). Personal structure analysis: A statistical method for investigating single personality. *Journal of Abnormal and Social Psychology, 37,* 163–183.

Baldwin, A. L., (1946). The study of individual pesonality by means of the intraindividual correlation. *Journal of Personality, 14,* 151–168.

Bem, D. (1983). Constructing a theory of the triple typology: Some (second) thoughts on nomothetic and idiographic approaches to personality. *Journal of Personality, 51,* 566–577.

Bem, D., & Allen, A. (1974). On predicting some of the people some of the time: The search for cross-situational consistencies in behavior. *Psychological Review, 81,* 506–520.

Bem, D., & Funder, D. C. (1978). Predicting more of the people more of the time: Assessing the personality of situations. *Psychological Review, 85,* 485–501.

Buss, A. (1978). The structure of psychological revolutions. *Journal of the History of the Behavioral Sciences, 14,* 57–64.

Cattell, R. B. (1944). Psychological measurement: Normative, ipsative, interactive. *Psychological Review, 51,* 292–303.

Dukes, W. F. (1965). $N = 1$. *Psychological Bulletin, 64(1)*, 74–79.

Duncker, K. (1945). On problem-solving. *Psychological Monographs, 58*(No. 5), 1–112.

Dymond, R. (1953). Can clinicians predict individual behavior? *Journal of Personality, 22,* 151–161.

Flanagan, O. (1981). Psychology, progress, and the problem of reflexivity: A study in the epistemological foundations of psychology. *Journal of the History of the Behavioral Sciences, 17,* 375–386.

Gärling, T., & Valsiner, J. (Eds.). (1985). *Children within environments: Towards psychology of accident prevention.* New York: Plenum Press.

Gergen, K. J. (1973). Social psychology as history. *Journal of Personality and Social Psychology, 26,* 309–320.

Gergen, K. J. (1982). *Toward transformation in social knowledge.* New York: Springer.

Holt, R. (1962). Individuality and generalization in the psychology of personality. *Journal of Personality, 30,* 377–404.

Köhler, W. (1976). *The mentality of apes.* New York: Liveright.

Lamiell, J. T. (1982). The case for an idiothetic psychology of personality: A conceptualization and empirical foundation. In B. A. & W. A. Maher (Eds.), *Progress in experimental personality research* (Vol. 11, pp. 1–64). New York: Academic Press.

Lamiell, J. T., Foss, M. A., Larsen, R. J., & Hempel, A. M. (1983). Studies in intuitive personology from an idiothetic point of view: Implications for personality theory. *Journal of Personality, 51(3),* 438–467.

Luria, A. R., & Artemyeva, E. Y. (1970). On two ways of reaching the validity of the psychological investigation. *Voprosy Psichologii,* No. 3, pp. 105–112.

Luria, A. R., & Majovski, L. V. (1977). Basic approaches used in American and Soviet clinical neuropsychology. *American Psychologist, 32,* 959–968.

Magnusson, D., & Allen, V. L. (1982). An interactional perspective for human development. *Reports from the Department of Psychology, the University of Stockholm* (Suppl. 56).

Magnusson, D., and Endler, N. (Eds.). (1977). *Personality at the crossroads: Current issues in interactional psychology.* Hillsdale, NJ: Erlbaum.

Mischel, W. (1968). *Personality and assessment.* New York: Wiley.

Mischel, W. (1979). On the interface of cognition and personality. *American Psychologist, 34,* 740–754.

Piaget, J. (1960). *The child's conception of physical causality.* Totowa, NJ: Littlefield, Adams.

Rogoff, B., & Wertsch, J. (Eds.). (1984). Children's learning in the "zone of proximal development." *New directions for child development* (No. 23). San Francisco: Jossey Bass.

Sarbin, T. (1942). A contribution to the study of actuarial and individual methods of prediction. *American Journal of Sociology, 48,* 593–602.

Simon, H. (1957). *Models of man.* New York: Wiley.

Skaggs, E. B. (1945). Personalistic psychology as science. *Psychological Review, 52,* 234–238.

Super, C. M., & Harkness, S. (1981). Figure, ground, and gestalt: The cultural context of the active individual. In R. Lerner & N. Busch-Rossnagel (Eds.), *Individuals as producers of their development* (pp. 69–86). New York: Academic Press.

Valsiner, J. (1984). Conceptualizing intelligence: From an internal static attribution to the study of the process structure of organism–environment relationships. *International Journal of Psychology, 19,* 363–389.

Van IJzendoorn, M. H., & Van der Veer, R. (1984). *Main currents of critical psychology.* New York: Irvington.

Vygotsky, L. S. (1962). *Thought and language.* Cambridge, MA: M.I.T. Press.

Vygotsky, L. S. (1978). *Mind in society.* Cambridge, MA: Harvard University Press.

Index